U0772008

第六届全国材料与热加工物理模拟及数值模拟学术会议论文集（2015）

周建新 赵春华 张 梅 曾建民 ◎主编

中国出版集团

世界图书出版公司

广州·上海·西安·北京

图书在版编目（ＣＩＰ）数据

第六届全国材料与热加工物理模拟及数值模拟学术会议论文集：2015 /
周建新等主编. ——广州：世界图书出版广东有限公司, 2015.8
　ISBN 978-7-5192-0173-9

　Ⅰ. ①第… Ⅱ. ①周… Ⅲ. ①材料–物理模拟–学术会议–文集
②热加工–物理模拟–学术会议–文集③材料–数值模拟–学术会议–
文集④热加工–数值模拟–学术会议–文集
Ⅳ. ①TB3-53②TG306-53

　中国版本图书馆 CIP 数据核字(2015)第 205942 号

第六届全国材料与热加工物理模拟及数值模拟学术会议论文集(2015)

策划编辑	杨力军
责任编辑	杨力军
封面设计	高艳秋
投稿邮箱	stxscb@163.com
出版发行	世界图书出版广东有限公司
地　　址	广州市新港西路大江冲 25 号
电　　话	020-84459702
印　　刷	虎彩印艺股份有限公司
规　　格	880mm × 1230mm　1/16
印　　张	20.25
字　　数	500 千
版　　次	2015 年 8 月第 1 版　2016 年 3 月第 2 次印刷
ISBN	978-7-5192-0173-9/O·0046
定　　价	78.00 元

前　言

牛济泰

　　材料在铸造、压力加工、焊接、热处理等热加工工艺过程中,以及在制成零部件后的实际服役过程中,会产生各种物理/力学行为,这些行为相当复杂,往往难以进行定量预测。伴随着先进的测试技术、计算机科学和控制理论的发展,材料现代物理模拟技术应运而生。对材料和热加工工艺来说,物理模拟通常是指利用小试样,借助某种试验装置再现材料在制备或热加工过程中的受热或同时受热、受力的物理过程,充分而精确地揭示材料在制备与热加工过程中组织与性能的变化规律,评定或预测材料在制备或热加工时可能会出现的问题,为研制新材料以及制定合理的加工工艺提供理论指导和技术依据。

　　数值模拟是利用数学模型来描述一个过程的基本参数的变化关系,采用数值方法求解,以获得过程的定量认识。随着数学科学和计算机软硬件的发展,目前已有了许多高效的数值计算方法。数值模拟技术的应用,不仅能预测某特定工艺的最终结果,而且能显示出工艺过程的变化情况,对加工过程的组织和性能变化规律能有更深入的了解,从而可使制定工艺不单凭经验,而是建立在更为科学、更为可靠的基础上,既可节省大量人力、物力,还可研究目前尚无法采用直接试验进行研究的复杂问题,并对工艺行为的结果进行评定和预测。

　　必须指出,数值模拟与物理模拟具有不同特点和应用范围,两者具有互补性,物理模拟是数值模拟的基础,数值模拟是物理模拟的归宿,只有将两者有机结合起来,才能更有效地解决材料科学与工程中的复杂问题,并获得符合实际的研究结果。

　　我国是在材料热加工领域应用物理模拟和数值模拟较早的国家,学术活动的开展也非常活跃。在中国机械工程学会材料分会下面设置有物理模拟与数值模拟专业委员会。从1990至2014年,以中国机械工程学会的名义共举办了7次(其中5次由国内国际合办)"材料与热加工物理模拟及数值模拟国际学术会议",现已成为国际上别具特色并具有较大影响力的系列学术会议。第一届材料与热加工物理模拟会议于1990年7月在哈尔滨召开。之后,考虑到物理模拟和数值模拟密不可分的学术关系和工程联系,将学术会议名称由"物理模拟"扩展为"物理模拟及数值模拟",于 1997 年 3 月在海南召开了第二届国际会议。以后分别在北京、上海、郑州、桂林等地举行会议。参会的国家和代表数目一届比一届多,论文的质量和水平也一届比一届高,为推动我国材料及热加工领域物理模拟与数值模拟技术的发展,增进与国外先进国家在本领域的相互交流、合作与接轨,起到了积极作用,也为建立国际学术组织打下了扎实的基础。2013 年 6 月第七届"材料与热加工物理模拟及数值模拟国际学术会议"在芬兰举行,经中国、美国、日本、俄罗斯、芬兰等国家代表的建议,来自 23 个国家的数百名代表经过充分协商,一致同意成立一个国际性的学术组织——"材料加工物理模拟及数值模拟联合会"。基于我国在材料物理模拟及数值模拟国际学术交流方面的突出贡献,一致同意此联合会总部设在中国,挂靠于中国机械工程学会,并选举我任联合会首任主席,美国、俄罗斯、日本的学者担任副主席。

这次在宜昌召开的第六届全国材料与热加工物理模拟及数值模拟学术会议,是近年来最新成果的展示。会议集中收录了自上次会议以来在物理模拟和数值模拟方面的国内最新研究成果。内容涉及连续铸造、半固态加工、轧制、锻造、挤压、拉拔、电弧焊和高能束焊、焊接热循环过程、热裂纹敏感性评价、冷裂纹敏感性评价、淬硬脆断倾向评价、焊接工艺优化热处理和粉末冶金;热压成型、高压成型以及新材料研制的许多新领域。

本论文集由华中科技大学周建新教授、三峡大学赵春华教授、上海大学张梅教授、广西大学曾建民教授主持编写。希望论文集的出版对于提高我国材料与热加工物理模拟与数值模拟水平和推进它的工程应用起到积极的作用。

感谢三峡大学机械与动力学院、华中科技大学材料成形与模具技术国家重点实验室、Dynamic Systems, Inc.、华铸软件中心以及国家数控重大专项 (2012ZX04012-011)、国家自然科学基金(No.51475181)、国家 863 课题(20130031003)对本次会议的资助。

2015 年 7 月 22 日 于哈尔滨

(牛济泰:材料加工物理模拟与数值模拟国际联合会主席,中国材料学会材料及热加工物理模拟与数值模拟专业委员会主任委员,俄罗斯自然科学院院士,哈尔滨工业大学教授、河南理工大学、西南科技大学特聘教授)

第六届全国材料与热加工
物理模拟及数值模拟学术会议

（2015 年 9 月 17-20 日　湖北宜昌）

主办单位

三峡大学

华中科技大学

上海大学

广西大学

中国机械工程学会材料分会

承办单位

三峡大学机械与动力学院

华中科技大学材料成形与模具技术国家重点实验室

上海大学材料科学与工程学院

广西大学材料科学与工程学院

中国机械工程学会材料分会物理模拟及数值模拟专业委员会

协办单位

Dynamic Systems，Inc.

北京科技大学高效轧制国家工程研究中心

宜昌市燕狮科技开发有限责任公司

华中科技大学华铸软件中心

湖北省机械工程学会

支持媒体

《机械工程材料》杂志社

大会顾问委员会

周 玉　牛济泰　涂善东　李德群　孙 军

大会组织委员会

主 任: 胡 军　牛济泰　周华民　赵春华

成 员: 周建新　张 屹　吴海华　张 梅
　　　　郭 锦　Wayne Chen　殷亚军
　　　　曾建民

大会学术委员会

主 任: 杨院生　雷永平

成 员: 曾建民　刘祖岩　陈 旭　周贤良
　　　　赵艳君　屈朝霞　刘春凤　王晓南
　　　　周旭东　周广涛　曲凤盛　王海燕
　　　　金 成　David Ferguson　张 梅
　　　　赵春华　周建新

大会执行主席

周建新　赵春华　张 梅　曾建民

目 录
Contents

材料物理模拟技术近年的发展、应用与前景

牛济泰

（哈尔滨工业大学、河南理工大学、西南科技大学教授）

摘　要：随着计算机科学与工程检测技术的迅速发展，现代物理模拟作为一门新兴技术引起世界各国科学界和工程界的广泛关注，应用的范围正迅速扩大，已成为21世纪材料研究的主要方法和手段。本文简要介绍了物理模拟技术的基本原理和特点，以及目前世界上应用的物理模拟试验装置，综述了我国在物理模拟试验装置方面的发展历程和在物理模拟技术应用与交流方面所开展的工作。最后，针对我国物理模拟技术领域存在的问题，对今后研究发展方向提出了建议。

关键词：物理模拟；应用

Developments of Materials Physical Simulation Technology and its Applications in the World and in China

NIU Ji-tai

(Professor of Harbin Institute of Technology, Henan Polytechnic University, and Southwest University of Science and Technology)

Abstract: With the rapid development of computer science and the engineering detection technique, modern physical simulation, as a novel technology, has attracted a wide attention of scientific and engineering circles throughout the world. Thanks to the widening applicable scope of physical simulation, it has been the primary methods and measures in the field of materials investigation. In this paper, fundamental principles and features of modern physical simulation technology was briefly introduced, and the experimental facilities were commented as well. Furthermore, the application and academic communication of the physical simulation in China and in the world were summarized. According to the problems of the modern physical simulation technology in China, we bring forward some suggestions in its research directions and prospects at last.

Key words: physical simulation；application

牛济泰,哈尔滨工业大学教授,博士生导师,俄罗斯自然科学院外籍院士,国际材料及热加工物理与数值模拟联合会主席。1941 年 10 月出生于河南省,1959 年考入哈尔滨工业大学焊接专业,1960 年加入中国共产党,1964 年哈工大毕业留校任教至今。在哈工大曾长期担任中层领导职务。现任哈工大材料工程研究所所长,河南理工大学、西南科技大学兼职教授,河南晶泰航空航天高新材料科技有限公司董事长。中国复合材料学会常务理事,中国机械工程学会材料分会常务理事,材料物理与数值模拟专业委员会主任,中国兵工焊接分会理事,中国汽车材料分会理事。担任《材料科学与工艺》《机械工程材料》《Ecology and industry of Russia》《Engineering Review》编委以及《焊接学报》英文编审等。在材料热加工物理模拟和数值模拟领域以及新材料焊接性和焊接新工艺等方面有突出造诣,开发了超高温物理模拟技术,完成了"碳/碳复合材料高温力学行为及性能预报"等高新技术研究成果。曾获国家及省部级科技进步一、二等奖 7 项,国家发明专利 20 余项,国内外刊物发表科技论文 240 余篇,专著 5 部,主持国家"863"、国家自然科学基金、解放军总装预研、国防科工委基金等科技项目多项。多次应邀到国外讲学或从事合作研究,培养国内外研究生数十名,在国内外材料模拟与焊接领域有较高知名度。

主要研究方向:1.金属材料焊接性及其先进连接技术。包括电子封装,超高强钢、高强铝合金、钛合金、形状记忆合金、超导材料、铝基复合材料以及异种金属材料连接新技术,汽车钢板与铝板连接新工艺;2.材料及其热加工物理模拟和数值模拟。包括低合金高强钢的控轧控冷物理模拟,超细钢焊接及其物理模拟,材料分子动力学模拟,航天宇宙飞船及空间站焊接接头安全可靠性评定和寿命预测等。

体积平均法凝固模型

吴孟怀 [1,2]，LUDWIG Andreas[2]，KHARICHA Abdellah [1,2]

（1. Christian-Doppler Laboratory for Advanced Process Simulation of Solidification and Melting, Dept. Metallurgy, Univ. of Leoben；2. Chair of Modeling and Simulation of Metallurgical Processes, Dept. Metallurgy, Univ. of Leoben）

摘　要： 建立一个理想的凝固模型所要面临的主要问题是怎样实现对跨尺度物理现象的耦合以及和对凝固过程中多相流体传输现象的处理。虽然有基于各种不同尺度上的凝固模型的尝试，但就目前的计算能力还无法想象出让一个单一模型实现从原子尺度到工程铸件尺度的跨越，所以体积平均法成为近期对凝固过程建模的趋势。其思想是立足于工程尺度，将铸件离散成足够多的微体积单元，对宏观传输现象如质量、动量、能量和溶质等进行直接求解，而对于那些发生在微观尺度上的如液-固界面处的质量交换，溶质再分配等现象进行在微体积单元内的体积平均处理，实现对发生在微观尺度上的凝固热力学、动力学和发生在宏观尺度上的多相流体传输现象的耦合。最终目的是对铸态组织和可能的铸造缺陷进行预测。本报告将对该领域的研究进展做简要综述，并提供相关模型算例。

关键词： 凝固；模型；铸件

Volume Average Modeling of Solidification

WU Menghuai[1,2]，LUDWIG Andreas[2]，KHARICHA Abdellah[1,2]

（1. Christian-Doppler Laboratory for Advanced Process Simulation of Solidification and Melting, Dept. Metallurgy, Univ. of Leoben；2. Chair of Modeling and Simulation of Metallurgical Processes, Dept. Metallurgy, Univ. of Leoben）

Abstract: A major scientific challenge in developing a solidification model lies in the requirement to bridge different length scales of physical phenomena, and to deal with the multiphase transport phenomena. No one single model is able to span the entire range of length scales from the process level of industrial interest down to the atomic scale due to the limitation of the current computational resources. Therefore, a trend for the process-modelling of solidification is to incorporate the interfacial phenomena occurring at the lower scale into the process scale of solidification with a so-called volume average approach. A casting of industrial interest can be discretized into large number of volume elements. The macroscopic transport phenomena （mass, momentum, enthalpy, species, etc.） are directly solved. All the microscopic phenomena occurring at the lower scale, e.g. the mass transfer and solute partitioning at the liquid-solid interface, will be volume-averaged in each volume element. Solidification thermodynamics, kinetics and multiphase hydrodynamics will be coupled. This presentation is to give an overview of the state-of-the-art of this field, and some modeling examples are demonstrated.

Key words: solidification; modeling; casting

项目资助：Christian-Doppler 基金会; FFG Bridge Early Stage(842441)，奥地利。

第一作者：吴孟怀，男。计算冶金学及凝固仿真。menghuai.wu@unileoben.ac.at。

　　吴孟怀，男，1963 年 7 月生，教授，博士生导师。1986 年获西北工业大学硕士，2000 年获德国 RWTH Aachen 亚琛工业大学工学博士，2008 年在奥地利 University of Leoben 莱奥本大学计算冶金学取得 Habilitation、教授、博导资格。现任奥地利莱奥本大学 Christian Doppler Laboratory － 先进凝固及熔化过程数值仿真实验室主任。先后主持欧洲共同体及奥地利多项重大研究课题，获得如 RFCS 欧盟钢铁研究基金、ESA 欧洲宇航局、FFG 奥地利研究促进会、FWF 奥地利科学基金、Christian Doppler 基金会，以及多家欧洲钢铁企业的资助。研究方向：多相流－体平均凝固过程数值仿真及其在冶金、铸造和其它材料加工过程中的应用，主要研究成果分别发表在该领域国际顶尖学术刊物上，例 *Metall. Mater. Trans. A*、*Acta Mater.*、*Comp. Mater. Sci.*、*Int. J. Heat and Mass Transfer* 等。共发表学术论文 200 余篇，SCI 收录 90 余篇，合作专著 10 部。鉴于吴孟怀教授在数值模拟领域的突出贡献，奥地利 Styria 州政府授予他 2010 年"Grundlagenforschung － 基础研究"年度荣誉奖(该奖项每年 1 次，每次只限 1 人)。

材料热成形模拟仿真技术在航天液体动力领域的应用需求及现状

李护林，杨欢庆，王　琳，郑　伟

（西安航天发动机厂）

摘　要：铸造、锻造、焊接、热处理等材料热成形技术是液体火箭发动机的主要工艺技术之一，其技术水平和产品质量直接关系到型号产品研制的先进性和可靠性。随着液体火箭发动机向高性能、轻质化的方向发展，大量的新材料和新结构应用于发动机上，给铸造、锻造、焊接及热处理等材料热成形工艺带来诸多挑战。传统的"经验+实验"材料成形工艺方法，由于周期长、成本高等缺点，已无法适应新型液体火箭发动机研制生产的要求。材料热成形模拟仿真技术通过对材料铸造、锻造、焊接、热处理过程中流场、温度场进行模拟，能够对成形过程中产品缺陷进行预测，达到优化成形工艺参数的目的，可大幅降低成本，缩短生产周期，在航天液体动力领域具有广泛的应用前景。本文分析了铸造、锻造、焊接、热处理等材料热成形模拟仿真技术在液体火箭发动机上的应用优势，综述了其在液体火箭发动机上的应用现状，并提出了上述技术在航天液体动力领域的后续应用需求。

关键词：材料热成形技术；模拟仿真技术；液体火箭发动机

Application Demands and Status of Material Thermal Forming Simulation Technology in Aerospace Liquid Propulsion Field

LI Hu-lin, YANG Huan-qing, WANG Lin, ZHENG Wei

（Xi'an Space Engine Factory）

Abstract: As one of the major technologies for liquid rocket engine manufacturing, the level of thermal forming technologies such as casting, forging, welding and thermal treatment and obtained product quality are directly related to the advancement and reliability of product development. Since high performance and light weight are the development trends of liquid rocket engines, a lot of new materials and new structures have been applied on liquid rocket engines, which brings many challenges to thermal forming technologies. Owing to long period and high costs, conventional "experience + experiment" material forming method could not satisfy the requirements of new liquid rocket engine development and manufacturing. Applying material thermal forming simulation technology, product defects could be predicted with the simulation of flow field and temperature field during casting, forging, welding and thermal treatment. With the advantages of forming parameters optimization, significant reduction of production costs and period, thermal forming simulation technology has

第一作者：李护林，男。液体火箭发动机制造工艺。date003@163.com。

通讯作者：杨欢庆，男。精密铸造及铸件后处理。18966827929@163.com。

a wide application prospect in aerospace liquid propulsion field. Based on the analysis of application advantages of simulation technology of thermal forming technologies such as casting, forging, welding and thermal treatment, the paper summarized its application status on liquid rocket engines and proposed further application demands in aerospace liquid propulsion field.

Key words: material thermal forming technologies; simulation technology; liquid rocket engine

李护林,研究员,我国液体火箭发动机专家,陕西省有突出贡献专家,中国航天科技集团公司学术技术带头人,集团公司首席工艺专家,集团公司特种加工工程技术中心副理事长、享受国务院政府特殊津贴,现任西安航天发动机厂总工程师兼总工艺师。长期从事载人航天、探月工程运载火箭发动机研制,在航天尖端技术创新方面取得了丰硕成果。作为液体火箭发动机研制的工艺技术领军人物,共参与或主持完成了 30 余项工艺攻关课题研究任务,为提高型号产品工艺技术水平及工艺稳定性做出了突出贡献。多项成果获得国防科学技术奖、军队科技进步奖、教育部技术发明奖等省部级以上奖项。2004 年度获集团公司优秀党员荣誉称号,2010 年获集团公司工艺工作先进个人荣誉称号。

三峡区域能源装备先进成型工艺与装备研究

赵春华[1]，吴海华[1]，丰　平[2]，陈从平[1]，石增敏[2]

(1.三峡大学 机械与动力学院；2.三峡大学 材料与化工学院)

摘　要：针对当前国内先进成形制造工艺和装备学科存在的基础研究不系统、不深入的共性状况，以国家、三峡区域和水电机械行业需求为目标，开展先进成形制造工艺与装备研究，以期提升石墨储能器件、胶结/大型焊接结构件、金属陶瓷刀具成形制造基础理论水平，并研制相关工艺装备。本论文方向主要围绕着增材制造及其工程应用技术、结构成形理论及控制技术和金属陶瓷刀具成型技术等三个方面开展研究。

关键词：石墨器件增材制造；金属陶瓷刀具成形制造；结构成形

Research on Advanced Molding Process and Equipment of Three Gorges Area Energy Equipment

ZHAO Chun-hua[1], WU Hai-hua[1], FENG Ping[2], CHEN Cong-ping[1], SHI Zeng-min[2]

(1. Department of Mechanics and Power Engineering, China Three Gorges University;
2. Department of Material and Chemistry, China Three Gorges University)

Abstract: In view of the common status of manufacturing technology and equipment research in basic research is not systematic and in-depth in current domestic, China Three Gorges University meet the region and hydro mechanical industry needs as the goal, to carry out advanced forming manufacturing technology and equipment research, in order to improve the graphite energy storage device, cementation and large welded structure, metal ceramic tool manufacturing level of basic theory, and the development of related technology and equipment. This paper mainly introduces some research work about the increase material production and its engineering application technology, structure forming theory and control technology and metal ceramic cutting tool technology and other three aspects to carry out research of our research groups.

Key words: graphite device manufacturing; metal ceramic cutting tool forming manufacturing; structure forming

基金项目：国家自然科学基金资助项目(51475266)。

第一作者：赵春华，女。机械制造。zhaochunhua@ctgu.edu.cn。

　　赵春华,教授,博士生导师,三峡大学"151"学术带头人。吉林省长春市人,1994年毕业于浙江大学机械制造及其自动化专业。2000年9月毕业于浙江大学CAD/CG专业,获工学硕士学位。2005年12月毕业于武汉理工大学载运工具运用工程专业,获工学博士学位。长期从事机械设计制造及其自动化的教学与科研工作,近5年来,主要从事故障诊断、润滑油分析、数字化设计及其制造等方面的研究工作。现任三峡大学机械与动力学院教授,机械设计制造及自动化专业硕士生导师。中国机械工程学会高级会员,IEEE高级会员,国家自然科学基金项目评审专家,*Chinese Science Bulletin*期刊审稿人。承担国家自然科学基金资助项目(Nature Science Fundation of China,资助号:51075234)"风电机组关键部件的失效机理与寿命预测研究";承担水电机械设备设计与维护湖北省重点实验室开放基金项目"基于WSN的塔式起重机安全预警技术研究"(编号2010KJX07);承担并完成湖北省自然基金项目"基于无线传感器网络的机群状态监测关键技术研究"。参加了国家级项目有国家自然科学基金资助项目"摩擦学系统状态特性的智能化描述方法的研究"(编号:5027511),国科金外资助中澳合作项目"基于监测信息的摩擦学和动力学的耦合关系研究"(第0311120207号),"十五"国家重大装备项目"300方首冲式挖泥船油液监测和故障诊断专家系统的研制"等项目的研究工作,并在其中担任主要负责人。在国内外刊物上发表论文100余篇,三大检索60篇次。

Gleeble 物理模拟系统及其新近进展

赵 奇，陈伟昌，David Ferguson

(Dynamic Systems Inc., P.O. Box 1234, Route 355 Poestenkill)

摘 要：简要回顾了物理模拟的发端及发展，介绍了 Gleeble 物理模拟系统及其在材料研发、性能测试以及工艺模拟方面的应用，特别是其对热成型技术进步的促进作用。总结了市场对物理模拟技术的新需求及 DSI 在物理模系统智能化、加强并拓宽对铸造和其他金属加工工艺模拟能力的一些新的尝试。最后，向大会隆重推出 DSI 的最新智能单机物理模拟器：Gleeble Welding Simulator(GWS)。

关键词：Gleeble；物理模拟；模拟器；GWS

Gleeble Physical Simulation and Recent Advances

ZHAO Qi, CHEN Wei-chang, David E. Ferguson

(Dynamic Systems Inc., P.O. Box 1234, Route 355, Poestenkill)

Abstract: After a brief review the history of physical simulation development various Gleeble physical simulator technologies and their capabilities are introduced. Applications in materials studies, materials testing, and processing simulation and the critical impacts of simulation on materials technology development and manufacturing technology advancements are presented. The new development of a highly intelligent physical simulation systems to meet the increasing demands of materials researchers in creating new technologies for simulating casting, deformation and other material processing processes is shown. Finally, the latest highly-intelligent Gleeble simulator: the standalone Gleeble Welding Simulator (GWS) is introduced and briefly described.

Key words: Gleeble; physical simulation; simulators; GWS

第一作者：赵 奇，男。粉末冶金。Jim.zhao@gleeble.com。

Qi Zhao, Ph.D.,
Senior Metallurgist
Dynamic Systems Inc.
P.O. Box 1234, Route 355
Poestenkill, NY 12140, USA
Jim.zhao@gleeble.com
+1 (518) 238-8180

Dr. Qi Zhao received a BS in Powder Metallurgy from Northeastern University, China in 1982, an MS in Materials Science from the Institute of Metal Research, Chinese Academy of Science in 1985, and Ph.D. in Metallurgy from the Massachusetts Institute of Technology in 1992. After working as a postdoctoral research fellow at MIT, he joined Hitchiner Manufacturing Company in 1993 as a project engineer. A year later, he joined Metal Casting Technology, a joint venture company between Hitchiner and General Motors, as a process engineer and later on as a member of the senior technical staff. In 2008, Dr. Zhao started Novarials Technology as cofounder and CEO. From 2010 to 2015, he worked as a senior casting metallurgist at GE's Global Research Center. In early 2015, he joined Dynamic Systems Inc. as a senior metallurgist focused on supporting and enhancing Gleeble sales in China and other Asian markets.

Through more than 20 years of experience in industrial R&D, manufacturing production, commercial development, and entrepreneurship, Dr. Zhao has continued to broaden his areas of technical expertiseand interest, including metal casting(IC, lost foam, sand, semi-solid, permanent mold, SX/DS), powder metallurgy(HIPing, MIM, PIM), diffusion bonding, 3D printing, welding, hot working (forging, rolling, extrusion, wire drawing), and continuous casting. He has received 28 US and international IP awards, coauthored 12 peer-reviewed technical papers. Other awards and recognition include the following:

1. As GE retiree nominated for GE GRC Whitney Award 2015 for Low Cryogenic Magnet MR System Development Program 2015

2. GE GRC High Temperature Alloy & Processing Lab ABOVE & BEYOND Award 2014

3. GE GRC Award to Inventors 2013

4. GE GRC Low Cryogenic Magnet MR System Development Program: Casting Expertise Award 2012

5. GE GRC PP20 Turbine Center Frame Fairing Casting Project: Casting Expertise Award 2012

6. NFS SBIR Phase II Award (Project# 1026642) ($500,000), 2010

7. NFS SBIR Phase I Award (Project#0910419) ($100,000), 2009

8. American Foundry Society's 2005 Howard F. Taylor Award recognizing the research paper with the greatest long term technical significance to the cast metals industry

9. American Foundry Society's Division 11 Best Paper Awards in 2003, 2005, and 2006

材料成形模拟仿真技术新进展

周华民，柳玉起，周建新

（华中科技大学 材料成形与模具技术国家重点实验室）

摘　要：材料成形模拟是华中科技大学材料成形与模具技术国家重点实验室的重要研究方向。本文简要介绍了该方向研究基地与团队的基本情况，重点阐述面向工程需求的该方向最新研究进展，包括：注射成形的多尺度模拟、铸造成形的全流程模拟、冲压成形的高精度模拟、焊接成形的高效率模拟，并分别给出了各类模拟新技术的应用案例与效果。

关键词：注射成形；铸造成形；冲压成形；焊接成形；模拟仿真

New Progress in Material Processing Simulation Technology

ZHOU Hua-min, LIU Yu-qi, ZHOU Jian-xin

（State Key Laboratory of Materials Processing and Die & Mold Technology,

Huazhong University of Science and Technology, Wuhan 430074, China）

Abstract: Material processing simulation is one of the main research fields of State Key Laboratory of Materials Processing and Die & Mold Technology, Huazhong University of Science and Technology. In this paper, firstly, basic information of the research centers and teams is introduced. Then, the new research progresses which are engineering demand oriented are expounded, including multi-scale simulation of injection molding, whole process simulation of cast forming, high precision simulation of stamping forming, and high efficiency simulation of welding forming. Last, application cases and its effects of each new simulation technology are also given, respectively.

Key words: injection molding; casting; stamping; welding; simulation

周华民，1974 年 7 月生。1992—1996 年华中理工大学本科（锻压专业），1996—2001 年华中科技大学直攻博（材料加工工程专业），获博士学位，留校工作。2004 年晋升教授，2007—2008 年美国 Wisconsin 大学高级研究学者。现任华中科技大学材料学院副院长、材料成形与模具技术国家重点实验室副主任，兼任国务院学位委员会学科评议组成员、中国模具工业协会数字化智能化技术部主任等，担任 Polymer-Plastics Technology and Engineering（美国 SCI 源刊）、Journal of Mechanical Engineering Science（英国 SCI 源刊）等期刊编委。主要从事塑料成形工艺、材料成形模拟、智能成形装备的研究。发表论文 200 余篇，其中 SCI 收录 80 余篇。2002 年获国家科技进步二等奖，2007 年获国家科技进步二等奖，2010 年获国家自然科学二等奖。2014 年获评教育部长江学者特聘教授，2013 年获中国青年科技奖，2011 年获国家杰出青年科学基金，2007 年获霍英东青年教师基金。

基金项目：国家杰出青年科学基金资助项目（51125021）。

第一作者：周华民，男。塑料注射成形模拟。hmzhou@hust.edu.cn。

第八届材料加工物理模拟与数值模拟国际会议和材料加工物理模拟与数值模拟国际联合会介绍

曾建民

(材料加工物理模拟与数值模拟国际联合会秘书长)

从 1990 至 2014 年,以中国机械工程学会的名义共举办了七次"材料与热加工物理模拟及数值模拟国际学术会议"。经过 20 多年的努力,本会议已成为国际上具有特色并有影响力的系列学术会议。第一次会议于 1990 年 7 月在哈尔滨召开,有 4 个国家的 80 多名代表参加。之后,考虑到物理模拟和数值模拟的密不可分的学术关系和工程联系,广泛征求国内外同行的建议,将学术会议名称由"物理模拟"扩展为"物理模拟及数值模拟",于 1997 年 3 月在海南召开了第二次国际会议,此次会议有 8 个国家的近 110 名代表参会;第三次会议于 1999 年 10 月在北京召开,有 16 个国家的 170 多名代表参加;第四次会议于 2004 年 5 月在上海召开,共有 22 个国家的 230 名代表参加了会议;第五次会议于 2007 年 10 月在郑州召开,有 35 个国家的 530 多名代表;第七次会议于 2013 年 6 月在芬兰举行,经中国、美国、芬兰、俄罗斯等国家代表的建议,来自 23 个国家的数百名代表经过充分协商,一致同意成立一个国际性的学术组织——"材料加工物理模拟及数值模拟联合会"。基于我国在材料物理模拟及数值模拟国际学术交流方面的突出贡献,一致同意此联合会总部设在中国,挂靠于中国机械工程学会材料分会,并选举中国牛济泰教授任联合会首任主席。美国 Ferguson David 博士、俄罗斯 Chumachenko E. N.教授、日本 Lino Hitoshi 博士担任副主席。会议决定建立国际性的物理模拟和数值模拟学术刊物,建立联合会网站,制定有关技术标准,使该领域的技术交流步入规范化和常态化。2014 年,专业委员会在北海召开年会,确定第八届"材料与热加工物理模拟及数值模拟国际学术会议"将于 2016 年 10 月在美国美丽的海滨城市西雅图举行。

第八届"材料与热加工物理模拟及数值模拟国际学术会议",由"材料与热加工物理模拟及数值模拟国际联合会"主办,会议主席为俄罗斯科学院院士,哈尔滨工业大学牛济泰教授。诺贝尔奖获得者 Shechtman 将在大会上做特邀报告。

国际联合会采用国际上通行的会员制。从事材料研究领域的人员,包括大专院校教师和研究生、研究院所和企业的科技人员,均可在志愿基础上成为联合会会员。联合会隶属中国科协管理,挂靠在中国机械工程学会材料分会。

联合会网站:http://www.ifpns.net

参加联合会的办法,登录上述网站后,在 newsletter 栏目下,下载中文或英文的申请入会表格和中英文的联合会章程。将填写好的表格通过电子邮件发给:ifpns130619@163.com

经联合会批准后,即发给会员证书。会员可以在 member 栏目中注册,下载联合会的文献资料。

联合会会员除了可以下载联合会的文件资料外,在联合会主办的国际会议、联合会举办的各种业务培训等方面还可享有注册费的优惠。

美国华盛顿的西雅图是美国太平洋沿岸西北部最大的城市,这里是波音飞机的故乡,也被叫作"飞机城"。微软公司也坐落在西雅图市。会议期间,将与美国的大学、研究院所和企业进行各种学术交流活动。

基于 CAD/CAE 集成的异型材挤出模头结构优化设计研究

李 力，唐红涛*，郭顺生，黄 浪，罗易彬

(武汉理工大学 机电工程学院)

摘 要：针对挤出模头三维参数化设计方法的不足以及实际应用中挤出模头积垢的问题，提出了整体式设计方案，基于整体型腔和整体模架两大模块进行设计。以 UG 为平台，以 VC 为开发工具，根挤出模头参数化设计流程，构建基于 UG 的异型材挤出模头 CAD 系统，实现应用于 CAE 的参数化模型快速建立。采用有限元数值分析方法，获得熔体流动的速度场，总结挤出模头结构参数如压缩比、平直段长度等对挤出流动的影响规律。引入最优化设计理论，以模头各功能段主要结构参数为设计变量，流道最大流量为约束变量，模头出口处型材截面上各个子区域平均流速的均方差为优化目标函数，建立优化数学模型。根据模型得到最优解并应用于 CAD 系统指导模型的建立。经优化设计后熔体流动的平衡性有显著提高，具有重要的理论意义和实际应用价值。

关键词：整体式设计；参数化模型；速度场；数学模型；优化

Study on Structural Design and Optimization of Profile Extrusion Die Based on CAD/CAE Integration

LI Li, TANG Hong-tao*, GUO Shun-sheng, HUANG Lang, LUO Yi-bin

(Wuhan University of Technology Mechanical and Electrical Engineering College)

Abstract: According to lack of parametric design methods for the design of extrusion die and the fouling problem in practical application, the integrated design method is put forward in this paper based on two modules of integral flow channel and integral die set. With VC as the development tool, the CAD system for design of extrusion dies is developed on the basis of the design process, which realizes the quick establishment of parametric model for CAE. Using finite element numerical analysis method, the velocity field of melt flow is obtained and the influence of die structure parameters such as compression ratio, length of flat section is summarized. Using main structural parameters of the functional section as design variables, maximum flow rate as the constraint variable, average flow velocity of each sub region as the optimization objective function, the optimization design theory is applied and the mathematical model is established. According to the model, the optimal solution is obtained, and the parameters are used to guide the establishment of structure through CAD system. After optimization design, the equilibrium of melt flow is significantly improved, which has important theoretical and practical value.

Key words: integrated design; parametric model; velocity field; mathematical model; optimization

基金项目：湖北省科技支撑计划项目(2014BAA032)。

第一作者：李 力(1992—)，男，硕士研究生。模具 CAD/CAE 研究。lilwut@163.com。

通讯作者：唐红涛(1987—)，男，讲师，博士。数字化模具设计研究。tanghongtaozc@163.com。

郭顺生(1963—)，男，教授，博导。CAD/CAM/PDM/ERP 基础理论及应用研究。

塑料异型材挤出模 CAD/CAE 技术及冷却系统优化设计研究

黄 浪，唐红涛*，郭顺生，李 力，郭 乔

(武汉理工大学 机电工程学院)

摘 要：针对塑料异型材结构复杂、分型繁琐、孔多易干涉、定型板块重复性设计等问题，设计了一套智能分型算法，自动创建分型板块。基于 UG 平台，开发了塑料异型材挤出定型模 CAD 系统。通过分析和计算定型模内型材冷却过程的数学模型、初始条件和边界条件，采用 ANSYS 有限元分析软件的瞬态热分析单元对塑料异型材定型冷却过程进行模拟，获得了定型模内型材的瞬态温度场。同时，以定型模冷却的均匀性为优化目标，通过对定型模温度场的分析，基于现有的 CAD 系统对定型模进行重构，实现 CAD/CAE 集成，达到对塑料异型材冷却水道优化设计的目的。

关键词：CAD/CAE；冷却系统；优化设计；定型模；温度场

Research on the CAE Technology of Extrusion Dies and Optimization Design of Cooling System for Plastic Profiles

HUANG Lang, TANG Hong-tao, GUO Shun-sheng, LI Li, GUO Qiao

(School of Mechanical and Electrical Engineering, Wuhan University of Technology)

Abstract: A method of intelligently typing is presented, by which sub-type plate is created automatically, solving the problems of complex structure, typing tedious, holes interference, plate repetitive design for the calibrators of plastic profile extrusion. Based on UG, a calibrator CAD system for plastic profiles is developed. With the mathematics models and the original and boundary conditions of the flowing process within runner and the cooling process in calibrators analyzed and computed, and the calibration and cooling process of an extrusion plastic profile simulated with the transient thermal analysis unit of ANSYS software in the combination of Newton's cooling law and the clamp force method, the transient temperature field of the profile in the calibrator is obtained. At the same time, in order to optimize the uniformity of the mold cooling, according to analysis of the temperature field of the profile, CAD/CAE integration is achieved with the reconstruction of the existing CAD system, realizing the purpose of optimizing the design of cooling system for the plastic profile.

Key words: CAD/CAE; cooling system; optimization design; calibrator; temperature field

---

基金项目：湖北省科技支撑计划项目(2014BAA032)。

第一作者：黄 浪(1991—)，男，硕士研究生。模具 CAD/CAE。hllwut@163.com。

通讯作者：唐红涛(1987—)，男，博士，讲师。数字化模具设计。tanghongtaozc@163.com。

郭顺生(1963—)，男，教授，博导。CAD/CAM/PDM/ERP 基础理论及应用研究。

钢材不稳态加热时间的研究与计算

王 东，首天成

(武汉纺织大学 机械学院)

摘 要：基于一维不稳态导热分析解，提出一个在第三类边界条件下的物体二维和三维不稳态加热时间的计算方法。其中，在求解超越方程 $\mu/B_i = ctg\mu$ 和 $J_0(\beta)/J_1(\beta) = \beta/\beta_i$ 的根时，应用插值法求解甚为方便。

关键词：钢材；不稳态加热；插值法；时间

Calculation of Heating Time under Transient Heat Conduction in Steel Plate

WANG Dong, SHOU Tian-cheng

(School of Mechanical Engineering, Wuhan Textile University)

Abstract: Based on the one dimensional analytical solution, a new calculation of two and three dimensional transient heating time is provided under the third boundary conditions. In solving of the root values with $\mu/B_i = ctg\mu$ for a large slab and $J_0(\beta)/J_1(\beta) = \beta/\beta_i$ for a cylinder, the Newton insert method may be more applicable for the solution.

Keyword: steel；transient heating；insert method；time

1 前言

关于物体材料的不稳态加热计算，一则，已知加热时间求解该时刻物体内的温度分布；另为，既定物体内某点(通常是中心点)温度时求解所需加热时间。姑且称前者为"正计算"，后者为"逆计算"。

在工程技术上，对于一维(例如板材加热)的正计算或逆计算，且不论计算的精度都可用其分析解的线算图方便求得。对于二维和三维问题的正计算，可用一维解的组合直接得到。但是对于逆计算加热，则较费周折，例如文献[1]介绍一种"试凑"的方法。然而，试凑法理解上直观，实际运作即有盲目性，尤其对三维问题很难经二、三次就能够试凑成功。也可以用数值法正计算来逆推计出加热时间，但是数值法运算涉及数据繁多而不便于手算，且每一步长的离散点温度分布只用作时间的过渡，并非是问题的直接答案。故本文提出一个应用一维不稳态分析解的组合来计算二维和三维问题加热时间的方法。

第一作者：王 东(1963—)，男，教授。机械工程材料。

2 基本方法

传热学已阐明,对于厚短方柱体(三维)和圆柱体(准二维)等规则物体其在第三边界条件下加热,且有均匀的初始温度分布时,可利用一维问题的组合求解[2]。一般地,对于三维问题其关系式为

$$H = H(x, y, z, \tau) = \theta_x(x, \tau) \cdot \theta_y(y, \tau) \cdot \theta_z(z, \tau) \tag{1}$$

式中 $H = \dfrac{t - t_f}{t_0 - t_f}$ ——物体内某点的相对过余温度;

$\theta_f = \theta_j(j, \tau) = \dfrac{t(j, \tau) - t_f}{t_0 - t_f}$ ——作为一维问题处理时,某坐标方向 $j(j = x, y, z, \tau)$ 的相对过余温度,其有式(2)或式(4)的分解关系式;

$t = t(x, y, z, \tau)$ ——在 τ 时刻物体坐标点 (x, y, z) 的温度;

$t(j, \tau)$ ——在 τ 时刻,作为一维问题处理时某坐标 $j(j = x, y, z)$ 的温度;

t_f ——加热介质温度;t_0 ——物体初始温度。

对于物体物性参量为常量(对于实际计算,通常取加热温度范围内的平均值)的一维加热,其在第三类边界条件下平壁的分析解为:

$$\theta_j = \sum_{n=1}^{\infty} \frac{2 \sin \mu_n}{\mu_n + \sin \mu_n \cos \mu_n} \cos(\mu_n \frac{j}{S}) e^{-\mu_n^2 F_0} \tag{2}$$

式中 S ——(加热方向)平壁厚度;对两侧对称加热时为 $1/2$ 厚度

$F_0 = \alpha \tau / s^2$ ——傅立叶准数(α 为导温系数);μ 为超越方程(3)的根

$$\mu / B_i = ctg\mu \tag{3}$$

式中 $B_i = \alpha s / \lambda$ ——毕欧准数;

α ——对流(或综合)换热系数;

λ ——物体材质的导热系数。

因为该方程的解有无穷多个根值,且每个根植对应于式(2)的和式,故在式(2)中 μ 的下标需冠以"n"以表示各个根植。圆柱体径向:

$$\theta_r = \theta_r(r, \tau) = \sum_{n=1}^{\infty} \frac{2J_1(\beta n)}{\beta n[J_1^2(\beta n) + J_0^2(\beta n)]} \cdot J_0(\beta n \frac{r}{R}) e^{-\beta_n^2 F_0} \tag{4}$$

式中 R——圆柱体半径;

$F_0 = \alpha \tau / R^2$ ——径向傅立叶准数;$B_i = \alpha R / \lambda$ ——径向毕欧准数。

β 是方程式(5)的根植,亦有无穷多个,并一一对应于式(4)和式中的 βn

$$\frac{J_0(\beta)}{J_1(\beta)} = \frac{\beta}{B_i} \tag{5}$$

式中 $J_0(\beta)$ ——零阶贝塞尔函数,其展开式为

$$J_0(\beta) = 1 - (\frac{\beta}{2})^2 + \frac{1}{(2!)^2}(\frac{\beta}{2})^4 - \frac{1}{(3!)^2}(\frac{\beta}{2})^6 + \cdots \tag{6}$$

$J_1(\beta)$ ——一阶贝塞尔函数,其展开式为

$$J_1(\beta) = \frac{\beta}{2}[1 - \frac{1}{2!}(\frac{\beta}{2})^2 + \frac{1}{2!3!}(\frac{\beta}{2})^4 - \frac{1}{3!4!}(\frac{\beta}{2})^6 + ...] \qquad (7)$$

对于实际问题的计算,可只取 $J_0(\beta)$ 和 $J_1(\beta)$ 展开式的前三项而略去其后各项,即有足够的精确性。

式(2)和式(4)都是迅速收敛的级数。若问题是当加热已扩散到物体中心处时,其内部的温度分布已摆脱初始温度分布的影响,即加热已进入第二期正规加热阶段,则此时,对平壁有 $F_0 \geq 0.3$,对圆柱体 $F_0 \geq 0.25$,在这种情况下,亦有式(2)或式(4)计算时,可取级数的第一项而略去其余各项使级数误差不超过 1% ;对于工程计算,一般地,当 $F_0 \geq 0.2$ 时,只取级数的第一项弃去其后各项已够精确。这样,事实上直接应用分析解计算时其困难处则在于有式(3)或式(5)解出 μ 和 β 值(尽管只取 1 个根植),它们通常采用图解法或迭代法求解。本文则应用前人研究所得的 μ 和 β 的数值表而进行插值计算求解,甚为简单方便。其具体方法见下节。μ 和 β 的数值表见文献[3]。

3 计算实例

一块断面尺寸 2($s_1 \times s_2 \times s_3$)=400mm×500mm×1500mm 初始温度 t_0=25$^o C$ 的钢坯料,置入炉温 t_f=1350 $^o C$ 的炉中各表面对称加热。该钢坯材质的平均导热系数 $\lambda = 26W/(m \cdot ^o C)$,导温系数 $a = 6.6 \times 10^{-6} m^2/s$ 。炉气对各表面的平均综合换热系数 $a = 265W/(m \cdot ^o C)$ 。试求当钢坯中心温度达 $t_c = 900 ^o C$ 时所需加热时间,以及此时表面上最高温度点的温度。

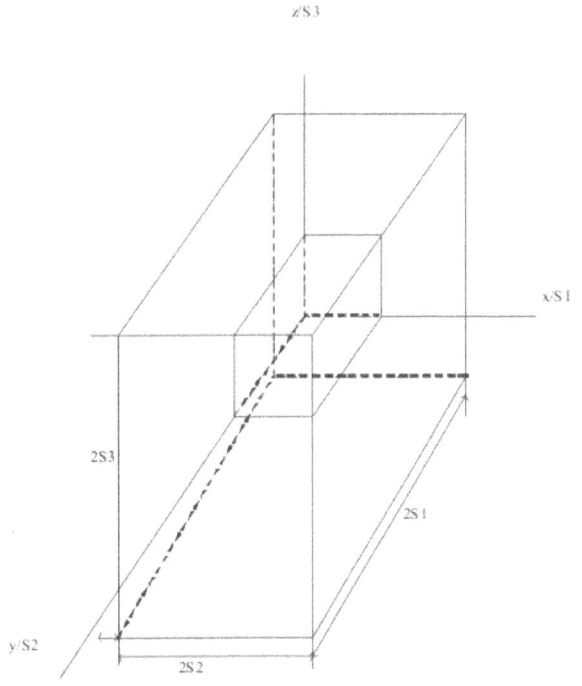

图 1 钢坯加热所取(无因次)坐标

本问题属于 3 个平壁(一维)问题的组合。由于对称加热故只研究钢坯的 1/8 部分便可,余 7/8 部分类推,其坐标系统示于图 1。

依题意,对中心点,根据式(1)有:$H_c = H(0,0,0,\tau) = \theta_x(0,\tau) \cdot \theta_y(0,\tau) \cdot \theta_z(0,\tau)$,而

$$H_c = \frac{t_c - t_f}{t_0 - t_f} = \frac{900 - 1350}{25 - 1350} = 0.3396 \quad \text{各个坐标方向的毕欧准数分别为:}$$

$$B_{i,x} = \alpha s_1 / \lambda = 265 \times 0.2 / 28 = 1.893 \quad, \quad B_{i,y} = \alpha s_2 / \lambda = 265 \times 0.25 / 28 = 2.366 \quad,$$

$$B_{i,z} = \alpha s_3 / \lambda = 265 \times 0.75 / 28 = 7.098$$

下面,应用 μ 的第一个根植的数值表,按照牛顿前插法(也可以采用后插法,视插值点的插值区间中靠"前"后靠"后"的位置而定),解出本问题的 3 个 μ 值。对 μ_x:从该数值表中对应于 $B_{i,x} = 1.893$ 的领近挑取出 4 个等距的 μ 值,列于表1;并将该 4 个数据的一阶,二阶和三阶差分值计算列在表右侧各列。

表1 4个等距 μ 值

序号	B_i	u	$\triangle u$	$\triangle^2 u$	$\triangle^3 u$
3	0.5	0.0533			
2	1.0	0.8603	0.2070		
1	1.5	0.9882	0.1279	−0.0791	
0	2.0	1.0769	0.0887	−0.0392	0.0399

参量 t 为: $t = \dfrac{B_{i,x} - B_{i,0}}{h} = \dfrac{1.893 - 2.0}{0.5} = -0.214$

式中 $h = 0.5$ —— B_i 的间距步长;代入牛顿后插公式:

$$\mu_x = \mu_0 + \left\{ \triangle \mu_0 + \left[\frac{\triangle^2 \mu_0}{2} + \frac{\triangle^3 \mu_0}{6}(1+t) \right](1+t) \right\} t$$

$$= 1.0769 + \left\{ 0.0887 + \left[\frac{-0.0392}{2} + \frac{0.0399}{6}(1-0.214) \right](1-0.214) \right\}(-0.214)$$

$$= 1.063 \text{ 弧度}(=60.75°)$$

将 $\mu_x = 1.0603$ 代入式(3)进行验算,其等号左端为 $\dfrac{\mu_x}{B_{i,x}} = \dfrac{1.0603}{1.893} = 0.5601$,

等号右端为 $ctg\mu_x = ctg60.75° = 0.5600$,验算表明,取三阶差分插值可使解具有满意的精确,故得 $\mu_x = 1.0603$ 。

用同样的方法得到: $\mu_y = 1.1120$, $\mu_z = 1.3789$ 。由于本问题是既定中心点,即 $j = x, y, z = 0$ 故 $\cos(\mu_j j / s) = 1 (s = s_1, s_2, s_3)$ 。

为书写方便,令:

$$U_j = \frac{2 \sin \mu_j}{\mu_j + \sin \mu_j \cos \mu_j} (j = x, y, z), \text{将 3 个 u 值分别代入上式,得到:}$$

$$U_x = 1.174, U_y = 1.188, U_z = 1.254 。$$

前已说明,本问题由 3 个一维问题的组合而求解,但必须是通过待求的加热时间来沟通并约束这一组合关系,即各个坐标方向的傅立叶准数 $F_{0,j}(j = x, y, z)$ 必定是由同一时间 τ 所组成的,即为 $F_{0,x} = a\tau / s_1^2$, $F_{0,y} = a\tau / s_2^2$ 及 $F_{0,z} = a\tau / s_3^2$,据此则得到等式:

$F_{0,x} \cdot s_1^2 = F_{0,y} \cdot s_2^2 = F_{0,z} \cdot s_3^2$ ，由此得：

$F_{0,x} / F_{0,z} = (s_3 / s_1)^2 = (750 / 200)^2 = 14.06$ ， $F_{0,y} / F_{0,z} = (s_3 / s_2)^2 = (750 / 250)^2 = 9$

故得到： $F_{0,x} = 14.06 F_{0,z}$ ， $F_{0,y} = 9 F_{0,z}$

将上述关系代入式（1）称为： $H_c = U_x U_y U_z \cdot \exp\left[-(14.06\mu_x^2 + 9\mu_y^2 + \mu_z^2)F_{0,z}\right]$

对上式两端取对数，并代入数值计算出：

$$F_{0,z} = \frac{In(U_x U_y U_z / H_c)}{14.06\mu_x^2 + 9\mu_y^2 + \mu_z^2} = 0.057$$

这样： $F_{0,x} = 14.06 \times 0.057 = 0.801$ ， $F_{0,y} = 9 \times 0.057 = 0.513$

三个坐标方向的 F_0 值中有二个的大于 0.3，足以说明，加热已扩散至物体中心处而进入正规加热期。据此，所求的加热时间 τ 为：

$\tau = F_{0,z} \cdot s_3^2 / a = 4858s \approx 1.35h$

如果本问题作为二维问题处理时，即图 1 示的 4 个侧面抽象为在 z 方向无限长，式（1）成为：

$H_c = \theta_x(0,\tau) \cdot \theta_y(0,\tau)$

经计算，加热时间 $\tau \approx 1.34h$ 。由此看出，本问题的 z 方向的两个端面的加热居次要作用，且当 s_3 大于 s_1 和 s_2 5 倍以上时可忽略 z 方向的传热作为二维问题处理。

关于本问题的后一问题，显然在物体的表面棱角处一无因次坐标点（1,1,1）具有最高的表面温度。对此，根据式（1）则写成该处的相对过余温度 H_w 为：

$H_w = \theta_x(1,\tau)\theta_y(1,\tau)\theta_z(1,\tau)$ ，因为： $j / s = 1(j = x, y, z; s = s_1 s_2 s_3)$

故： $\cos(\mu_x \cdot x / s_1) = \cos 60.75^o = 0.4886$ ， $\cos(\mu_y \cdot x / s_2) = \cos 63.71^o = 0.4429$ ，

$\cos(\mu_z \cdot x / s_3) = \cos 79.00^o = 0.1908$

这样： $H_w = U_x U_y U_z \cos\mu_x \cos\mu_y \cos\mu_z \cdot \exp[-(14.06\mu_x^2 + 9\mu_y^2 + \mu_z^2)F_{0,z}] = 0.014$

由关系式： $H_w = (t_w - t_f)(t_0 - t_f)$ ，

从而得到该点温度： $t_w = t_f + (t_0 - t_f)H_w = 1331^o C$

由此可以看出，事实上当求出加热时间后，再由正计算可计算出在该时刻物体任何部位处温度。

4 结论

以上计算方法，通过对应用数值法计算来对照，其结果极为接近，从而证明这种方法的正确性。所需指出的是，所述方法是建立在加热已进入正规阶段这以基点之上的；而由加热扩散至物体中心处时的问题本身便可直观直接判断加热已进入正规阶段，由此而决定应用所述的方法来计算加热时间。

对于第一类边界条件，其边界温度为常量的加热（冷却）的时间计算，例如在大容器溶液介质中

淬火问题,这一方法亦适用,且无需求解超越方程式(3)或式(5),较第三类边界条件更容易些。本文所提出的计算过程脉络清晰,便于分析,这一方法颇适合于制定加热工艺时使用。

参考文献:

[1] 王补宣. 工程传热传质学[M]. 北京: 人民教育出版社,1993,87—89.

[2] B. 萨琴科. 传热学[M]. 北京: 高等教育出版社,1997,234—236.

[3] J. 威尔缔. 工程传热学[M]. 北京: 人民教育出版社,1993,116—117.

[4] 周筠清. 传热学[M]. 北京: 北京科技大学出版社,1995,45—47.

钢材热加工中 Newton 插值法的分析与应用

王 东[1]，黄俊鸣[1]，王春霖[2]

（1.武汉纺织大学 机械学院；2.武钢公司）

摘 要：阐述了应用 Newton 插值法求函数表所列数据以外的数据，建立插值曲线方程求解方程近似根等问题，文中所举 2 个实例都是钢材热加工技术中的一些较为典型的问题。

关键词：钢材；Newton 插值法；热加工；温度

Application of Newton Interpolation Method in Steel Heating Technology

WANG Dong[1], HUANG Jun-ming[1], WANG Chun-ling[2]

（1.School of Mechanical Engineering, Wuhan Textile University; 2.Wuhan Iron & Steel Co）

Abstract：This paper describes how to seek data exclusive of data listed in the function table and establish the interpolation curve equation in order to solve the approximate root of the equation. The two heating examples shown in the paper are all representative problems in the Steel heating technology.

Keywords：steel; newton interpolation; heating; temperature

1 前言

插值计算方法的基本思想是：构造某个简单函数用以逼近原函数，通过计算接近的关系式从而得到研究对象的近似值。逼近函数的类型有多种选法，但其基本上是代数多项式。这是因为数学中业已阐明，有相当广泛的函数可以用代数多项式逼近[1]。建立代数多项式也有多种方法，其中 Newton（牛顿）插值的法官法具有递推性，其组成很有规律性，方便于实际运算，并能应用较少的已知数据达到应用的精度。

包括牛顿插值法的诸多插值方法，在一般教本中有详细的论证。本文避开数学上的逻辑推理论证过程，就几个钢材热加工程技术上较典型的实际问题综合地提出牛顿插值法的应用。

2 Newton 插值法

已知函数 $y = f(x)$，如果它是给出的函数表，其 n+1 个节点 $x_0, x_1 \cdots x_n$ 处的函数值是 $y_0, y_1 \cdots y_n$；

第一作者：王 东（1963— ），男，教授。机械工程材料。

如果给出的是 $y = f(x)$ 的解析式,则在上述节点处取值 $y_0, y_1 \ldots y_n$,要求建立一个次数不超过 n 的多项式 $N_n(x)$,使 $N_n(x) = y_i$(i=0,1…n)。为此,只要在 n+1 个已知节点 x_i 以外再给一个节点 x,此时将点 x 也看作一个节点,推出逼近原函数 $f(x)$ 的牛顿插值多项式 $N_n(x)$ 为:

$$N_n(x) = f(x_0) + f[x_0, x_1](x - x_0) + f[x_0, x_1, x_2](x - x_0)(x - x_1) + \ldots$$
$$+ f[x_0, x_1, \ldots, x_n](x - x_0)(x - x_1)\ldots(x - x_{n-1}) \tag{1}$$

$N_n(x)$ 与原函数 $f(x)$ 之间的关系是:

$$f(x) = N_n(x) + R_n(x) \tag{2}$$

Rn(x)称为余项[2],造成余项的原因是用差商代替了微商。通过余项的计算可得到原函数 $f(x)$ 与逼近函数 $N_n(x)$ 之间的误差。但是实际计算余项颇为困难,尤其是在只给出函数表的情况下。然而,通过对余项的分析说明,余项的大小与插值节点的取法有关,所得的结论是:取与(欲插入的) x 距离最近的几个节点作为插值区间会使余项最小。此外,由差分误差传播的分析—稳定性问题的分析表明,对实际问题盲目追求高阶差分(或差商)并不可取。

大多数给出的函数表,或是全区间是等距的,或虽然全区间不等距而子区间是等距的。式(1)适用于等距和不等据节点的计算。当节点等距分布时,用差分代替差商从而可以避免多次除法便于计算,因而导出了牛顿前插公式和后插公式。在实际运作时,究竟要采用哪一个公式,应视插值点在插值区间的位置而定。

设给定的节点是由小到大排序,即 x₀<x₁<…<xₙ,并有等距步长 $h = x_i - x_{i-1}$ (i=1,2…n)。如果靠近 x₀ 处插值(前插),按照前述使余项为最小的思想,则应挑选邻近 x₀ 的节点 x₁,x₂…作为插值点。为方便计算,做变换 $x = x_0 + ht$,参量 t 为 $t = \dfrac{x - x_0}{h}$ (t>0),这样

$$N_n(x) = y_0 + \frac{t}{1!}\Delta y_0 + \frac{t(t-1)}{2!}\Delta^2 y_0 + \ldots + \frac{t(t-1)\ldots(t-n+1)}{n!}\Delta^n y \tag{3}$$

在计算各阶差分(或差商)时列出如例1(例2)示的表格甚为清晰方便。

3　冶金工程中的应用

3.1　曲线方程

在工程技术中,通常用最小平方和的方法拟合曲线。如前述、插值法是构照逼近函数,故可以将所有的已知函数点连接起来而构成"插值曲线"用以表述物理量间的定量关系。插值曲线与拟合曲线的功效相同,但是所建立的方法却不一样,即后者所构造的曲线不要求通过所有的已知点,只要求它与已知点的偏差平方和最小,故拟合曲线是交替靠近已知点。

据此,一般而言,当实验数据较少或没有适宜的函数曲线可以拟合时,采用插值曲线比较妥当。

例1:为研究转炉炉衬在高温下被侵蚀的情况,需查明转炉炉衬工作层在作业时的温度沿其厚度的分布。在150t转炉上约位于钢水面下30mm沿炉衬工作层15、75、132、172和600mm处埋设5只热电偶测温。对本问题而言,所关心的是由测知温度范围而推断耐火材料软化带的厚度。由于工作层只

有 600mm 厚,以及测温条件的恶劣,故没有必要也不可能埋设跟读的热电偶以获得较多点的温度。然而,既已获得 5 个实验数据,除满足问题需要外,不妨用它们开发另一个功用,即由此建立表述炉衬工作层的热积蓄量(即焓值),用以转炉热平衡。

该转炉第 6 次的出钢前,由上述部位所测得的 5 个温度读数,并取炉衬工作层表面 x=0 出的温度为 1600℃(等于钢液温度)作为补充读数得到 6 个数据,见表 1。按前述稳定性的思路,拟分成二段各建立起牛顿插值函数如下:

表 1　温度读数

x/m	t/℃	一阶差商	二阶差商	三阶差商
0	1600			
0.015	1356	16266.67		
0.075	1182	−2900	178222.27	
0.132	777	−7105.28		
172	763	−350	69641.86	
0.6	600	−380.84	−65.9	−132776.69

解:将上表中上斜线行上的值带入式(1),经整理,得到炉衬二段各自温度分布 $t = t(x)$ 为:

$$t_1 = N_2(x) = 1600 - 18940x + 178222.27x^2 \quad (0 \leqslant x \leqslant 0.172)$$

$$t_2 = N_2(x) = 2630.1 - 27562.46x + 119964.22x^2 - 13277.96x^3 \quad (0.172 \leqslant x \leqslant 0.6)$$

这样,工作层在出钢前单位体积的焓值 Q 为:

$$Q = \rho \left\{ \int_0^{0.075} c_1 t_1 dx + \int_{0.075}^{0.6} c_2 t_2 dx \right\} \quad \text{,kJ.m}^{-3}$$

式中 ρ —工作层耐火材料的密度,kg.m^{-3};

c_1、c_2—工作层耐火材料的(所在温度范围内)平均比热,kJ.kg^{-1}.℃$^{-1}$

3.2　牛顿—爱尔米特插值

例 2:物体的导热系数 a ,是表述物体在加热时温度在物体内扩散的性能,故 a 亦称热扩散率,它是温度的函数。文献[3]载有中碳钢 a =f(t)的实验曲线,示于图 1。

从图 1 可以看出,它是由 7 个实验数据的点连成的曲线;a 随温度的升高而降低,在 1000℃ 时为最低值,以后稍有回升。图 1 未给出曲线方程。

现在的问题是:如果对该中碳钢材进行加热计算时,例如对板材的加热,可按一维不稳定态处理,则要解偏微分

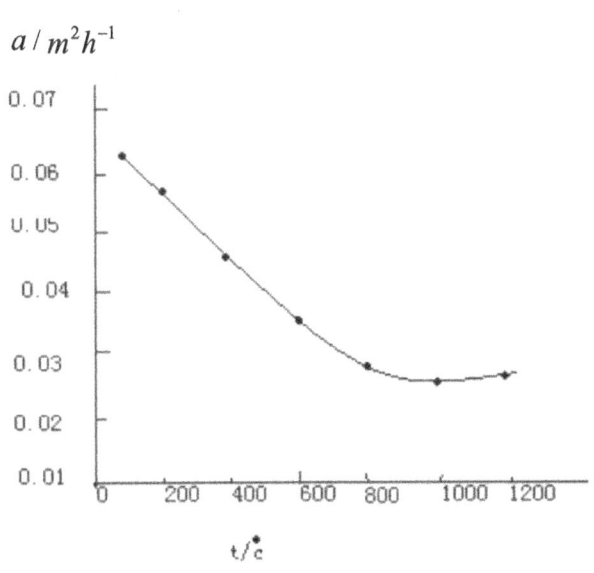

图 1　中碳钢的热扩散率 a 随温度 t 的变化

方程:$\frac{\partial t}{\partial \tau} = a\frac{\partial^2 t}{\partial x^2}$(其中 τ 为时间);如若得知钢板加热所在范围内的 a 的平均值 a_m,则用它作为常量取代微分方程中的变量 a,会使解该微分方程变得容易(用数值法求解也是这样)。因此,本问题为:根据图 1 的曲线点求其插值曲线方程,进而求出 a 的平均值 a_m。由图 1 读出的 7 个数据见表 2 如下:

表 2　温度读数

t /℃	125	200	400	600	800	1000	1200
a/ m^2h^{-1}	0.0525	0.04875	0.04	0.03	0.02625	0.0225	0.023

解:分析这 7 个数据的规律,结合前述稳定性的思想,宜采用较低等次的插值多项式,故拟分成 $125℃ \sim 400℃$,$400℃ \sim 800℃$ 和 $800℃ \sim 1200℃$ 的三段插值。但在 t=1000℃ 时为插值曲线的点,从而由直观直接判断除该处 a 的一阶微商为零,在建立插值曲线方程时理应考虑利用这一有利条件,其方法就是爱尔米特插值法。因此,所采用的方法是:对于第一段和第二段曲线各建立二次抛物线方程,而对第三段则应用牛顿—爱尔米特插值法建立三次函数曲线方程。关于第一段(a_1)和第二段(a_2),应用和例 1 相同的方法得出其 $a_1=N_2(t')$ 和 $a_2=N_2(t')$ 各为:

$a_1=0.02773t'^2-0.05939t'+0.05932$　$(0.125 \leqslant t' \leqslant 0.4)$

$a_2=0.07813t'^2-0.1281t'+0.07875$　$(0.4 \leqslant t' \leqslant 0.8)$

以上二式中的 $t'=10^3t$

爱尔米特插值多项式是带有微商的插值多项式,其建立方法是:在原有原有条件基础上,增加多少条件就增加多少项,以增加的条件来确定这些项的内容。牛顿-爱尔米特插值法,就是用式(1)或式(2)构造牛顿插值多项式,然后再加上所述的补充条件。

在计算方法[2]中给出了 m 个($m \leqslant n$)节点一阶微商值的牛顿-爱尔米特插值多项式的一般建立的方法。但是,本问题有些特殊,即第三段的三个节点中,只有一节点的一阶微商为已知,余两个节点的一阶微商为未知,因此不能直接套用该方法,需依据以下给出的 4 个条件,推出插值多项式,该 4 个条件见表 3 为:

表 3　条件读数

$t'(=10^3t)$	0.8	1.0	1.2
a	0.02625	0.0225	0.0230
$a'(=da/dt)$		0	

设插值多项式记为 $H_3(t')$,使它满足 $H_3(t_i')=a_i$(i=0,1,2)和 $H'_3(t'_1)=0$。按前述的思路,$H_3(t')$ 可表示为:

$H_3(t')=N_2(t')+P(t'-0.8)(t'-1)(t'-1.2)$

其中,$N_2(t')$ 是应用式(1)构造的牛顿二次抛物线方程,即为:

$N_2(t')=0.04688t'^2-0.10131t'+0.07875$

为确定系数 P,对 $H_3(t')$ 求导,并注意到条件 $H'_3(1)=0$,即有:

$H_3'(1)=N_2'(1.0)+P[(1.0-0.8)(1.0-1.2)]=0$

即：$-9.375 \times 10^{-3} - 0.04p = 0$ ，故有 P=-0.2344

代入 $H_3(t')$，便得牛顿–爱尔米特三次插值曲线方程，即：

$$a_3 = H_3(t') = -02344\ t'^3 + 0.750\ t'^2 - 0.7969\ t' + 0.3038 \qquad (0.8 \leq t' \leq 1.2)$$

据此，所求在 125—1200℃温度区间内的中碳钢导热系数的平均值 a_m 为：

$$a_m 10^3 \int_{125}^{1200} a\,dt = [\int_{0.125}^{0.4} a_1\,dt' + \int_{0.4}^{0.8} a_2\,dt' + \int_{0.8}^{1.2} a_3\,dt']$$

计算出：$a_m = 0.03188 m^2/h$

4 结论

牛顿插值法是根据函数表的离散数据建立代数多项式的逼近原函数。据此可以得出结论:凡要求通过原函数用以表达事物之间的关系的,均可用构造出的插值多项式来加以近似描述。故除本文所述的几个实际应用的典型问题外,还有如数值微分等其它方面的用途以及派生出新的数值方法。

参考文献:

[1] 中科院数学研究所数理统计组. 常用数理统计方法 [M]. 北京:科学出版社,2004 年.

[2] 白玉山. 计算方法 [M]. 辽宁:辽宁出版社,2004 年.

[3] 蔡乔方. 加热炉 [M]. 北京:冶金出版社,2002 年.

[4] 江宏俊. 流体力学 [M]. 北京:高教出版社,2005 年.

DP 工艺 GH4169 合金热变形组织演变研究

司家勇[1,2]，廖晓航[1]，刘　娜[1]，钟利萍[1]

(1.中南林业科技大学 机电工程学院；2.中南大学 粉末冶金研究院,中南大学 粉末冶金国家重点实验室)

摘　要：针对 DP 工艺 GH4169 合金,研究了变形温度为 900℃~1060℃、应变速率为 $0.001~0.5s^{-1}$ 的热变形过程中的微观组织演变特征及δ相形貌变化，分析了δ相对 GH4169 合金高温变形行为和变形机制的影响。结果表明:微观组织中大量片层状/长针状δ相在高温压缩变形过程中由于变形断裂和溶解断裂的综合作用,发生弯曲、扭折直至球化;随着变形温度的升高或应变速率的降低,δ相含量逐渐减少,合金动态再结晶晶粒尺寸和动态再结晶体积分数逐渐增大。

关键词：GH4169 合金；DP 工艺；δ相；热变形；动态再结晶

Study on Hot Deformation Microstructure of Delta-processed GH4169 Alloy

SI Jia-yong[1,2]，LIAO Xiao-hang[1]，LIU Na[1]，ZHONG Li-ping[1]

(1. College of Mechanical and Electrical Engineering, Central South University of Forestry and Technology;
2. State Key Laboratory of Powder Metallurgy, Central South University)

Abstract: The microstructure evolution of delta-processed GH4169 alloy was studied at temperature range of 900-1060℃ and strain rate range of 0.001-0.5 s^{-1}. The effects of δ phase on the mechanism of dynamic recrystallization were discussed. The results show that the bending, kinking and spheroidization of plate-like or needle-like δ phase take place due to the effects of deformation and dissolution breakages during hot deformation, and needle-like δ phase transferrs to spherical or rod-shaped δ phase. With increasing deformation temperature and decreasing strain rate, the content of δ phase decreases, but the volume fraction and grain size of dynamic recrystallization increase.

Key words: GH4169; delta-processed; δ phase; hot deformation; dynamic recrystallization

基金项目:国家"863"计划项目(2012AA03A514)。

第一作者及通讯作者:司家勇,男。高温合金成形工艺研究。sjy98106@163.com。

TC4 钛合金微弧氧化膜的高温曝露行为研究

陈泉志，唐仕光，郭宇明，黄祖江，李伟洲 *，曾建民

（广西大学 材料科学与工程学院）

摘　要：利用微弧氧化技术，分别以相同浓度的 Na_2SiO_3、Na_3PO_4、$NaAlO_2$ 为电解液对 TC4 钛合金进行表面处理。对处理后获得的陶瓷膜样品进行恒温氧化和热冲击实验，分析氧化及热冲击前后样品的物相结构、形貌、膜层厚度和元素成分。结果表明：陶瓷膜样品经过 750℃/30h 的氧化增重符合抛物线规律，在 Na_3PO_4 溶液中形成的膜层有较好的抗氧化性，$NaAlO_2$ 的次之。经过 100 次 650℃水冷至 28℃ 的循环热冲击后，三种电解液中获得的氧化膜都出现了不同程度的开裂，在 Na_3PO_4 溶液中形成的氧化膜并无剥落，而在 $NaAlO_2$、Na_2SiO_3 溶液中的氧化膜都出现剥落现象，其中以 Na_2SiO_3 溶液中形成的氧化膜剥落最为严重，说明涂层的抗氧化性及抗热震能力与陶瓷层的表面形貌、厚度以及相组成密切相关。

关键词：微弧氧化；钛合金；电解液；高温氧化；抗热震性

Study on High Temperature Behavior of Micro-arc Oxidation Coating on TC4 Titanium Alloy

CHEN Quan-zhi, TANG Shi-guang, GUO Yu-ming, HUANG Zu-jiang, LI Wei-zhou, ZENG Jian-min

(School of Materials Science and Engineering, Guangxi University, Guangxi)

Abstract: Ceramic coatings were prepared on the surface of TC4 titanium alloy by means of micro-arc oxidation (MAO) in the same concentration of Na_2SiO_3、Na_3PO_4、$NaAlO_2$ solutions, respectively. The constant temperature oxidation and thermal shock behavior of samples with and without coatings were studied in this paper, analysis the phase composition, morphology, thickness and elemental composition of coated samples, which before and after the constant temperature oxidation and thermal shock. The high oxidation resistance and thermal shock resistance properties were investigated by the oxidation weight gain tests. The results indicated that the curves of weight gains of the coated samples exposed at 700℃/30h showed a shape of parabola, the coating formed in Na_3PO_4 solution had the best oxidation resistance property, and the coating formed in $NaAlO_2$ solution were in the second place. After the 100 times thermal shock experiment set 650℃ to 28℃, the coatings had suffered different degrees of cracking in the three kinds of electrolyte, and the coating formed in the Na_3PO_4 solution maintained a relatively nice morphology, and had not spall, the coating formed in the Na_2SiO_3 solution and $NaAlO_2$ solution were spall and broken, and the coating formed in the Na_2SiO_3 solution was the worst. It showed that the oxidation resistance and thermal shock resistance properties were closely related to the morphology, thickness and phase of the coatings.

Keywords: micro-arc oxidation; titanium alloy; electrolyte; high-temperature oxidation; thermal shock

---

基金项目：广西自然科学基金(2010GXNSFD013006，2014GXNSFCA118013)；广西高等学校高水平创新团队项目(第二批)；
　　　　　广西自然科学基金创新研究团队项目(2011GXNSFF018001)。

第一作者：陈泉志，男。金属材料表面处理与防护。quanzhi_cheni@163.com。
通讯作者：李伟洲，男。材料的表面防护。liwz2008@hotmail.com。

铌合金表面渗镀复合涂层及其抗氧化研究

杨　阳，黄祖江，唐仕光，李伟洲，张修海，曾建民

(广西大学 材料科学与工程学院)

摘　要：利用包埋渗结合化学镀技术在铌合金表面制备了复合涂层，研究涂层在退火过程中的元素扩散行为及涂层的高温抗氧化性能。结果表明:复合涂层以晶态 Al_3Nb 和非晶态 Ni 相为主;退火过程中渗层中的 Al 元素向外扩散,涂层转变为晶态,形成了 $NiAl_3$、Al_3Nb、NiAl 相。对退火前后的涂层进行 1000℃恒温氧化实验,20h 后沉积态涂层的增重为 $7.7mg/cm^2$, 表面主要含 NiO、Al_2O_3、NiAl 相; 退火态涂层样品的增重为 $4.9mg/cm^2$,表面生成了 Al_2O_3、$NiNb_2O_6$、$NiAl_2O_4$ 等相。氧化后涂层与基体结合良好。退火态涂层表面由于富 Al 元素,氧化后形成较多的 Al_2O_3,比沉积态的涂层能更有效地减缓氧化进程,提高铌合金的抗氧化性。

关键词：包埋渗；化学镀；复合涂层；高温抗氧化性

Oxidation Resistance of Duplex Coatings Prepared by Pack Cementation Combined with Electroless Plating on Nb-based Alloy

YANG Yang, HUANG Zu-jiang, TANG Shi-guang, LI Wei-zhou, ZHANG Xiu-hai, ZENG Jian-min

(School of Materials Science and Engineering, Guangxi University)

Abstract: In order to improve the oxidation resistance of Nb-based alloy C103, the methods of pack cementation and electroless plating were used to prepare duplex coatings on the surface of C103. The element diffusion behavior and oxidation resistance of the coatings before and after annealed were investigated. The results indicated that the as-deposited Ni-P/Al riched coating was composed of crystalline Al_3Nb and amorphous Ni. After vacuum heat treatment, the Al element diffused from the inner to the outer coating. $NiAl_3$、Al_3Nb and NiAl phases were formed. After high temperature of 1000℃/20h oxidation experiments, the deposition state was composed by NiO, Al_2O_3, NiAl phases, the weight gain was $7.7mg/cm^2$; Annealing state coating was given priority to with Al_2O_3, $NiNb_2O_6$, $NiAl_2O_4$ phases, weight increment was $4.9mg/cm^2$; Both the coatings were well combined with matrix. The Al was riched on the surface of annealing state coating. Oxidation process was slow down because of the formation of richer Al_2O_3 coating, which effectively improve the oxidation resistance of C103 alloy.

Key words: embedding permeability; electroless; composite coating; high temperature oxidation resistance

基金项目:国家自然科学基金 (51371059,51001032);广西自然科学基金(2014GXNSFCA118013,0731013);广西高等学校高水平创新团队 项目;广西自然科学基金创新研究团队项目(2011GXNSFF018001)。
第一作者:杨 阳,女。金属材料表面处理。Cathrynyang0210@163.com。
通讯作者:李伟洲,男。金属材料表面处理 liwz2008@hotmail.com。。

LGB38MnV 钢热压缩本构模型的建立

刘升旭 [1]，蒙秋红 [1]，何奥平 [2]，段亚菲 [2]，曾建民 [2] *

（1.广西柳工机械股份有限公司；2.广西大学 材料科学与工程学院）

摘　要：LGB38MnV 非调质钢是由一种自主开发的用于装载机半轴用的新钢种。本工作利用 Gleeble3500 热力学模拟机对非调质钢 LGB38MnV 进行单道次高温压缩，研究了 LGB38MnV 钢在变形温度为 850℃~1000℃、应变速率为 0.01/s~10/s 时的流变应力。采用双曲正弦形式的 Arrhenius 方程，研究了该材料在变形过程中材料参数随应变的变化规律，并最终建立本构方程。同时，根据所得应力-应变曲线，初步研究了不同变形条件下动态回复和动态再结晶过程，为该钢种的锻造提供了基本工艺数据。

关键词：LGB38MnV 钢；本构方程；动态回复；动态再结晶

Establishment of a Constitutive Model to Predict Hot Forming in 38MnV Steel

LIU Sheng-xu[1], HE Ao-ping[2], DUAN Ya-fei[2], ZENG Jian-min[2]*

（1.Guangxi Liugong Machinery Co., LTD; 2.School of Materials Science and Engineering, Guangxi University）

Abstract：LGB38MnV is a new kind of self-developed steel that is used for the half axle in shovel loader. In the present work, high temperature compression test of LGB38MnV non-quenched and tempered steel was conducted on Gleeble3500 machine in the temperature range of 850℃ -1000℃ and the strain rate range of 0.01/s-10/s, the flow stress curves were studied in this paper. The variation of material contents with different strain values were studied and constitutive equation was built according to an Arrhenius type equation. The dynamic recovery and recrystallization processes during hot compression were studied on the basis of stress-strain curves, which has provided basic process parameters for the new steel.

Key words: LGB38MnV steel；constitutive equation；dynamic recovery；dynamic recrystallization

基金项目：桂科攻 1348010-1。

第一作者：刘升旭，男。黑色金属新型材料。282291525 @qq.com。

通讯作者：曾建民，男。黑色金属新型材料。zjmg@gxu.edu.cn。

Mn 对 6061 铝合金再结晶行为的影响

王友彬[1]，曾建民[2]

(1.西北工业大学 凝固技术国家重点实验室；2.广西大学 有色金属及材料加工新技术教育部重点实验室)

摘　要：采用硬度测试、金相显微镜、EBSD 及投射电镜观察等分析方法，研究了微量元素 Mn 的添加量对 25%、50%及 75%冷轧变形量的 6061 铝合金的再结晶行为的影响。实验结果表明，在相同的冷轧变形量下，6061 铝合金的再结晶温度和再结晶激活能随着 Mn 含量的增加而提高；加入 Mn 后形成的细小的弥散相可以阻碍晶粒的长大和晶界的迁移，抑制 6061 铝合金的再结晶。

关键词：再结晶；冷轧；激活能；6061 铝合金

Effects of Mn Addition on Recrystallization of 6061 Al Alloy

WANG You-bin[1], ZENG Jian-min[2]

(State Key Laboratory of Solidification Processing, Northwestern Polytechnical University, China;
Key Laboratory of Education Ministry in Non-ferrous Metals and New Processing Technology, Guangxi University)

Abstract: Effects of Mn addition on recrystallization of 6061 Al alloy after cold-rolling with 25%, 50%, 75% deformation were studied by hardness test, optical microscopy, EBSD and TEM. The results show that the activation energy of recrystallization increases with the rising of Mn content and Mn can boost the recrystallization temperature during the same cold-rolling deformation. The dispersed phase induced by Mn addition can inhibit the grain growth and impede the movement of boundaries, hence, Mn addittion can inhibit the recrystallization.

Key word: recrystallization; cold rolled; activation energy; 6061 Al alloy

基金项目：广西有色金属及特色材料加工重点实验室开放基金(GXKFJ14-03)；国家"973"计划前期研究专项项目(2012CB722804)。

第一作者：王友彬，男。铝合金热处理工艺。wangyoubin114@126.com。

通讯作者：曾建民，男。铝合金的铸造及加工工艺、金属表面处理。zjmg@gxu.edu.cn。

时效对 A7N01 铝合金 PMIG 焊接接头组织与性能影响研究

梁天权，覃秋梅，莫小梅，李伟洲，曾建民

（广西大学 材料科学与工程学院）

摘　要：研究了 120℃、24h 时效处理对 A7N01 铝合金脉冲熔化极气体保护焊（PMIG）焊接接头金相组织、室温拉伸力学和耐腐蚀性能的影响。结果表明：低温时效处理可细化 PMIG 焊接接头的焊缝金属晶粒度和强化相，增加组织均匀性和强化相数量。时效焊接接头的抗拉强度由时效前的 227.83MPa 提高到了 247.77MPa，提高了焊接接头的力学性能。时效处理使接头的自腐蚀电位向正方向移动，自腐蚀电流减小，提高了焊接接头的耐腐蚀性能。详细分析了时效改善接头组织与性能的主要因素。

关键词：A7N01 铝合金；PMIG；时效；组织；性能

Influence of the Microstructure and Performance of the PMIG Welding Joint for A7N01 Aluminum Alloy by Aging Treatment

LIANG Tian-quan, QIN Qiu-mei, MO Xiao-mei, LI Wei-zhou, ZENG Jian-min

（School of Materials Science and Engineering, Guangxi University）

Abstract: The influence of the microstructure and performance in tensile strength and corrosion resistance for the welding joint of A7N01 aluminum alloy by aging with parameter of 120℃, 24h was investigated in this paper. The results show that the grain size of the welding metal and strengthening phase of the joint can be fined by low temperature aging treatment, and it can also promote the microstructure become homogeneous and increase the strengthening phase. Then the aging treatment can enhance the mechanical property of the welding joint from 227.83MPa to 247.77MPa in tensile strength. It can make the self-corrosion potential move positively, and reduce the self-corrosion current as well, improving the corrosion resistance of the welding joint. The main factors improving the microstructure and performance by aging are analyzed in detail.

Keywords: A7N01 aluminum alloy; PMIG; aging treatment; microstructure; performance

基金项目：国家自然科学基金（51361003）；教育部高等学校博士学科点专项基金（20134501120002）；
广西自治区"八桂学者"专项经费资助和广西创新团队项目（2011GXNSFF018001）；
广西高等学校高水平创新团队及卓越学者计划（第二批）。
第一作者：梁天权，男。腐蚀与防护、先进材料连接技术。liangtianquan@126.com。

Gleeble3500 用于开发低成本钛合金的研究

王同波，李伯龙，袁 杰

(北京工业大学 材料科学与工程学院)

摘 要：本文主要借助 Gleeble 热模拟技术拟定 Ti-xFe-3Al 的锻造工艺，以开发新型超高强低成本钛合金。利用扫描电子显微镜(SEM)、金相显微镜(OM)和电子背散射取向分析技术(EBSD)表征材料的微观结构，利用万能试验机测定合金的力学性能。结果表明：基于 Gleeble 热模拟得到力学曲线，可确定 Ti-7Fe-3Al、Ti-10Fe-3Al 和 Ti-12Fe-3Al 合金最佳精锻温度为 850℃±20℃。该温度用于实际锻造工艺，获得强度高达 1200MPa、无剪切组织的 Ti-xFe-3Al。

关键词：低成本；超高强；Ti-Fe-Al

Study on Development of Low-cost Titanium Alloy by Gleeble3500

WANG Tong-bo, LI Bo-long, YUAN Jie

(Beijing University of Technology, Material Science and Engineering)

Abstract: In this paper, forging process of Ti-xFe-3Al was worked out by Gleeble thermal simulation technique, to develop a new low cost Titanium alloy with super high strength. The microstructure of alloy was characterized by Scanning Electron Microscope (SEM), Optical Micoscope (OM) and Electron Backscatter Diffraction (EBSD). Mechanical properties' testing were carried out in universal testing machine. It was found that, based on stress-strain curve, the best forging temperatures of Ti-7Fe-3Al, Ti-10Fe-3Al and Ti-12Fe-3Al were determined to be 850℃±20℃. The temperature was used to forging process to obtain Ti-xFe-3Al alloy whose strength could be up to 1200 MPa and microstructure had no shearing zone.

Key words: low-cost; super high strength; Ti-Fe-Al

基金项目：国家自然科技基金项目(NO.51371013)。
第一作者：王同波，男。轻合金的相变与形变。wtb@emails.bjut.edu.cn。
通讯作者：李伯龙，男。轻合金的相变与形变。blli@ bjut.edu.cn。

粘性流体微滴堆积－固化过程动力学数值模拟

陈从平，黄杰光，王小云

（三峡大学 机械与动力学院）

摘　要：在微滴喷射式 3D 打印过程中，粘性材料在自身内应力场、表面张力场及温度场的耦合作用下堆积－固化，其固化形态难以用准确的物理模型来描述。本文在分析固化过程力学特性的基础上，利用 FLOW-3D 软件对粘性材料微滴堆积-固化过程进行数值建模与多场耦合动态力学行为进行仿真，并基于 VOF 法追踪材料固化过程的形态变化，获得材料特性、固化率、堆积速度、微滴及基板初始温度等多物理因素对材料最终固化形态的耦合作用规律，其结论可为相关理论研究及实际应用提供指导。

关键词：微滴喷射；堆积-固化；多场耦合；动力学；数值模拟

Dynamics Numerical Simulation on Deposition and Solidification Processes of Viscous Material Micro-droplet

CHEN Cong-ping, HUANG Jie-guang, WANG Xiao-yun

（College of Mechanical & Dynamics, China Three Gorges University）

Abstract: on the processes of micro-droplet injection 3D printing, viscous material pile-up and freeze due to the coupling impact of its internal stress field, surface tension field and temperature field. It's difficult to represent its solidification front morphology with accuracy physical model. A FLOW-3D software is applied to model the processes of deposition and solidification of viscous material droplet and simulate the dynamics behavior of the processes which coupled varieties of physical fields base on analysing the dynamics properties of solidification in this article. And the VOF method is applied to track the variation of form during the process of solidification. That acquires the law of the coupling impact of material behavior, pile-up velocity, the initial temperature of micro-droplet and substrate, solid fraction and the like physical factors. The conclusion may instruct the research of correlation theory and practical application.

Keywords: micro-droplet jetting; pile-up; the coupling of varieties of physical fields; dynamics; numerical simulation

基金项目：国家自然科学基金(51475266,51005134)；三峡大学研究生科研创新基金(2015CX039)。
第一作者：陈从平，男。微流体动力学、3D 打印。mechencp@163.com。
通讯作者：黄杰光，男。微流体动力学、3D 打印。422975820@qq.com。

DH32 船板钢埋弧焊接头力学性能分析

梁国俐，齐铁力，苑少强，杨跃辉，李　敬，张晓娟

（唐山学院 机电工程系）

摘　要：本文以 DH32 高强度船板钢为研究对象，通过拉伸、冲击试验手段，对 DH32 船板钢埋弧焊接头进行了力学性能测试。拉伸试验结果发现所有断裂均发生在拉伸试样的母材区，DH32 船板钢在大热输入焊接条件下，焊缝和焊接热影响区的强度好于母材，并没有出现热影响区软化现象。冲击试验发现，焊缝平均冲击功 55J，熔合线平均冲击功 58J，HAZ 平均冲击功 75J，冲击数值均满足船级社冲击功不小于 34J 的要求。

关键词：DH32 船板钢；埋弧焊；拉伸强度；冲击韧性

Analysis on Mechanical Properties of Joint by Submerged Arc Welding for DH32 Shipping Plate Steel

LIANG Guo-li, QI Tie-li, YUAN Shao-qiang, YANG Yue-hui, LI Jing, ZHANG Xiao-juan

(Department of Electro-Mechanical Engineering, Tangshan College)

Abstract: Based on the DH32 high-strength shipping plate steel this paper investigated the mechanical properties of joints by the submerged arc welding (SAW) using the tensile, impact test. The results of tensile test found that all the fracture occurred in the base area of tensile specimens of DH32 steel, the strength of the weld and heat affected zone (HAZ) is better than base material in the SAW process, and the softened zone in HAZ didn't appeared. It was found that the worst zone of toughness of joint is the weld zone, the impact value is 55 J, which caused by slag, the toughness of fusion line is 58J, the toughness of HAZ is 75J. The impact value are all above the 34J which is the register of shipping requirements.

Key words: shipping plate steel; SAW; tensile strength; impact toughness

基金项目：唐山市科技计划项目(14130237B)，河北省钢铁联合研究基金项目(E2015105052)。
第一作者：梁国俐(1974—)，女，副教授，工学博士。新型钢铁材料焊接性能研究与开发。guoliliang428@163.com。

Ni-Mg-Al-LDH 的制备及其对甲基橙的吸附作用

谢襄漓[1]，徐国永[1]，王林江[2]

(1.桂林理工大学 化学与生物工程学院；2.广西有色金属及特色材料加工国家重点实验室培育基地)

摘　要：本文用水热法合成了 Ni:Mg:Al 比值分别为 2:1:1 和 4:1:1 的碳酸根型层状双氢氧化物 (Ni-Mg-Al-LDH)，考察了合成条件、结构、焙烧产物特征及其对甲基橙的吸附作用。在晶化温度 140℃、晶化时间 14 小时的条件下，得到结构完整、结晶特性好的 Ni 参杂 Mg-Al 层状双氢氧化物。将合成的 Ni-Mg-Al 层状双氢氧化物在 410℃条件下焙烧 2 小时，得到层状双金属氧化物。将该焙烧产物用于吸附甲基橙溶液，针对吸附剂的用量、吸附溶液的 pH 值以及吸附温度对吸附作用的影响进行研究。实验结果表明：对于浓度为 200mg/L 的甲基橙溶液，在吸附剂的用量为 0.1g/100mL，pH 值为 8.00，吸附反应温度为 50℃的条件下，吸附率高达 96%。该层状双金属氧化物可以作为处理印染废水中甲基橙的吸附剂。

关键词：Ni-Mg-Al-LDH；水热合成；焙烧；吸附；甲基橙

Preparation and Adsorption Properties of Ni-Mg-Al-LDH on Methyl Orange

XIE Xiang-li[1], XU Guo-yong[1], WANG Lin-jiang[2]

(1.College of Chemistry and Bioengineering,Guilin University of Technology; 2. Ministry-province Jointly-constructed Cultivation Base for State Key Laboratory of Processing for Non-ferrous Metal and Featured Materials)

Abstract：In this paper, layered double hydroxides　(LDHs) with the ratio of Ni/ Mg/Al of 2:1:1 and 4:1:1 were synthesized by hydrothermal method. The results shown that the structure and crystalline characteristics of 2:1:1 LDH is very good at the conditions of the crystallization temperature of 140℃, crystallization time of 14 hours. When LDH was calcined at 410℃ for two hours, the original features of LDH peaks disappeared and became layered double oxide (LDO). The calcined product LDO was used as adsorbent for methyl orange in water solution. The results shown that the adsorption rate of LDO for methyl orange is 96% at the methyl orange concentration of 200mg/L, the amount of the adsorbent LDO of 0.1g per 100mL solution, pH value of 8.00 and the adsorption temperature of 50℃. Ni-Mg-Al-LDH can be used as an adsorbent of methyl orange dyeing wastewater treatment.

Keywords: Ni-Mg-Al-LDH; hydrothermal synthesis; calcination; adsorption; methyl orange

基金项目:国家自然科学基金(41272064)。

第一作者:谢襄漓,女。分析化学。xiexl@glut.edu.cn。

通讯作者:王林江,男。矿物材料。wlinjiang@163.com。

矿物混合料球团斜盘造粒的模拟

曾晓乐，曾建民

（广西大学 材料科学与工程学院）

摘　要：基于湿法造粒，以聚乙烯醇(PVA)—水系统为喷雾溶液，研究斜盘造粒过程中粘结剂含量对混合粉料造粒颗粒性能(包括颗粒的强度、堆积性能、流动性等)的影响。并以吸水率、体积密度和气孔率为指标，研究该造粒条件下所得颗粒的干压成型性能。实验结果表明：随粘结剂浓度增加，粉体成球率可达 90%以上，0.5~2mm 筛分率逐渐降低，且颗粒的平均尺寸增大，形状趋于不规则；PVA 浓度越高，颗粒初始含水率越低，颗粒强度越大；比较各浓度下干颗粒的堆积密度、休止角和漏斗流出时间，我们发现该实验条件下 15vol%的 PVA 含量造粒效果最好。此外，对干压成型过程来说，造粒后的颗粒相对于粉体，能够得到更加致密的坯体结构。

关键词：混合粉料；斜盘造粒；粘结剂；流动性

Simulation of Inclined-pan Granulation for Pelletizing Mineral Mixtures

ZENG Xiao-le, ZENG Jian-min

（School of materials science and engineering, guangxi university）

Abstract: Based on wet granulation, we conduct the research in the process of inclined-pan granulation with which respect to the effect of polyvinyl alcohol (PVA) additives content on the pelletizing properties (include the strength, bulk performance, fluidity etc) which are relate to aggregates of mixed powders via PVA-H2O system as the spray solution. Dry pressing molding performance of the granules obtained with this method also be considered by employing the water absorption, bulk density and porosity of sintered compact as indicators. The results show that more than 90% powders turned into pellets, while screening percent between 0.5mm and 2mm square mesh sieve decrease gradually with the increasing of PVA concentration, moreover, the average size of the particles increased and the shape became irregular. The higher concentration of PVA means lower moisture content and greater strength to wet grains. Comparing the bulk density, dwell angle and flow time of dried conglomerations granulated with different PVA content, respectively, we found that the best best content of PVA effect granulation under our laboratory conditions is 15vol%. In addition, we will get more densified sintered compacts in dry-pressing process if suppressing with granules.

Key words：mixed powder; inclined-pan granulation; binder agent; fluxility

---

基金项目:广西八桂学者建设工程专项经费资助项目。
第一作者:曾晓乐,男,研究生。plpzxl@163.com。
通讯作者:曾建民,男,教授。zjmg@gxu.edu.cn。

ZL205A 铝合金铸件的偏析行为及机理研究

胡　武[1]，曾建民[1]，陈燕飞[2]，吴继峰[2]

(1.广西大学 材料科学与工程学院；2.南昌洪都航空集团公司机电国际公司)

摘　要：利用 X 射线探伤、扫描电子显微镜、能谱仪和光学显微镜，对具有线性偏析组织的 ZL205A 合金铸件进行分析研究。结果表明：ZL205A 合金铸件粗大的宏观偏析组织从边缘以枝杈状或者相互平行的曲线条向铸件内部延伸，与热裂组织相似，并伴随有缩松或者微裂纹；ZL205A 合金铸件中 Cu 原子处于晶内的能量比处在晶界的能量高，而且 Cu 元素在晶界偏聚可以降低界面能，使得 Cu 原子有自发地向晶界偏聚的趋势，发生共晶偏析的元素为 Cu 元素并以 θ(Al₂Cu) 共晶相的形式存在，晶界上的偏析组织由 θ(Al₂Cu) 相和 α(Al) 相组成；线性偏析与热裂之间的相互关系可用凝固收缩补偿理论进行解释。热裂纹的产生促进了线性偏析的形成，消除铸件线性偏析组织可从防止铸件的热裂缺陷形成入手。

关键词：ZL205A 合金；偏析；热裂

Study on Behavior and Mechanism of Segregation in ZL205A Alloy Casting

HU Wu[1], ZENG Jian-min[1], CHEN Yan-fei[2], WU Ji-feng[2]

(1. School of Materials Science and Engineering, Guangxi University;
2. The Electromechanical International Company of Nanchang Hongdu Aviation Group)

Abstract: The specimens produced by linear segregation ZL205A alloy castings were studied by X-ray inspection, OM, SEM and EDS. The study suggested that the macro-segregation appeared in branches or parallel curves at grain boundaries and extended from the edges to the interiors of specimens, similar to the hot cracking behavior, and is accompanied by some shrinkage porosities or micro-cracks. In the intracrystalline Cu atomic energy is higher than in grain boundary energy and the Cu elements concentrated in the grain boundaries can reduce interfacial energy, so Cu atom has a spontaneous tendency to concentrate to the grain boundary. The eutectic segregation elements as Cu element and form eutectic θ (Al₂Cu) phase, the organizations of grain boundary segregated are composed of θ(Al₂Cu)phase and α (Al) phase. The relationship between linear segregation and hot cracking can be explained by solidification shrinkage-repairing theory. Hot crack produced to promote the formation of linear segregation and to eliminate the casting's linear segregation organization can prevent from the formation of cracking defects of castings.

Key words: ZL205A alloy; segregation; hot cracking

基金项目：国家自然科学基金 (51064003)；广西自然科学基金(2011GXNSD018009)。
第一作者：胡　武,男。铸造铝合金缺陷及性能研究。huwunh@163.com。
通讯作者：曾建民,男。铸造铝合金缺陷及性能研究。zjmg@gxu.edu.cn。

0 引言

ZL205A 合金是一种可固溶强化的高强度铸造铝合金，在 T6 热处理状态下其抗拉强度σ_b可达 500Mpa 以上[1-3]，是目前强度最高的铸造铝合金之一。ZL205A 是 Al-Cu-Mn 系合金，并添加有 Ti、Cd、V、Zr 和 B 等微量元素，因此凝固时结晶温度范围宽，热裂倾向大，使得铸件极易产生宏观偏析[4-6]，造成铸件报废。在航空铸件中出现严重的偏析往往是灾难性的[7]。铸件中宏观的偏析可用 X 射线探伤进行检测，在探伤底片上呈现白色。国内的学者对此进行了一定的研究，如李玉胜、史晓平等人[4,8-10]的研究表明铸件中的带状偏析、云雾状偏析和线性偏析的偏析产物是 Al$_2$Cu，属于共晶偏析，白点偏析则是属于是富集大比重金属混合物的比重偏析。贤福超[6,11]等人研究的块状偏析也属于比重偏析，发现预制锭中粗大的块状组织对铸件块状偏析的形成具有一定的形态遗传性。但目前对偏析机理的研究还欠深入，企业生产中偏析导致产品报废的现象时有发生。本文以出现有线性偏析的 ZL205A 合金铸件为对象，分析和探讨铸件的偏析行为及其与裂纹之间的相互关系。

1 实验方法

实验所用的为模拟实际生产而铸造出来的带有线性偏析的 ZL205A 合金铸件。偏析的取样位置及对应的探伤底片如图 1 所示，偏析位置附近为铸件壁厚的过渡处，底片中圈内的线条，即所谓的线性偏析。

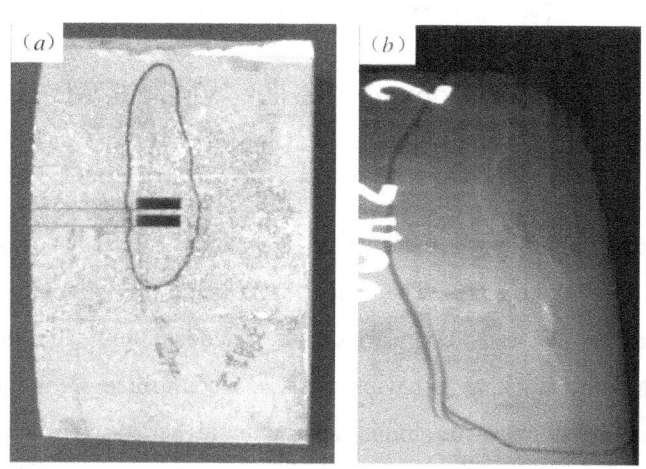

图 1　铸件取样位置及对应的探伤底片:(a)取样位置(b)对应的探伤底片
Fig.1　Sampling positions of the casting and X- ray picture of the segregation:
(a)the sampling positions,(b)X- ray image.

对取得的样品经磨制、抛光和 0.5%HF 腐蚀后，利用 Leica 金相显微镜和日本日立的场发射扫描电子显微镜观察偏析组织的微观形貌，以及用 EDS 进行点分析、线扫描和面扫描分析偏析组织的成分及分布。

2 实验结果及分析

2.1 ZL205A 合金铸件的晶界偏析行为

结合 X 光探伤底片、金相照片及能谱分析，得试样的宏观偏析位置及相应的金相照片如图 2 所示。图 2(a)中的 a 曲线所围的区域为试样宏观偏析区域，此区域与 X 光探伤底片中发现的宏观偏析位置相吻合,由图可看出偏析区域的长度超过了铸件壁厚的 2/3;图 2(b)为图 2(a)中 b 箭头所指位置的金相照片, 可以看到粗大的宏观偏析组织以类似枝杈状或者近似相互平行的曲线条分布在裂缝的边沿或者晶界上并向基体内部延伸,此现象与热裂行为相似。另外,晶界上还分布有许多细小的枝晶状或网状的偏析组织,深黑色的孔洞为缩松或者微裂纹。由于经过热处理时,细小的偏析组织可以被 α(Al)相充分固溶,故对铸件性能产生影响的主要是那些粗大的宏观偏析组织。

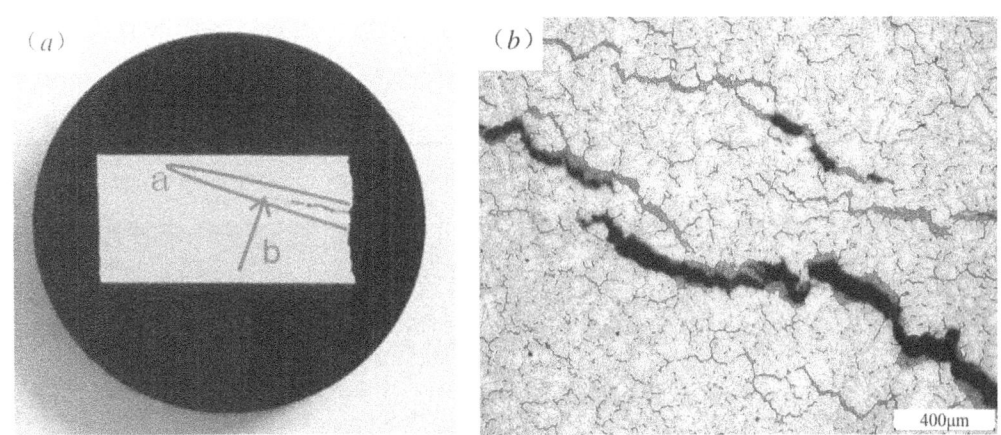

图 2　铸件金相试样偏析区域及相应位置的金相照片:(a)偏析区域(b)金相照片

Fig.2　The segregation area and OM image of casting specimen:(a) segregation area, (b) OM image.

对宏观偏析区域及基体组织进行更微观的 OM 和 SEM 照片观察(如图 3 所示)。图 3(a)中的灰色枝晶状组织为偏析组织,亮白色并且分布有小灰点的组织为基体组织;对图 3(b)中的基体组织和偏析组织进行 EDS 成分分析,分析结果如表 1 所示,对比数据可以看出,Cu 元素在晶界处大量偏析,其质量含量大约是基体的 12.3 倍,未发现 Mn 或者其他元素在晶界处偏析,O 主要在制备样品过程中,样品被氧化所致。为了进一步了解 Cu 元素在偏析样品中的分布情况,对样品的 Al、Cu 元素进行了线扫描和面扫描,其扫描图像如图 4 所示。从图中可以更加直观地观察到 Cu 元素在晶界偏析的情况,发现在宏观的晶界偏析组织中聚集了大量的 Cu 元素。此现象与文献[12]发现的"晶界高于晶内,一次晶界高于二次晶界"的 Cu 元素分布特点类似。由以上分析可知,ZL205A 合金铸件往晶界的偏析元素为 Cu 元素。根据表 1 偏析组织中的 Al 和 Cu 原子百分比 Al/Cu≈ 2.38,Al-Cu-Mn 三元相图[13]和文献[6]中的物像分析,得 Cu 元素往晶界偏析是以 θ(Al2Cu)共晶相存在的,而偏析组织则由 θ(Al2Cu)相和 α(Al)相组成。

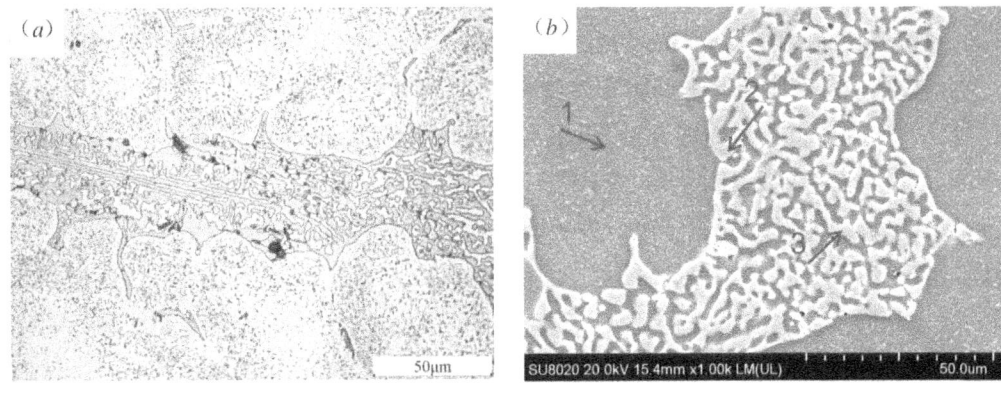

图 3　ZL205A 合金偏析组织的 OM 和 SEM 照片:(a)偏析组织的 OM 照片(b)偏析组织的 SEM 照片

Fig.3　OM and SEM images of segregation structure in ZL205A alloy: (a) OM image, (b) SEM image.

表 1　ZL205A 合金基体及偏析组织成分分析结果

Tab.1　The composition of matrix and segregation in ZL205A

元素	基体		偏析组织					
	1		2		3		平均	
	wt%	at%	wt%	at%	wt%	at%	wt%	at%
Al	94.88	96.87	49.84	69.12	49.91	69.23	49.88	69.18
Cu	3.99	1.73	49.39	29.08	49.37	29.08	49.38	29.08
Mn	0.44	0.22	–	–	–	–	–	–
O	0.68	1.18	0.77	1.8	0.72	1.69	0.75	1.75

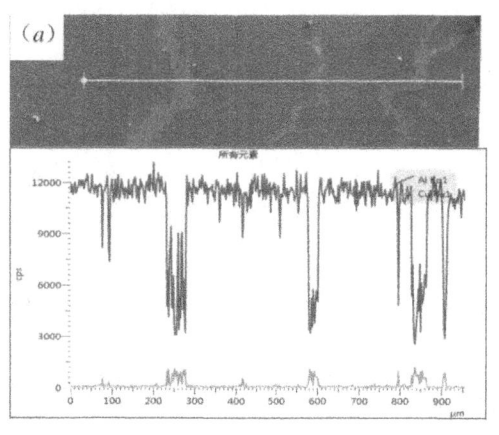 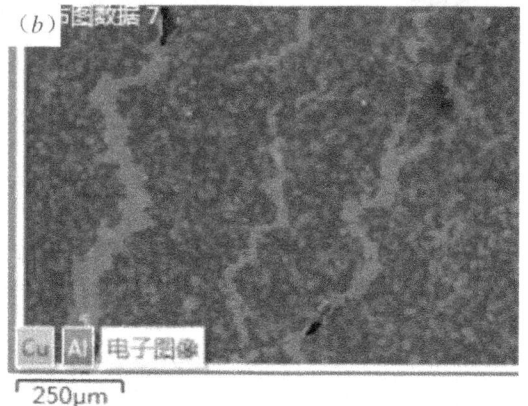

图 4　ZL205A 合金偏析元素线扫描与面扫描分析图:(a)线扫描图(b)Cu、Al 分布面扫描图

Fig.4　Line and surface scanning analysis of ZL205A alloy:
(a) line scanning image, (b) surface scanning image of Cu and Al.

2.2 ZL205A 合金铸件的晶界偏析机理

晶界处 Cu 元素含量高于晶内的现象,可用相应的偏析理论进行解释:从能量角度看,根据二元系恒温吸附方程[9]:

$$\Gamma_i = -\frac{1}{RT}\frac{\partial\gamma}{\partial\ln x} = -\frac{x}{RT}\left[\frac{\partial\gamma}{\partial x}\right]$$

式中:Γ_i 为单位表面吸附 i 组元的量,或单位表面积上溶质浓度和晶体内部平均浓度之差;γ 为比界面能;x 为溶质原子在晶体中的平衡体积浓度;R 为气体常数。当 Γ_i 增大,即单位表面吸附 i 组元的量增大时,有 $\partial\gamma/\partial x<0$,使得比界面能降低,得出当晶界偏聚溶质原子时,可降低界面能。对于 ZL205A 合金,由于晶界缺陷比晶内多,使得 Cu 原子处于晶内的能量比处在晶界的能量高,而且 Cu 元素在晶界偏聚还可以降低界面能,故 Cu 原子有自发地向晶界偏聚的趋势。

从凝固过程的溶质分布看:在凝固过程中,随着温度的降低,合金以枝晶状生长,枝晶的形成开始于不稳定的平面状固液界面的破开扰动得以增强,一直到被扰动界面的尖端和凹谷处的生长出现明显差别为止,尖端还能往横向排出溶质,而凹谷处则积累由尖端排出的多余溶质[12]。这样随着温度的继续降低,Cu 元素不断地在晶界处积累,降到共晶温度时 Cu 在晶界的偏聚程度达到最大。这一过程的溶质分布可用著名的 Scheil 方程[14]进行解释:

$$\omega_L = \omega_{C0} \varphi_L{}^{-(1-k)}$$

式中：ω_L 为凝固过程中某一时刻 t 时液相溶质质量分数，ω_{C0} 为初始溶质质量分数，φ_L 为液相体积分数，k 为溶质分配因数（$k=\omega^*_S / \omega^*_L$，$\omega^*_S$、$\omega^*_L$ 分别为凝固界面上固相和液相的溶质质量分数）。由公式可知 k 值的大小将影响着溶质偏聚程度的大小，当 k>1 时则表现为负偏析，而当 k<1 时，表现为正偏析。对于 ZL205A 合金铸件，根据实验现象可知 Cu 元素发生正偏析，即 k<1，此时在凝固过程中液相体积分数 φ_L 不断减小，剩余的 Cu 溶质不断被排入凝固界面前的液相，使得剩余的溶液产生 Cu 溶质富集。

由于 ZL205A 合金铸件结晶温度范围宽，在壁厚较大的大部位，凝固以糊状方式进行，凝固时先形成的网状骨架之间的溶液由于凝固收缩产生孔隙造成负压对液体有抽吸作用，随后这些孔隙部位即被剩余的富 Cu 溶质的低熔点共晶液填充，当达到共晶温度时 Al 与 Cu 发生共晶反应生成 θ（Al2Cu）相并且分布在晶界上，最终产生晶界共晶偏析。

3 线性偏析与热裂之间的相互关系

线性偏析与热裂之间的相互关系可由凝固收缩补偿理论[15,16]进行解释，对于 Al-Cu 系铸造合金，当 Cu 含量处于 4%~5% 时热裂倾向性到达最大[3]。ZL205A 的铜含量为 4.6%~5.3%，在铸造凝固过程中极易在铸件壁厚变化的热节部位附近处产生热裂纹，使得未凝固的富 Cu 合金熔液通过枝晶通道往热裂纹处进行填充，当达到共晶温度（548℃）时发生二元共晶反应[13]：$L \rightarrow \alpha(Al) + \theta(Al_2Cu)$。当补缩通道被堵或者合金熔液不足时，共晶反应产生的凝固收缩就得不到补充或者热裂纹处未被填满，当收缩继续进行，如果应力过大，组织会再次被撕裂，凝固后最终得到的组织形貌则表现出在偏析组织处出现缩孔甚至裂纹。当应力相对较小时，产生的热裂纹不大，此时合金液可以完全填充，凝固后的形貌只有粗大的线性偏析组织和少量的缩孔，并不出现裂纹；当应力相对较大时，产生的一次热裂纹较大，合金液难以完全填充，此外随着凝固收缩的继续进行，应力进一步增大，部分合金液凝固后产生的共晶偏析组织被再次撕裂产生二次裂纹，最终得到的组织形貌不仅有粗大的线性偏析组织和少量的缩孔，而且还出现了裂纹，如图 2 所示。由此分析可推断，晶界粗大的低熔点线性偏析组织的形成与热裂具有极大的相关性，热裂纹的产生促进了线性偏析的形成。

故要消除铸件宏观的线性偏析组织可从防止铸件的热裂缺陷形成入手。另外，根据宏观偏析的判断式可知，当 $v/u=-\beta/(1-\beta)$ 时无宏观偏析，因此要解决宏观偏析还可以从液体流动速度和凝固速度的关系入手，即[17]：一是 v 与 u 两者方向相反；二是 v/u 的绝对值要小，即 v 要小，而 u 要大。如：加大冷却速度，缩短固 - 液两相区的凝固时间，尽量使 u 值增大。而注温太高、注速太快，均会延缓铸件冷却，从而使宏观偏析加剧。

4 结论

（1）ZL205A 合金铸件粗大的线性偏析组织以枝杈状或者相互平行的曲线条分布在裂缝的边沿或者晶界上并向基体内部延伸，与热裂行为相似，并伴随有缩松或者微裂纹。

（2）ZL205A 合金铸件发生共晶偏析的元素为 Cu 元素，Cu 原子有自发地向晶界偏聚的趋势，并以

$\theta(Al_2Cu)$相的形式大量分布在晶界上。

(3)线性偏析与热裂之间的相互关系可由凝固收缩补偿理论进行解释,消除铸件宏观线性偏析组织可从防止铸件的热裂缺陷形成入手。

参考文献:

[1] 刘伯操. HZL_205高强度铸造铝合金及其应用[J]. 航空材料,1982,(02): 12–14.

[2] 刘伯操,向启尧,钱景新. ZL_205高强度铸造铝合金[J]. 航空材料,1978,(5): 11–17.

[3] 黄恢元主编. 铸造手册(第3卷)[M]. 北京: 机械工业出版社,1993: 86.

[4] 李玉胜, 翟虎, 王涛等. ZL205A合金大型铸件带状偏析组织对力学性能的影响 [J]. 铸造,2008,(03): 219–222.

[5] 张瑛洁,冯志军,李巨文等. 大型ZL205A铸件的成分偏析和过烧组织研究[J]. 铸造,2003,(08): 545–546+628.

[6] 贤福超,郝启堂,李新雷等. ZL205A合金晶界偏析行为研究[J]. 铸造技术,2012,(12):1391–1393.

[7] 贾宝仟,柳百成. 偏析对铸件缩松、应力应变及热裂的影响[J]. 大型铸锻件,1996,(04): 1–4.

[8] 史晓平,李玉胜. ZL205A合金偏析缺陷研[J]. 铸造,2011,60(10): 1022–1026.

[9] 李玉胜,翟虎,闫卫平等. ZL205A合金壳体铸件线性偏析缺陷形成机理研究[J]. 铸造,006,(11):1170–1173.

[10] 李玉胜,翟虎,马宝民等. 高强度ZL205A合金大型铸件"白点"偏析研究[J]. 铸造,2007,56(2): 185–187+191.

[11] 贤福超,郝启堂,范理. ZL205A合金块状偏析形成机理[J]. 稀有金属材料与工程,2014,43(4): 941–945.

[12] 董晟全,周敬恩,严文等. 高强度铸造铝合金凝固过程中的元素偏析[J]. 铸造技术,2002,(01): 50–52.

[13] 龚磊清,金长庚,刘发信等. 铸造铝合金金相图谱[M]. 长沙: 中南工业大学出版社,1987: 39–41.

[14] 周尧和,胡壮麟,介万奇. 凝固技术[M]. 北京: 机械工业出版社,1998: 64–70.

[15] 丁浩,傅恒志,刘忠元等. 凝固收缩补偿与合金的热裂倾向[J]. 金属学报,1997(09).

[16] 王业双,王渠东,丁文江等. 合金的热裂机理及其研究进展[J]. 特种铸造及有色合金,2000(02).

[17] 胡汉起. 金属凝固[M]. 北京: 冶金工业出版社,1985: 326–334.

模压变形 5052 铝合金的组织结构与室温拉伸性能

杨开怀 [1,2]，曾建民 [1]，邹泽昌 [2]，陈文哲 [3]

(1.广西大学 广西有色金属及特色材料加工重点实验室；2.福建船政交通职业学院 机械工程系；3.厦门理工学院)

摘　要：采用模压变形(Groove Pressing,GP)法变形 5052 铝合金板材,系统研究了 GP 变形对 5052 铝合金组织结构和室温拉伸性能的影响规律。结果表明：GP 变形能够显著细化 5052 铝合金,平均晶粒尺寸随变形道次的增加而减小；变形 4 道次试样平均晶粒尺寸约为 $1.1\mu m$。GP 变形能够显著强化 5052 铝合金, 抗拉强度随变形道次的增加而单调递增；其伸长率随变形道次的增加基本保持不变且在变形 3 道次后缓慢提高,这与 GP 变形中的晶粒细化微观过程和试样的位错运动微观机制紧密相关。

关键词：模压变形；5052 铝合金；组织；拉伸性能

Microstructures and Tensile Properties of 5052 Aluminum Alloy Processed by Groove Pressing

YANG Kai-huai[1,2], ZENG Jian-min[1], ZOU Ze-chang[2], CHEN Wen-zhe[3]

(1. Ministry-province Jointly-constructed Cultivation Base for State Key Laboratory of Processing for Non-ferrous Metal and Featured Materials; 2. Department of Mechanical Engineering, Fujian Chuanzheng Communications College; 3. Xiamen University of Technology)

Abstract: Microstructures and tensile properties of commercial 5052 aluminum alloy sheets processed by groove pressing （GP）, using a novel self-designed groove pressing die, were investigated. The results show that GP can effectively refine the coarse-grained structure of 5052 aluminum alloy to an ultra fine grain range. The average grain size gets decreased with the increasing of the number of pressings, and decreased to 1.1 μ m after 4 passes. GP are quite good for strengthening the 5052 Al alloy, and the ultimate tensile strength increases with the increasing of pass number. The elongation of the sample decreases firstly and then keeps constant approximately, and increases slightly after the third pass. The tensile properties are closely related to grain refinement process and the mechanism of dislocation movement during GP.

Key words: groove pressing; 5052 aluminum alloy; microstructures; tensile properties

基金项目：广西大学广西有色金属及特色材料加工重点实验室开放基金资助(合同编号：GXKFJ14-02)；
福建省高校杰出青年科研人才培育计划资助(JA14371)；福建省交通厅重点项目资助(201210)。
第一作者：杨开怀(1980—)，男，博士，副教授。先进结构材料研究。ykh1347@sina.com。
通讯作者：曾建民，男，教授，博导。clxy3693@126.com；陈文哲，教授，博导。chenwz@fzu.edu.cn。

等温锻整体铝合金汽车轮圈轮辐结构优化数值模拟研究

王家宣，曾　欣，周弘庆，余明喜

(南昌航空大学 航空制造工程学院)

摘　要：采用 DEFORM-3D 数值模拟软件对整体式铝合金汽车轮圈等温锻造进行模拟，得到不同的轮辐结构在等温锻造下的流动应力—应变曲线，并分析不同轮辐结构对轮圈等温锻造过程中的等效应力场、等效应变场分布情况的影响。根据模拟结果得到：整体式铝合金汽车轮圈在 450℃、变形速率为 2mm/s 以及摩擦系数为 0.3 的条件下进行等温锻造，轮辐孔连皮体积越大，锻件的锻造载荷越小，等效应变和等效应力的分布情况更为良好，等效应变的极限差也越低，高应力值的区域面积较小；轮辐孔数量的多少对锻造载荷的影响不是很大，无轮辐孔结构的锻件等效应变场分布最为均匀，内部组织均匀，锻件出现缺陷的概率最小，但在后续加工轮辐孔时容易将锻件内部的纤维流线切断，对于有轮辐孔结构的锻件来说，轮辐孔数量越多，等效应变场分布越均匀，锻件出现缺陷的概率越小。

关键词：数值模拟；铝合金；轮圈；等温锻造

The Numerical Simulation Study of Unitary Aluminum Alloy Wheel Spokes Structure Optimization by Isothermal Forging

WANG Jia-xuan, ZENG Xin, ZHOU Hong-qing, YU Ming-xi

(School of Aeronautical Manufacturing Engineering Nanchang Hangkong University)

Abstract: Using the DEFORM-3D numerical integral of automotive aluminum wheel isothermal forging simulation software, to obtain the flow of spoke structure in isothermal forging under different stress strain curve, and the analysis of different structures on equivalent wheel spokes in isothermal forging process of stress field, strain field distribution. According to the simulation results: integral aluminum alloy car wheels under the condition of the forging temperature at 450℃, the deformation rate for isothermal forging at 2mm/s and the friction coefficient is 0.3, the forging load is smaller and equivalent strain and equivalent should stress distribution is better, the equivalent strain limit is lower, high stress value area is smaller by the spoke holes even skin volume turn large; influence by the number of spokes hole to forging load is not obviously, no spoke hole structure forging strain field has the most uniform distribution, uniform internal organization, forging defect probability is smallest, but in the subsequent processing spoke holes easy to cut off the forgings with inner fiber flow line, for the spoke hole structure of forgings, the spoke hole number, effect becomes more uniform field distribution, the probabilityof forging defect is smaller.

Key words: numerical simulation; aluminum alloy; wheel; isothermal forging

---

第一作者：王家宣，男。材料加工特种精密成形。wjiax@sohu.com。

脉冲电流作用下偏晶合金连续凝固过程研究

江鸿翔，赵九洲

（中国科学院 金属研究所）

摘　要：本文以 Cu-Bi-Sn 偏晶合金为对象，实验与模拟相结合，研究了脉冲电流对偏晶合金连续凝固组织演变的影响。结果表明，脉冲电流主要通过改变液-液相变过程中弥散相液滴的形核能垒来影响偏晶合金凝固过程及组织。当弥散相液滴电导率大于基体熔体电导率时，脉冲电流促使液-液相变过程中弥散相液滴的形核能垒降低，随着电流密度峰值的增加，偏晶合金凝固组织中弥散相粒子的尺寸减小。当弥散相液滴电导率小于基体熔体电导率时，脉冲电流促使液-液相变过程中弥散相液滴的形核能垒升高，随着峰值电流密度的增加，偏晶合金凝固组织中弥散相粒子的尺寸增加。

关键词：偏晶合金；脉冲电流；连续凝固；形核率

Study of Continuous Solidification of Cu-Bi-Sn Alloys Under Effect of Electric Current Pulses

JIANG Hong-xiang, ZHAO Jiu-zhou

(Institute of Metal Research, Chinese Academy of Sciences)

Abstract：Continuous solidification experiments were carried out with Cu-Bi-Sn alloys under the effect of electric current pulses （ECPs）. A model was developed to describe the microstructure formation of immiscible alloys under effect of ECPs. The experimental and numerical results demonstrate that ECPs mainly affect the microstructure formation through changing the nucleation behaviors of the precipitated phase droplets （PPDs）. When the PPDs have a higher electrical conductivity compared to the matrix, ECPs enhance the nucleation rate and promote the formation of a well dispersed microstructure, whereas ECPs cause a decrease in the nucleation rate and promote the formation of a phase segregated microstructure.

Keywords: immiscible alloy; electric current pulses; solidification; nucleation rate

---

基金项目：国家自然科学基金项目(51271185)，(51471173)以及中国载人航天工程项目资助。
第一作者：江鸿翔，男，博士。合金凝固过程研究。
通讯作者：赵九洲，男，博导。合金凝固和固态相变方面研究。jzzhao@imr.ac.cn。

焊接峰值温度对 X80 大变形管线钢热影响区软化的影响

由宗彬 [1,2]，李烨铮 [1,2]，刘 宇 [1,2]

(1.油气管道输送安全国家工程实验室；2.中国石油天然气管道科学研究院)

摘 要：本论文研究了焊接峰值温度对 X80 大变形钢热影响区软化的影响，利用 Gleeble 3500 热模拟试验机模拟热输入为 20kJ/cm、峰值温度为 500℃至 1300℃ 的单次焊接热循环，针对模拟试样，进行了室温拉伸试验、示波冲击试验、维氏硬度试验，采用扫描电子显微镜、金相显微镜进行了显微组织的分析，研究不同峰值温度下 X80 大变形管线钢单道焊热影响区的力学性能和组织特征。试验结果表明：峰值温度为 800℃~1000℃时，X80 大变形钢焊接热影响区出现了软化现象，其中屈服强度和维氏硬度的下降较为显著，抗拉强度的变化较小，金相组织以铁素体和贝氏体为主；峰值温度为 1200℃~1300℃时，屈服强度和抗拉强度明显增大，均匀延伸率显著减小。在 X80 大变形管线钢自保护半自动焊的现场焊接热输入条件下，单道焊临界热影响区出现了显著的软化现象。

关键词：X80 大变形钢；焊接峰值温度；热影响区软化；力学性能；显微组织

Effect of the Peak of Temperature on Softening of Heat-affected Zone of X80 Pipeline Steel with Excellent Deformability

YOU Zong-bin[1,2], LI Ye-zheng[1,2], LIU Yu[1,2]

(1. National Engineering Laboratory for Pipeline Safety, China; 2. Pipeline Research Institute of CNPC)

Abstract: The effect of the peak of temperature on Softening of heat-affected zone of X80 pipeline steel with excellent deformability were studied for this paper, the welding heat input for 20 kJ/cm and the peak temperature of 500℃ to 1300℃ of single welding thermal cycle were simulated by the Gleeble-3500 thermal simulated test machine. For simulated samples, the room temperature tensile test, impact test and vickers hardness test were haved, the microstructure was analyzed by scanning electron microscope and metallographic microscope. Under different peak temperature, the mechanical properties and organizational characteristics of single pass welding heat affected zone of X80 pipeline steel with excellent deformability were studied. The results show that, when peak temperature was 800℃~1000℃, the softening was finded for welding heat affected zone of the X80 pipeline steel with excellent deformability, the drop of the yield strength and vickers hardness was significant, but the change of tensile strength is small, the microstructure was given priority to with ferrite and bainite; when peak temperature was 1200℃~1300℃,with the yield strength and tensile strength increased obviously, the percentage totle extension at maximum force decreased significantly. Under the condition of the welding heat input of self-protection semi-automatic welding for X80 pipeline steel with excellent deformability, the critical heat affected zone of single pass welding appeared significant softening phenomenon.

Key words: X80 pipeline steel with excellent deformability; peak temperature; the softening of heat-affected zone; mechanical properties; microstructure

基金项目:油气管道输送安全国家工程实验室资助项目;中国石油天然气管道局项目。

第一作者:由宗彬,男。管道工程材料。youzongbin@126.com。

工艺参数对 TiAl 基合金熔体凝固行为的影响

叶喜葱，赵光伟

（三峡大学 机械与动力学院）

摘　要：本文利用金属型底浇式真空吸铸方法制备 TiAl 合金叶片铸件，研究吸铸工艺参数对合金熔体凝固行为的影响。利用 ProCAST 软件，计算 TiAl 基合金的物性参数，并分别对不同的铸型温度、浇注温度、浇注速度等工艺参数对合金熔体的凝固行为影响进行了数值模拟研究，得到了 TiAl 基合金金属型底浇式真空吸铸的优化工艺参数。根据优化的参数，获得了形状完整、铸造缺陷少的叶片。金属型的强制冷却和 Si 元素的共同作用细化了铸件组织，平均晶粒尺寸小于 20um。

关键词：真空吸铸；TiAl 基合金；凝固行为；缩孔

The Influence of Process Parameters on the Solidification Behavior of TiAl Based Alloy Melt

YE Xi-cong，ZHAO Guang-wei

（School of Mechanical and Power Engineering, China Three Gorges University）

Abstract: In this paper, the TiAl alloy blade casting was obtained by metal-bottom pouring vacuum suction casting, and the influence of suction casting process parameters on the solidification behavior of TiAl based alloy melt. The ProCAST software was used to calculate the TiAl based alloy physical parameters, and the influence of mold temperature, pouring temperature, pouring velocity parameters on the solidification behavior of the molten alloy was simulated study. The optimization process parameters of TiAl based alloy metal-bottom pouring vacuum suction casting was obtained. On the optimization process parameters condition, the blade casting was obtained with complete shape and less casting defects. The effect of permanent mold and Si element refine the structure of casting, and the average grain size is less than 20um.

Key words: suction xasting; TiAl based alloy; solidification behavior; shrinkage

---

第一作者：叶喜葱，男。新型结构精密材料成形技术。yexc@ctgu.edu.cn。

镁合金铸轧板坯中心线偏析形成机制

翁文凭[1]，盛敏奇[1]，屈天鹏[1]，许继芳[1]，王 璟[1]，陈 琦[2]

(1.苏州大学 沙钢钢铁学院；2.苏州有色金属研究院有限公司)

摘 要：采用金相显微镜、扫描电镜等手段，分析镁合金铸轧板坯中心线偏析的组织、形貌与微区成分；利用数值模拟方法，计算机仿真镁合金铸轧过程中凝固传热与流动过程；结合铸坯中心线偏析组织、形貌及成分特征，分析镁合金铸轧板坯中心线偏析形成过程。结果表明：中心线偏析由富含Al、Zn溶质元素的金属间化合物组成，Mn元素在偏析区域内贫化，偏析区域主要由细小等轴晶粒组成；铸轧速度对镁合金铸轧板坯中心线偏析形成过程产生重要的影响，当控制铸轧速度在合理范围时，可利用过热金属液冲刷凝固前沿引起富含溶质金属液强迫对流，改善溶质元素扩散条件，抑制中心线偏析形成。

关键词：镁合金；板带铸轧；中心线偏析；数值模拟

Formation of Central-line Segregations during Twin-roll Casting of Magnesium Alloys

WENG Wen-ping[1], SHENG Min-qi[1], QU Tian-peng, XU JI-Fang-[1], WANG Jing[1], CHEN Qi[2]

(1. Shagang Scholl of Iron and Steel, Soochow University;2. Department of Materials Science and Technology, Suzhou Nonferrous Metals Research Institute)

Abstract: Microstructure and micro-area chemical analysis of central-line segregations of magnesium alloys strip processed by twin-roll casting were investigated by OM, SEM and WDS. Solidification process of TRC strip was analyzed by numerical simulation under different conditions. The results indicate that central-line segregations were compromised by solute rich intermetalics of Al, Zn elements, the percentage composition of Mn element in central-line segregations was decreased, furthermore the structure of central-line segregations changes to equiaxed grain with fine grain size. The central-line segregations could be decreased or eliminated by controlling the speed in a proper location because the overheated melts supplied by feed tip could washouts the solidification front, which caused the re-melt of the solidification front and the forced convection, and it played an important role on the solute elements diffusion around the solidification front.

Key words: magnesium alloys; twin-roll casting strip; central-line segregations; numerical simulation

---

基金项目：国家自然科学基金(No.51204115)。

第一作者：翁文凭，男。轻合金材料制备及其液态成形。wenpingweng@suda.edu.cn。

通讯作者：盛敏奇，男。金属材料及界面工程。shengminqi@suda.edu.cn。

C-Mn 钢等温热处理的力学性能研究

周旭东[1]，刘香茹[2]，李 俊[3]，王 健[3]

(1.河南科技大学 材料科学与工程学院；2.河南科技大学 物理工程学院；3.宝山钢铁股份有限公司研究院)

摘 要：采用物理模拟方法，在 Gleeble1500D 热力学模拟试验机上对碳含量为 0.16% 的 C-Mn 钢冷轧板进行不同温度和不同时间的等温热处理试验和力学性能测试，探索工艺参数对力学性能的影响规律，并建立力学性能数学模型。结果表明：600.℃~650.℃范围内力学性能变化随温度最迅速，随时间最快速，这一区间是铁素体再结晶 S 曲线最陡的区间。依据加权方法建立的分段力学性能数学模型具有较高的精度，延伸率的绝对误差为 1.82%，屈服强度绝对误差为 21.87MPa，抗拉强度绝对误差为 21.00MPa。

关键词：C-Mn 钢；等温热处理；力学性能

Mechanical Properties of Isothermal Heat Treatment of C-Mn Steel

ZHOU Xu-dong[1], LIU Xiang-ru[2], LI Jun[3], WANG Jian[3]

(1. School of Materials Science and Engineering, Henan University of Science and Technology; 2. School of Physics and Engineering, Henan University of Science and Technology; 3. Baosteel Institute, Baoshan Iron & Steel Co)

Abstract: Both the isothermal heat treatment experiment of C-Mn cold-rolled sheet steel with carbon content 0.16% and mechanical properties test are carried out with physical simulation method on Gleeble1500D thermal mechanical simulation testing machine, at different temperatures and at different duriation of heat treatment, in order to explore the influence of isothermal heat treatment process parameters on mechanical properties, and finally to establish the mathematical models of the mechanical properties of the steel. The results show that: on the range of 600 ~ 650℃, the mechanical properties changes with the most rapid steep as temperature elevates, and with the fastest speed as time increases. Because the ferrite recrystallization S shape curve at this range is of steepest. The segmented mathematic models based on weighting method have better accuracy. The elongation absolute error is 1.82%, the yield strength absolute error is 21.87MPa, and the ultimate strength absolute error is 21.00MPa.

Key words: C-Mn steel; isothermal heat treatment; mechanical properties

第一作者：周旭东，男。材料成形模拟技术。syuuzhou@163.com。

热处理温度和冷速对 DP980 钢烘烤硬化性能的影响

刘香茹[1]，周旭东[2]，李　俊[3]，王　健[3]

(1.河南科技大学 物理工程学院；2.河南科技大学 材料科学与工程学院；3.宝山钢铁股份有限公司研究院)

摘　要： 高强双相钢 DP980 钢是铁素体马氏体双相钢,其抗拉强度达 980MPa 等级,多用于汽车领域。本文采用 Gleeble1500D 热力模拟试验机为手段,对 DP980 的烘烤硬化性能进行研究,选用的热处理温度分别为 750℃、850℃和 950℃,保温时间为 2min,冷却速度分别为 10℃/s、20℃/s 和 50℃/s,预变形 2%后进行 170℃下 20min 烘烤硬化处理, 并对试验样品进行微观组织观察。研究结果表明：DP980 钢具有烘烤硬化性能,烘烤处理后,铁素体相中析出很多大小在 10~20nm 的相；热处理温度及冷速对 DP980 烘烤硬化值都有很大的影响,850℃、冷速 50℃/s 时 DP980 的烘烤硬化值最为理想,其组织为细小的马氏体板条(板条长度最长 10μm 左右)和铁素体组织。

关键词： DP980 钢；烘烤硬化；热处理

Effect of Heat Treatment Temperature and Cooling Rate of DP980 Steel on Bake-hardening Property

LIU Xiang-ru[1], ZHOU Xu-dong[2], LI Jun[3], WANG Jian[3]

(1. School of Physics and Engineering, Henan University of Science and Technology; 2. School of Materials Science and Engineering, Henan University of Science and Technology; 3. Baosteel Institute, Baoshan Iron & Steel Co)

Abstract: High-strength dual-phase steel DP980, ferritic martensitic dual-phase steel, has the tensile strength of 980MPa grade, used for the automotive parts. In this paper, with Gleeble1500D thermal simulation test machine, the bake-hardening properties of DP980 were studied, the choice of the heat treatment temperature was 750℃, 850℃ and 950℃, holding time of 2 min, the cooling rate was 10℃/s, 20℃/s and 50℃/s, 2% of pre-deformation, the bake-hardening aging process at 170℃ 20min, and the test sample microstructure was observed. The results show that:

DP980 steel has good bake-hardening property. After the baking process, there are a lot of precipitates particle in size 10~20nm in the ferrite phase; DP980 bake hardening value is affected greatly by the heat treatment temperature and cooling rate. Finally the heat treatment parameter of 850℃ and cooling rate 50℃/s is the best one for the highest bake hardening value of DP980 steel with the microstructure of fine martensite laths (lath longest length about 10μ m) and ferrite.

Keywords: DP980 steel, bake hardening, heat treatment

第一作者:刘香茹,女。材料计算设计与模拟。*xiangruliu314@126.com。*

从赤泥还原铁合金的试验模拟

何奥平，曾晓乐，曾建民 *

(广西大学 有色金属及加工新技术教育部重点实验室)

摘　要：本工作采用小型电弧炉进行从赤泥中还原铁合金的冶金过程模拟。为丰富铁合金成分，在配料时，赤泥中加入了红土镍矿。结合还原产品的元素组成和分布，主要研究了碳含量对原料中铁和其他有价金属元素还原效率的影响。实验结果表明：随着碳含量增加，铁和镍、铬、钛等有价金属都能得到还原，并且回收率显著提高，有价元素镍、铬、钛均能较好的进入到合金中。综合各金属元素在合金中的品位和回收率，在赤泥和红土镍矿质量配比为 1:1 时，最佳焦粉配比为 25%，铁合金总回收率可达到 82.55%，铁、镍、铬、钛的品位分别为 88.47%、1.08%、2.08% 和 2.78%。可见，电弧冶金技术有可能成为赤泥和红土镍矿还原制备铁合金的新工艺，具有使用前景。

关键词：赤泥；红土镍矿；铁合金；还原

Experimental Simulation of Reduction of Fe Alloy from Red Mud

HE Ao-ping, ZENG Xiao-le, ZENG Jian-min

(Key Lab. of Nonferrous Materials and New Processing Technology, Ministry of Education, Guangxi University)

Abstract: This work aims at simulating the metallurgical process of reduction of Fe alloy from red mud using small arc furnace. In order to enrich the composition of the Fe alloy, laterite nickel ore was added in the raw materials. The effect of reduction rate of Fe and other valuable metals was studied by combining element composition with distribution. The results show that the reduction rate of Fe, Ni, Cr and Ti increases significantly with the increasing of carbon content in the raw materials. Ni, Cr and Ti can be all dissolved into the alloy. The optimum addition of coke is 25% by mass percentage for the best metal grade and recovery in the case of 1:1 by mass ratio between red mud and laterite nickel ore. The total recovery of the Fe alloy reaches 82.55%. The grade of Fe, Ni, Cr and Ti is 88.47%、1.08%、2.08% and 2.78%, respectively. Therefore, the arc metallurgy may become a potential technique for reduction of Fe alloy from red mud and laterite nickel ore.

Key words: red mud; laterite nickel ore; Fe alloy; reduction

基金项目：广西自治区"铝合金材料先进加工技术"八桂学者专项经费资助；广西大学科研基金项目(XJZ140258)。

第一作者：何奥平(1979—)，女，讲师，在读博士。

通讯作者：曾建民(1955—)，男，教授，博士。

Ni 基高温合金中合金化元素对 γ/γ' 相界面错配位错运动影响

樊沁娜 [1,2]，**王崇愚** [1,3]，**于 涛** [1]

(1.钢铁研究总院功能材料所；2.东北大学 理学院；3.清华大学 物理系)

摘 要: Ni 基高温合金中的 γ/γ' 界面错配位错网对于抑制位错运动、提高合金蠕变性能有重要的作用。本文采用分子动力学和嵌入原子势方法研究了在剪切形变下合金化元素(Re、Co 和 W)对错配位错在 γ/γ' 相界面运动的影响。研究发现在剪切形变过程中,错配位错运动通过扭折的形核和扭折沿着位错线的迁移来实现。合金化元素 W 和 Re 可以阻碍错配位错的运动而 Co 不能,W 阻碍错配位错运动的能力比 Re 强,与元素对错配位错钉扎作用的强弱一致。

关键词：Ni 基高温合金；相界面；错配位错；分子动力学

The Effect of Alloying Elements on the γ/γ' Interface Misfit Dislocation Motion in Ni-based Superalloys

FAN Qin-na [1,2]，**WANG Chong-yu** [1,3]，**YU Tao** [1]

(1.Division of Function Materials, Central Iron and Steel Research Institute;
2.College of Science, Northeastern University; 3. Department of Physics, Tsinghua University)

Abstract: γ/γ'interface misfit dislocation network plays an important role in preventing the motion of dislocations and improving the creep resistance in Ni-based superalloys. The alloying effect of elements (Re、Co and W) on the motion of misfit dislocations in γ/γ' interface under applied shear stress in Ni-based superalloys is investigated by employing Molecular Dynamics (MD) simulation using Embedded Atom Method (EAM) potential. During the shearing process, it is found that the misfit dislocation move through the nucleation of kinks and migration along the dislocation line. Both the alloying element Re and W prevent the motion of the misfit dislocation but Co don't. The ability of preventing the motion of the misfit dislocation of W is stronger than that of Re, which is consistent with the order of pinging effect of misfit dislocation..

Key words: Ni-based superalloy; interface；misfit dislocation; molecular dynamics

0 引言

Ni 基高温合金具有出色的高温力学性能,被广泛用作航空发动机叶片的重要结构材料[1,2]。Ni基单

基金项目:国家重点基础研究发展计划资助项目(2011CB606402)。

第一作者:樊沁娜,女,钢铁研究总院、东北大学联合培养博士生。fqn528@163.com。

晶高温合金具有独特的"马赛克"共格结构：$L1_2$ 有序结构的高体积分数的强化析出相 γ′–Ni3Al 嵌入在具有 FCC 结构的 γ–Ni 基体相中。两相不同的晶格常数会产生晶格应力，而这种晶格应力可以通过错配位错网络的形成得到释放[3,4]。这些错配位错网络可以有效阻止 γ 基体相中的位错切割进入 γ′ 相，从而提高体系的蠕变强度[5,6]。为了提高材料的力学性能，Ni 基高温合金中加入了大量的固溶强化元素来[7,8]，探索这些元素对错配位错网络的运动具有重大意义。

合金化元素对错配位错稳定性的影响已被广泛研究。Geng[9]利用第一原理方法 DMol 研究发现，Ta 和 W 都可以提高 a/2 [110] (001)刃型错配位错的稳定性。Chen 等[10]用 DMol 方法研究发现，合金化元素可以显著提高界面结合能和剪切强度，Re 和 Ru 的协同强化效应要强于单独的 Re 和 Ru[11]。对于 Co、Re 和 W，元素对错配位错的钉扎作用随着原子半径的增大而增大[12]。

由于实验很难观察在原子层次的位错运动，分子动力学(MD，Molecular dynamics)方法作为一个准确有效的方法，可以从原子机制上来研究固溶元素对高温合金的力学性能的影响。MD 方法需要势函数来提供原子间的相互作用，然而由于缺乏多元势函数，文献中仅有目前只有 H 杂质对 γ/γ′ 相界面错配位错运动的影响[13]，目前还没有合金化元素对错配位错运动的相关研究。因此我们采用本组建立的系列三元势[14,15,16] (Ni-Al-X，X 代表 Re、Co 和 W)对合金化元素对错配位错运动的影响进行了研究。

1 模型

我们构造了(001)面上 γ(Ni)/γ′(Ni3Al)相界面的错配位错模型。如图 1 所示，原子模型的取向为 X=[110]、Y=[110]及 Z=[001]，模型在 Z 方向分为上、下两部分。上半部分为 γ(Ni)相，尺寸为(n+1)[110]、(n+1)及 15[001]，下半部分为 γ′(Ni3Al)相，尺寸为 n[110]、n 及 15[001]。n 的取值通过方程 $(n+1)a_\gamma\sqrt{2} = na_{\gamma'}\sqrt{2}$ 求解确定 n=75，γ(Ni)及 γ′(Ni3Al)相的晶格常数分别为 a_γ(3.52Å)、$a_{\gamma'}$(3.567Å)。在 X、Y 方向上采用周期边界条件，在 Z 方向采用自由边界条件，γ(Ni)相及 γ′(Ni3Al)相之间定义为相界面和滑移面，模型中包含 1,368,120 个原子。我们使用分子动力学模拟软件 LAMMPS[17]弛豫体系得到错配位错模型，所用的时间步长为 5×10^{-15}s，并且维持体系的温度为 5K。弛豫后得到 γ(Ni)/γ′(Ni3Al)相界面的错配位错模型错配位错的伯格斯矢量为 a/2<110>(a 为晶格常数，3.567Å)。使用可视化软件 OVITO[18]观察体系的结构变化。OVITO 软件中的 CNA(Common Neighbor Analysis)[19]算法可以通过区分体系中不同结构(fcc，bcc，hcp 等)来识别位错。

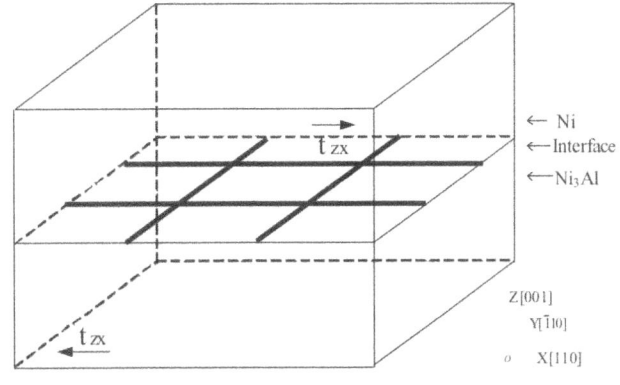

图 1 错配位错模型

Fig.1 Diagrammatic sketch of interface misfit dislocation model.

在高温合金中,Re、W、Co 偏向分布在 γ(Ni)相,因此我们分别在 γ(Ni)相中随机掺入了 3at.%浓度的 Re、W、Co。采用本组构造的系列三元 Ni-Al-X(X 为 Re、Co、W)[17,18,19]研究合金化元素对错配位错运动的影响,在得到体系的错配位错模型之后,我们以恒定的速率对体系施加均匀剪切形变。加载过程中采用速度标定方法保持体系的温度不变。

2 结果与讨论

2.1 随机掺杂合金化元素 Re、W、Co 对错配位错构型的影响

为了研究合金化元素 Re、W、Co 对错配位错构型的影响,在 γ(Ni)相中分别随机掺入了 3%浓度的 Re、W、Co,然后对体系进行弛豫。图 2 是利用 OVITO 软件中 CNA 功能给出弛豫后的界面位错构型,所示平面是 XY 平面。

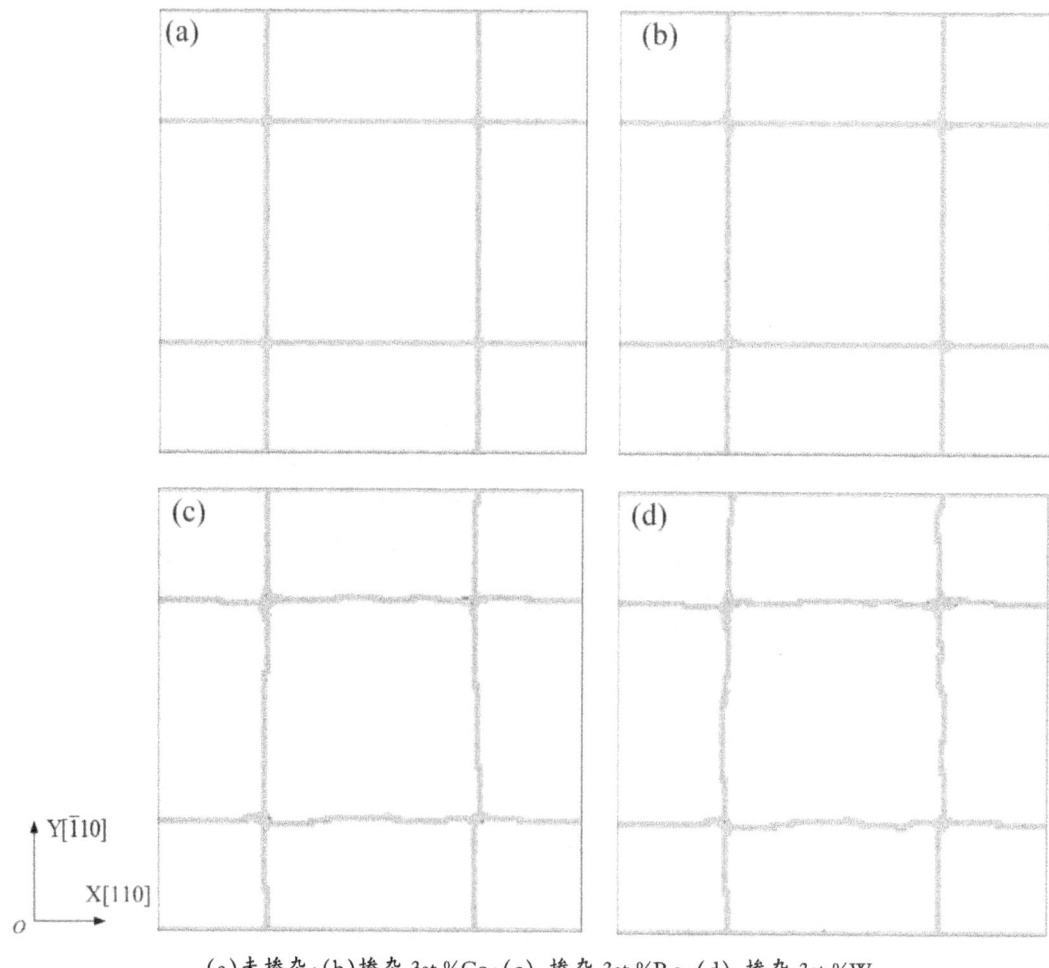

(a)未掺杂;(b)掺杂 3at.%Co;(c) 掺杂 3at.%Re;(d) 掺杂 3at.%W

图 2　弛豫后的 γ/γ′ 相界面错配位错网

Fig.2　γ/γ′ interface misfit dislocation network after relaxing

由图 2(b)可以看出,Co 掺杂后对 γ/γ′ 相界面错配位错结构影响最小,只有在位错交叉处略微不同。图 2(c)、(d)分别为在错配位错模型中掺杂 3%的 Re 和 W 后的错配网格结构。可以看出错配位错

上产生了扭折,而且掺杂 W 之后产生的扭折数量比 Re 更多。产生扭折的原因是由于掺杂元素的原子半径较大,从而引起了两相的错配度的变化[20]。在有掺杂原子分布的区域的错配度变化比较大,这样就引起了错配位错的扭折。同时也可以看出,Co、Re 和 W 引起的错配位错的产生扭折的数量也随着原子半径的增大而增多。

2.2 错配位错运动的扭折机制

本节以随机掺杂 3 at.%的 W 元素模型为例研究错配位错移动的过程。为了使错配位错开始运动,在 XZ 面施加剪切形变。剪切形变的定义 $\varepsilon = (\partial / \partial z)u_x$,$u_x$ 为 X 方向的位移。我们对体系施加平均速度为 $1.0 \times 10^{-2}/ps$ 的均匀剪切形变,观察错配位错的运动过程。

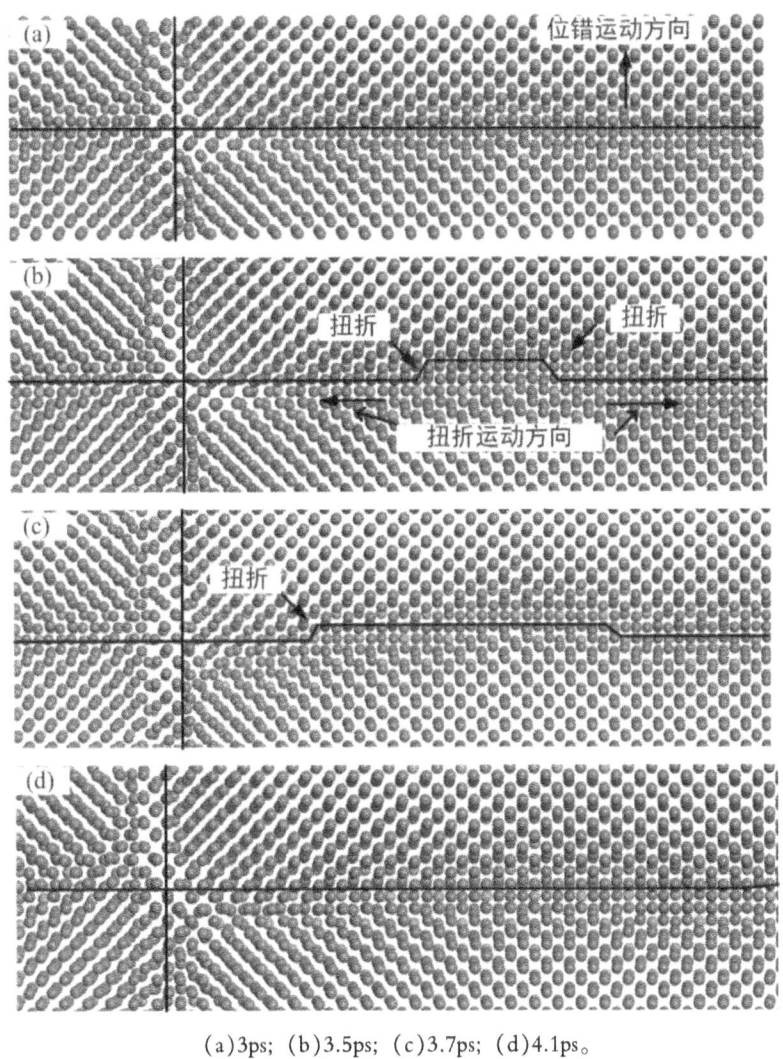

(a)3ps;　(b)3.5ps;　(c)3.7ps;　(d)4.1ps。

图 3　一段错配位错在不同时刻的位错结构图

Fig.3　Structure of a piece of moving misfit dislocation at different time

由于弛豫后的错配位错出现了扭折的结构,我们取一段没有扭折的平整位错线来观察它的运动。图 3 给出了所取位错段的运动情况,子图为不同时刻的位错结构图。在时间为 3.0ps 时,位错线平直的躺在 Peierls 能谷中;3.5ps 时剪切应变达到 0.035,有一节位错线越过了 Peierls 势垒到达了下一个能谷,形成了一个原子尺度的双扭折;随后在 3.7ps 时,扭折沿着位错线继续迁移;最终,在 4.1ps 时,和符号相反的扭折湮没,使得位错线向前移动。这表明,在掺入合金元素后,错配位错的运动依然是通过扭折的形核和迁移来实现。根据位错理论[21],扭折的形成能小于整段位错运动的 Peierls 势能,所以扭折首先形核逐渐移动并带动整段位错的移动。

2.3 随机掺杂合金化元素 Re、W、Co 对错配位错运动的影响

为了研究合金化元素对错配位错运动的影响,我们分别在 γ/γ' 相界面错配位错模型中随机掺杂了 3 at.%的 Co、Re 和 W。弛豫得到稳定构型后,对体系在 XZ 面施加平均速度为 1.0×10-2/ps 的均匀剪切形变,位错沿 X 方向移动。图 4 所示为在剪切作用下,相界面体系未掺杂第三元素和分别随机掺入 Co、Re 和 W 之后,体系运动了 5ps 后的界面错配位错网结构。由图 4 可见,四种情况下错配位错交叉处的位错运动均超前于其他部位,说明位错交叉点的移动带动了位错其他部分的移动。对于以上四种情况,施加剪切形变之后 5ps 位错交叉部位的位置基本相同,说明掺杂第三元素后基本不影响位错节点处的运动。这是由于位错交叉处的势能较高,对应的原子键和能力较弱,所以掺杂元素对位错的影响也较弱。由图 4 可见,当随机掺入 3 at.%的 W 元素位错运动 5ps 后位错线上的扭折最多,位错宽度也最大,说明 W 元素对错配位错运动的影响最大。

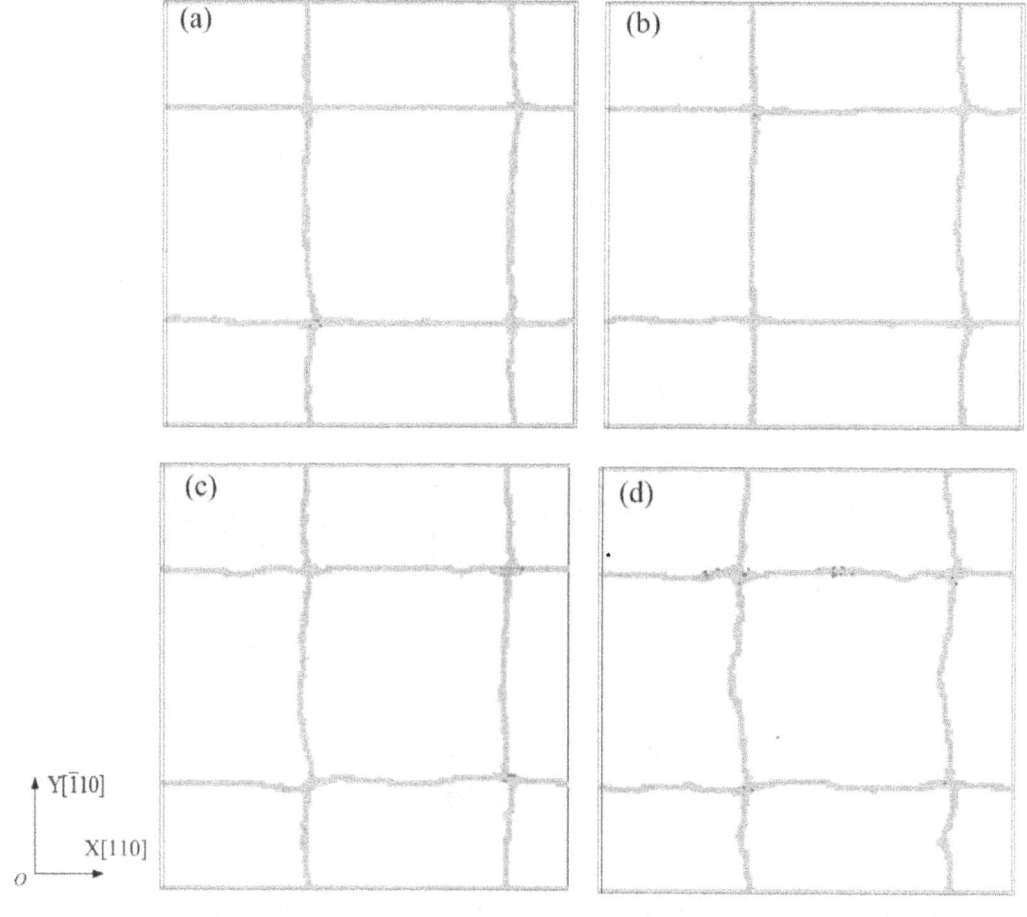

(a)未掺杂;(b)掺杂3 at.%Co;(c) 掺杂 3 at.%Re; (d) 掺杂 3 at.%W。

图 4 施加均匀剪切应变 5ps 后 γ/γ' 相界面错配位错网

Fig.4 Structure of γ/γ' misfit dislocation network under homogeneous shear strain at 5ps.

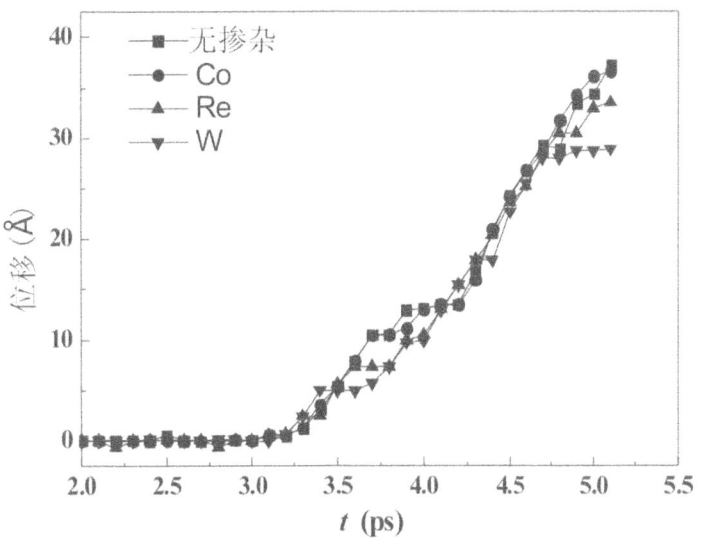

图 5 右侧错配位错的最小位移

Fig.5 Minimum displacement of the right side misfit dislocation

施加均匀剪切形变后,整条位错不同部分的移动速度不同。由于加入不同合金化元素后,位错交叉部位移动最快且位置基本相同,为了定量观察并比较加入不同合金化元素后错配位错的变化,选取位错最慢处的位移进行比较。我们统计了 XY 平面上右侧伯格斯矢量为 a/2<110> (a 为晶格常数,3.567Å)的错配位错线沿 X 方向的最小位移,如图 5 所示。由图 5 可见在 3.0ps 之前,剪切形变量小于 0.03 之前,位移为 0,位错线没有运动。当 3.2ps 时,剪切形变量达到 0.032,位错运动开动,随后位移上升达到一个平台去,之后再次上升达到平台,说明位错的移动是阶梯式的。每一个阶梯约为 2.5Å,相当于一个伯格斯 a/2<110>。由上节分析可知,这是由于位错发生了扭折,当扭折形核并带动整个位错线运动到一个伯格斯矢量后,整段位错从一个 Peierls 势垒到达了下一个能谷。当 4.2ps 时,剪切形变达到 0.042,位错的位移达到约 15Å 时,错配位错开始快速移动,说明错配位错的移动比较均匀,错配位错呈现整体移动的趋势。随后,达到 4.7ps 后掺杂 W 的体系再次出现了平台,而其他模型的位移持续增大,表明位错继续移动。由图 5 可见,随机掺入 3 at.% 的 Co 后位错的移动距离和未掺杂时基本一致,而掺杂 3 at.% 的 Re 和 W 后错配位错的位移始终比未掺杂第三元素时小,表明掺入 Re 和 W 之后会使错配位错的运动速度变慢。W 阻碍错配位错运动的能力比 Re 强。在研究这三种合金化元素对界面错配位错的钉扎作用表明[19],Co、Re 和 W 对错配位错的钉扎作用强弱顺序 Co<Re<W。这三种元素对错配位错运动的阻碍作用与钉扎作用强弱顺序一致。

3 结论

采用分子动力学和嵌入原子势的方法模拟了合金化元素 Co、Re 和 W 对错配位错运动的影响,研究表明:

(1)合金化元素会使错配位错的构型发生变化,Co、Re 和 W 引起的错配位错的产生扭折的数量也随着原子半径的增大而增多。

（2）位错的运动通过扭折的形核和移动实现。

（3）元素 Co 对界面错配位错运动影响不大，Re 和 W 可以阻碍位错的运动，W 阻碍位错错配位错的能力强于 Re。表明合金化元素对错配位错运动的阻碍程度和对错配位错钉扎作用的强弱顺序一致。

参考文献：

[1] Ross E W, Sims C T. Nickel-base alloys [J]. Wiley-Interscience, John Wiley and Sons, Superalloys II--High Temperature Materials for Aerospace and Industrial Power, 1987: 97-133.

[2] Reed R C. The superalloys: fundamentals and applications [M]. New York: Cambridge university press, 2006.

[3] Lahrman D F, Field R D, Darolia R, et al. Investigation of techniques for measuring lattice mismatch in a rhenium containing nickel base superalloy [J]. Acta Metallurgica, 1988, 36(5): 1309-1320.

[4] Devincre B, Kubin L P, Lemarchand C, et al. Mesoscopic simulations of plastic deformation [J]. Materials Science and Engineering: A, 2001, 309: 211-219.

[5] J. X. Zhang, T. Murakumo, Y. Koizumi, et al. Interfacial dislocation networks strengthening a fourth-generation single-crystal TMS-138 superalloy[J]. Metallurgical and Materials Transactions A, 2002, 33 (12): 3741-3746.

[6] N. Clément, M. Benyoucef, M. Legros, et al. In situ deformation at 850 C of standard and rafted microstructures of nickel base superalloys [C]//Materials science forum. 2006, 509: 57-62.

[7] Giamei A F, Anton D L. Rhenium additions to a Ni-base superalloy: effects on microstructure [J]. Metallurgical transactions A, 1985, 16(11): 1997-2005.

[8] Harris K, Erickson G L, Sikkenga S L, et al. Development of the rhenium containing superalloys CMSX-4 and CM186 LC for single crystal blade and directionally solidified vane applications in advanced turbine engines [J]. Superalloys 1992, 1992, 297-306.

[9] Geng C Y, Wang C Y, Yu T. First-principles investigation of the alloying effect of refractory elements Ta and W in the misfit dislocation core of γ/γ' (001) interface [J]. Physica B: Condensed Matter, 2005, 358(1): 314-322.

[10] Chen K, Zhao L R, John S T. A first-principles survey of γ/γ' interface strengthening by alloying elements in single crystal Ni-base superalloys [J]. Materials Science and Engineering: A, 2004, 365(1): 80-84.

[11] Chen K, Zhao L R, John S T. Synergetic effect of Re and Ru on γ/γ' interface strengthening of Ni-base single crystal superalloys [J]. Materials Science and Engineering: A, 2003, 360(1): 197-201.

[12] Fan Q N, Wang C Y, Yu T, Du J P, Physica B: Condensed Matter, 2015, 456: 283-292.

[13] Yu Tao, Xie Hong-Xian, Wang Chong-Yu. Effect of H impurity on misfit dislocation in Ni-based single-crystal superalloy: molecular dynamic simulations [J]. Chinese Physics B, 2012, 21 (2): 026104-026109.

[14] Du J P, Wang C Y, Yu T. Construction and application of multi-element EAM potential (Ni-Al-Re)

in γ/γ′ Ni–based single crystal superalloys [J]. Modelling and Simulation in Materials Science and Engineering, 2013, 21(1): 015007–27.

[15] Du J P, Wang C Y, Yu T. The ternary Ni–A–Co embedded–atom–method potential for γ/γ′ Ni–basedsingle–crystal superalloys: Construction and application [J]. Chinese Physics B, 2014, 23(3): 033401–8.

[16] Fan Q N, Wang C Y, Yu T, et al. A ternary Ni－Al－W EAM potential for Ni–based single crystal superalloys [J]. Physica B: Condensed Matter, 2015, 456: 283–292.

[17] http://lammps.sandia.gov/

[18] Stukowski A. Visualization and analysis of atomistic simulation data with OVITO － the Open Visualization Tool [J]. Modelling and Simulation in Materials Science and Engineering, 2010, 18(1): 015012.

[19] Cleveland C L, Luedtke W D, Landman U. Melting of gold clusters [J]. Physical Review B, 1999, 60(7): 5065–5077.

[20] 郭建亭.高温合金材料学(应用理论基础) [M], 北京: 科学出版社, 2008.

[21] Hirth J P, Lothe J. Theory of dislocations [M]. New York: Wiley, 1982.

钢锭冒口感应加热技术基础研究

赵红昌，顾　涛，周　宣，张美玲，王明家 *

（燕山大学 材料学院；燕山大学 亚稳材料制备技术与科学国家重点实验室）

摘　要：开发了一种钢锭冒口感应加热新技术，系统研究了冒口感应加热技术对铸锭凝固过程的影响规律。采用有限元模拟方法对冒口感应加热条件下的铸锭凝固过程进行模拟计算，得到冒口区域的电磁场、温度场分布并对缩孔的深度及形状进行预测。通过 200kg 铸锭中间试验，测温试验和实际钢锭解剖对数值模拟结果进行了验证，两者吻合度较高。在自然凝固条件下冒口凝固时间为 995s，缩孔深度为 150mm；感应加热输入功率为 13kw 时，冒口凝固时间延长至 1410s，缩孔深度减少至95mm。结果表明：感应加热技术输入的外热源可以控制铸锭冒口凝固进程，延缓冒口凝固时间，减少缩孔深度，显著减小冒口尺寸，提高铸锭利用率。

关键词：冒口感应加热；铸锭；缩孔；数值模拟

Fundamental Research of Induction Heating Technology on Ingot Feeder

ZHAO Hong–chang, GU Tao, ZHOU Xuan, ZHANG Mei–ling, WANG Ming–jia*

(State Key Laboratory of Metastable Materials Science and Technology, Yanshan University)

Abstract：In this paper, a new induction heating technology on ingot feeder was developed, and the effects on solidification process were studied systematically. The solidification process of ingots in different induction heating conditions were simulated by adopting the Finite Element Method (FEM), also the temperature and magnetic field distribution at the feeder, together with the position and shape of the shrinkage cavity were predicted. In order to validate the calculation results, a 200kg ingot semi-plant test was conducted. From the results of cooling curves at characteristic positions and analyzing the distribution and size of the shrinkage cavity by cutting the semi-plant ingot, experimental results were perfectly matched with the numerical analysis. As the solidification time of feeder is 995s and the depth of shrinkage cavity is about 150mm in natural condition, while, when the induction heating power turn into 13kw, the results were shown 1410s and 95mm respectively. The result shows that solidification process of feeder can be controlled by outer source of heat from induction heating, and also the depth of shrinkage cavity can be greatly reduced to improve the utilization rate of ingot by using this new technology.

Key words：feeder induction heating; ingot; shrinkage porosity; numerical simulation

第一作者:赵红昌,男。大型铸锭凝固技术及数值模拟。hczhaoysu@126.com。
通讯作者:王明家,男。大型铸锻件及热处理。mingka@ysu.edu.cn。

深冷处理对中碳低合金钢 ZG30CrSiMnMo 冲击磨料磨损性能的影响

刘　英[1,2]，夏一龙[1]，郭　红[1]，杨　凯[1]，李　卫[1,2]

（1.暨南大学 材料系；2.广东省耐磨及特种功能材料工程技术研究开发中心）

摘　要：对不含 Ni 及含 1.29wt.% Ni 的中碳低合金耐磨钢 ZG30CrSiMnMo 进行 1050℃×2h 水淬+250℃×2h 回火热处理和 1050℃×2h 水淬+液氮(-196℃)深冷处理×8h+250℃×2h)回火热处理,研究深冷处理对材料的显微组织、硬度及冲击磨料磨损性能影响。研究结果表明:不含 Ni 及含 1.29wt.% Ni 的 ZG30Cr2iMnMo 钢显微组织为回火马氏体;不含 Ni 及含 1.29wt.% Ni 的 ZG30CrSiMnMo 钢淬火+回火后的硬度分别为 HRC48.5、HRC49.2；淬火+深冷处理+回火后硬度明显提高，分别为 HRC52.8、HRC54.5；在冲击功为 4.5J 的冲击磨料磨损条件下,淬火+深冷处理+回火钢的耐磨性显著优于淬火+回火钢的耐磨性。深冷处理使钢的硬度与耐磨性显著提高,Ni 的加入也使耐磨性明显提高。ZG30CrSiMnMo 钢主要磨损机制为疲劳剥落磨损和切削磨损。

关键词：ZG30CrSiMnMo 钢；热处理；深冷处理；硬度；冲击磨料磨损

Effect of Cryogenic Treatment on the Impact-abrasion Behavior of ZG30CrSiMnMo Steel

LIU Ying[1,2], XIA Yi-Long[1], GUO Hong[1],YANG Kai[1], LI Wei[1,2]

(1. Department of Materials Science and Engineering, Jinan University; 2. Guangdong Province Engineering Technical R&D Center of Wear Resistant and Special Functional Materials)

Abstract: The excluding Ni and containing 1.29wt.%Ni steel ZG30CrSiMnMo were performed to 1050℃×2h water quenching+250℃×2h tempering heat-treatment and 1050×2h water quenching + liquid nitrogen(-196℃) cryogenic treatment×8h+250℃×2h) tempering heat treatment, The effects of cryogenic treatment on the microstructure, hardness and impact-abrasion resistance of the steel ZG30CrSiMnMo were investigated. The results show that the microstructure of excluding Ni and containing 1.29 wt..% Ni steel ZG30CrSiMnMo is tempered martensite. The hardness of excluding Ni and containing 1.29 wt. %Ni steel ZG30CrSiMnMo after water quenching + tempering heat-treatment is HRC48.5, HRC49. 2, respectively; and after water quenching + cryogenic treatment + tempering heat-treatment, the hardness is HRC52.8, HRC54.5, respectively. As the impact work of 4.5J is adopted in the impact-abrasive wear test, the impact-abrasive wear resistance of ZG30CrSiMnMo steel after water quenching + cryogenic treatment + tempering heat-treatment were superior to the material after water quenching + tempering heat-treatment. Cryogenic treatment has obviously improved the hardness and wear resistance of steel, the addition of Ni and obviously improve the wear resistance. The impact-abrasive wear mechanism is mainly fatigue peeling wear and micro-cutting.

Key words: low-alloyed cast steel; heat treatment; cryogenic treatment; hardness; the impact-abrasion behavior

---

第一作者:刘　英,男,副研究员。liuying2000ly@163.com。

Ce 元素对铸造态 Mg-Li-Al-Ca 系合金
在 NaCl 溶液中的腐蚀行为的影响

杨云龙 [1,2]

(1.吉林工业职业技术学院; 2.吉林大学 材料科学与工程学院)

摘　要：用电化学方法研究 Mg-11Li-3Al-1Ca-1Ce 合金和 Mg-11Li-3Al-1Ca 合金在 3.5%Na-Cl 溶液中的腐蚀行为,利用扫描电镜观察腐蚀后的表面形貌,利用 X 射线衍射仪分析了腐蚀产物的组成,通过失重法测试腐蚀速率。结果表明：Mg-11Li-3Al-1Ca 合金在添加稀土元素 Ce 后,在试样表面形成更为复杂[Mg(OH)2、Ca(OH)2、Al(OH)3]的腐蚀产物,从而使 Mg-11Li-3Al-1Ca-1Ce 合金的抗腐蚀性优于 Mg-11Li-3Al-1Ca 合金。

关键词：镁合金; 抗腐蚀性; 稀土元素

Effects of Ce on Corrosion Behaviour of
Mg-Li-Al-Ca Alloy in NaCl

YANG Yun-long[1,2]

(1. Jilin Vocational College of Industry and Technology; 2. Department of Materials Science and Engineering, Jilin University)

Abstract: Corrosion behaviours of Mg-11Li-3Al-1Ca-1Ce alloy and Mg-11Li-3Al-1Ca in 3.5% NaCl solution were studied by electrochemistry. And surface prographs were observed by SEM after corrosion tests. And corrosion products were analysed by XRD. And corrosion ratio was determined by weight loss method. These results showed that corrosion products of Mg-11Li-3Al-1Ca alloys were made of Mg (OH)2 and Ca (OH)2 and Al (OH)3 after adding Ce, and the corrosivity of Mg-11Li-3Al-1Ca-1Ce alloys is better than that of Mg-11Li-3Al-1Ca alloys.

Kewords: magnesium alloy; corrosion resistance; rare earth element

---

第一作者:杨云龙,男。化工机械。Yyl1976@126.com。

基于 Fluent 的空气射流冷却 BNbRE 钢轨模拟

王海燕 [1,2]，郑梦珠 [1,2]，高雪云 [3]，陈树明 [1,2]，吴志峰 [1]，邹存哲 [1]

(1. 内蒙古自治区白云鄂博矿多金属资源综合利用重点实验室；2. 内蒙古科技大学 材料与冶金学院；]
3.中冶东方工程技术有限公司长材事业部)

摘　要：本文采用 Fluent 软件，基于标准 k-ε 湍流模型对 BNbRE 钢轨在空气射流冷却过程的流场和温度场进行计算,讨论不同冷却风速下淬火钢轨二维与三维模型的温度场及传热特性变化规律。结果表明:随着风速提高,钢轨表面传热系数增大,换热效果增强。从钢轨流场速度矢量变化图可以看出,对于入口风速为 100m/s 的情况,当冷却时间达到 80s 时,上部射流与壁面冲击后产生的贴壁气流分散到两侧,会影响下部射流方向。与前者相比,当入口速度为 200m/s 时,当冷却时间为 40s 时,贴壁气流已明显影响了两侧下部射流。钢轨冷却过程的温度分布情况显示,将入口风速从 100m/s 提高到 200m/s 后,钢轨的冷却速度明显加快,终冷温度从 543℃降至 358℃,平均冷却速度也由 2.55℃/s 增加至 4.42℃/s。本模型对于合理配置喷嘴部位和与调节出口风速,满足钢轨淬火冷却工艺的要求,从而控制最终组织具有重要指导意义。

关键词：BnbRE；空气射流；有限元

Simulation for Air-jet Cooling BNbRE Rail Based on Fluent

WANG Hai-yan[1,2], ZHENG Meng-zhu[1,2], GAO Xue-yun[3],

CHEN Shu-ming[1,2], WU Zhi-feng[1], ZOU Cun-zhe[1]

(1. Key Laboratory of Integrated Exploitation of Bayan Obo Multi-Metal Resource; 2. School of Material and Metallurgy, Inner Mongolia University of Science and Technology; 3. Beris Engineering and Research Corporation)

Abstract: In this paper, flow and temperature fields in the air jet cooling process of BNbRE rail was calculated by Fluent software based on the standard k-ε turbulence model, and the 2D and 3D models temperature field and the heat transfer characteristics change under different wind speed quench cooling rail law was discussed. The results showed that the rail surface heat transfer coefficient increases, and heat transfer enhances with the wind speed increase. For the inlet velocity of 100m/s the case in the rail flow field velocity vector maps, when the cooling time reaches 80s, the gas dispersion produced by the upper jet impact with the wall will affect the lower part jet direction. Compared with the former, when the inlet velocity is 200m/ s and the cooling time is 40s, the adherent air flow has significantly affected both sides of the lower part of the jet. The temperature distribution of the rail cooling process indicate that the cooling rate of rail is significantly increased when the inlet velocity increased from 100m/s to 200m/s, the final cooling temperature decrease from 543℃ to 358℃, the average cooling rate increased from the 2.55℃ /s to 4.42℃ /s. The model has important significance for the rational allocation of parts and a nozzle outlet velocity adjustment, to meet the requirements of rail quenching process, thereby controlling the final microstructure.

Key words: BnbRE; air jets; finite element

基金项目：国家自然科学基金(51361021,51101083)；内蒙古自然科学基金(2013MS0813)。

第一作者：王海燕(1975—)，女,副教授。windflower126@163.com。

大功率热沉的散热仿真与优化

段亚菲[1]，曾建民[1]，刘永祯[2]，梁丽华[3]，陈 平[3]

(1.广西大学 材料科学与工程学院；2.摩比天线技术(深圳)有限公司；3.南南铝业股份公司)

摘 要：微电子技术发展迅速，微细化和高密度是电子器件的发展方向，散热因素引起的电子器件可靠性问题变得更加突出。本文利用ANSYS有限元软件对大功率热沉的传热学问题进行了数值模拟，研究了热沉的运行环境、结构参数及表面传热系数对热沉的影响，并与实验结果进行了对比分析。利用正交试验方法分析了不同因素水平对热沉散热性能的影响，对热沉的结构进行了优化。

关键字：ANSYS；散热器；数值模拟；正交试验

The Thermal Simulation and Optimization of The High Power Heat Sink

DUAN Ya-fei[1], ZENG Jian-min[2], LIU Yong-zhen[2], LIANG Li-hua[3], CHEN Ping[3]

(1. School of Materials Science and Engineering, Guangxi University;
2.Mobi-antenna Technology (shenzhen) co., LTD; 3. Alnan Aluminum co., LTD)

Abstract: Ultra-micronization and high density have becoming the developing direction of electronic devices with the rapid development in microelectronics technology. Reliability problems caused by thermal factors become a puzzling problem. By means of the finite element software ANSYS, the heat transfer of high-power heat sink has been analyzed, which deals with the effect of the operating environment, structural parameters and the surface heat transfer coefficient of the heat sink. Comparison between simulation and experiment results has been made. The influence of different factors on heat dissipating performance of the heat sink has been studied using the orthogonal experiment method.

Key words: ANSYS; radiator; numerical simulation; orthogonal experiment

0 引言

随着微电子技术、大规模集成电路技术的迅速发展，热因素引起的可靠性问题变得更加突出。热流自芯片流向外部环境会受到阻碍，称为热阻。

对热沉而言，Yue-Tzu Yang[1]等人研究了旋转和固定式直翅型热沉冲击冷却的数值模拟和优化。Bladimir Ramos-Alvarado[2]等人对配置在电子产品、燃料电池和太阳能电池的带有微流道的液体冷却

基金项目：国家自然科学基金 (51064003)；广西自然科学基金(2011GXNSD018009)；广西科学研究与技术开发(1348001-1)。

第一作者：段亚菲，女。数值仿真。609944582@qq.com。

通讯作者：曾建民，男。数值仿真。zjmg@gxu.edu.cn。

热沉进行了计算流体动力学研究。Vitor A.F.Costa[3]等人对在自然对流条件下 LED 灯径向热沉冷却进行了研究。陈明、胡安[4]等人对绝缘栅双极型晶体管传热模型进行了建模分析。王金兰[5]等人研究了一种多芯片封装(MCP)的热仿真设计。韩冬和何闻[5]研究了基于 ANSYS 的大功率晶体管热沉,并进一步提出改进热沉的方案。刘永贞、梁丽华等[6-7]研究了 CPU 散热器的散热和接触热阻之间的关系。

到目前为止,国内外主要从热沉形状和散热片的加工工艺来提高散热片的散热性能。本文将通过研究热沉的运行环境、结构参数及表面传热系数对热沉性能的影响,并对其进行优化设计,获得质量轻、热阻小,并使热沉具有良好的散热性能。

1 研究方法

1.1 实验仪器

为了模拟芯片的发热过程,采用自行设计的 XGYF-200A 功率模拟仪,如图 1 所示。包括恒功率发生器,测温热电偶。将散热片放置在功率模拟仪的发热体上,两者之间涂以导热硅脂,以减小界面热阻。热电偶可以设置在散热器的任何位置,实现任意点的温度监控。

图 1 XGYF-200A 型仪器

Fig.1 XGYF-200A instrument

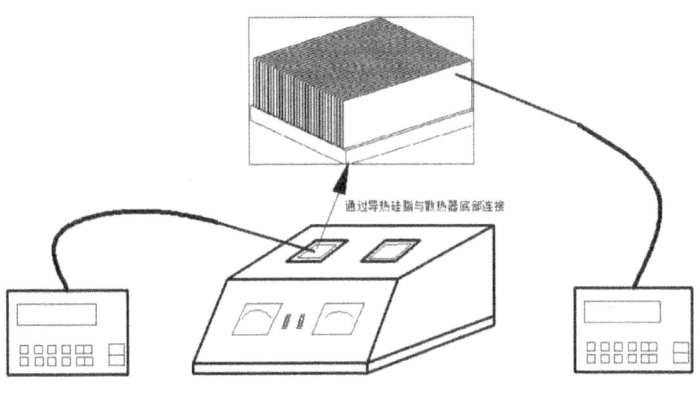

图 2 测试系统示意图

Fig.2 Schematic diagram of testing system

为便于和仿真进行比较,采用 Ti32 红外热像仪记录温度场。它是接收被测目标发出的红外辐射,依靠接收目标自身辐射的红外信号,并将这种热量转化为带有温度数据的可视化图像。说明:由于高于绝对零度(-237℃)的物体都会发出红外辐射。

试验的测试示意图如图 2 所示,将实验设备按示意图连接起来。试验系统包括两个部分:加热部分和测试部分。加热部分主要通过 XGYF-200A 型仪器,在给定的功率下,接通直流电源,使整体的加热部分相当于一个纯电阻电路,把所消耗的电能全部转化为仪器上方铜块的热能;测试部分主要包括数据采集仪、热电偶,热电偶采集的温度信号通过数据采集仪进行信号转换,在显示界面显示出来,通过记录,跟踪测试点温度随时间变化的过程。散热器与热源之间涂有导热硅脂,主要用来减少接触面之间的接触热阻。试验过程主要通过调节直流电源的电压和电流来改变对热源的加热功率。

1.2 建立物理模型和数学模型

利用 UG 三维绘图软件绘制热沉的结构,根据实际安装状态,与热源相接处的发热区域为一矩形区域,如图 3 所示。

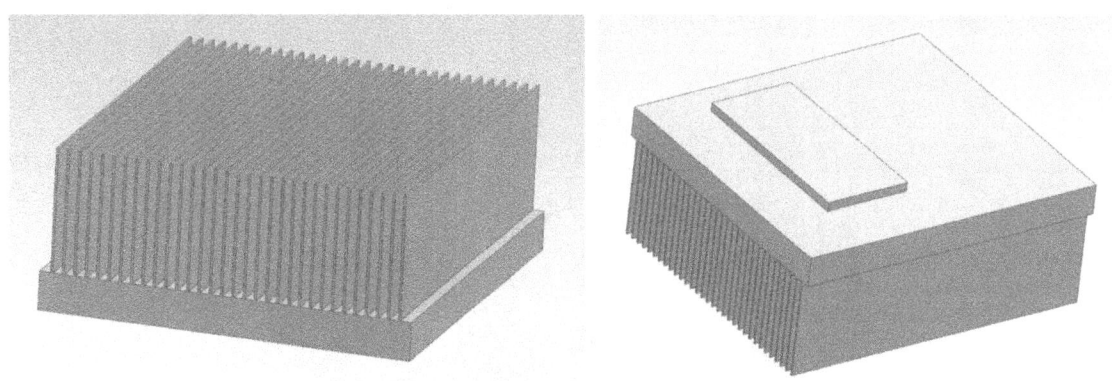

图 3 热沉的几何实体模拟

Fig.3 Geometric entity simulation of heat sink

数值仿真的数学模型采用三维非稳态导热微分方程[8]。

$$\rho c \frac{\partial t}{\partial \tau} = \frac{\partial}{\partial x}\left(\lambda \frac{\partial t}{\partial x}\right) + \frac{\partial}{\partial y}\left(\lambda \frac{\partial t}{\partial y}\right) + \frac{\partial}{\partial z}\left(\lambda \frac{\partial t}{\partial z}\right) \quad (1\text{-}1)$$

其中:λ ——热沉材料的导热系数　　　ρ ——热沉材料的密度

c ——热沉材料的比热容

这是三维非稳态导热微分方程的一般形式。

2 热沉的数值模拟

2.1 边界条件的设置

对热沉进行网格划分之后,根据热沉在实际环境中的运行及简化后的计算模型,对计算模型进行数值计算的边界条件设置。

(1)模型的物性

模型包括热沉和简化后的 IGBT 晶体管。热沉的材料为铝合金;由于在进行实验数据的收集时,采用的发热源材料为紫铜,为了实验与模拟结果相符合,所以在模拟发热源的材料物性时,所选取的也是紫铜。

表 1　材料的物性参数

Tab.1　Material　physical　parameters

	热导率 W/m·℃	密度 Kg/m³
铝合金	209	2700
紫铜	398	8930

在模拟计算散热器的热传导过程中,一方面散热器置于空气中,因为空气没有辐射与吸收的能力,辐射散热与其工作的室内环境有关;另一方面,根据斯忒藩-玻尔兹曼定律,物体对外辐射能量的总能力 E 与其绝对温度的四次方成正比,故在物体处于低温时热辐射时,往往可以忽略。

(2)热源功率

根据 IGBT 晶体管在正常工作时,它的额定功率为 400 W。在 ANSYS 热分析中,可以施加五种边界条件(温度、热流率、对流、热流、热生成率),根据发热源的额定功率可以计算出其上的热生成率 Q,即:

$$Q = \frac{400W}{110mm \times 50mm \times 5mm} = 14.5 \times 10^6 J/(m^3 \cdot s) \tag{1-2}$$

(3)对流换热系数

当热沉处于自然对流的状态,热沉与周围外界环境之间的对流换热系数为

$$h = 0.59 \times (1.5 \times 10^7 \times 0.699)^{1/4} \times \frac{0.029}{5.5 \times 10^{-2}} = 17.7W/(m^2 \cdot ℃) \tag{1-3}$$

2.2 时间步长的确定

时间步长是影响非线性求解精度和效率的最大因素。综合考虑该实验条件,选择时间步长为 10S。

3 实验数据与模拟数据的对比分析

为了验证上述数值仿真的可靠性,对设计的热沉进行模拟仿真,并与实验数据进行对比分析。

本次试验的主要目的是分析散热器的散热性能是否满足散热要求,对散热器底部、散热器翅片的温度进行监测,得到监测点的温度随时间的变化曲线。试验数据的分析如下:

(1)散热器的底部温度的分析

模拟 IGBT 晶体管芯片的热量主要通过纯电阻电路把电能转换为导热体的热能,然后通过导热硅

脂把热能传递到散热器中,由散热器把芯片产生的热量导出。本次试验在封闭的环境条件下进行,分别对散热器底部的温度和散热翅片的温度进行检测,其结果如图4所示：

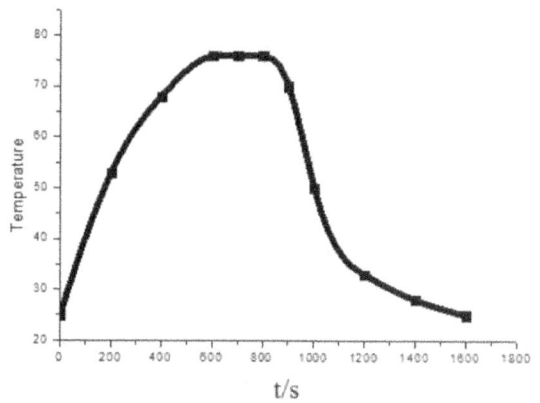

图4 散热器底部温度分布

Fig.4 Distribution of temperature at the bottom of the heat sink

图5所示,是通过数据采集仪,一定时间间隔内测量散热器底部的温度分布。散热器底部的最高温度为78℃,其温度值低于90℃,说明所设计的散热器满足正常工作条件下的要求。

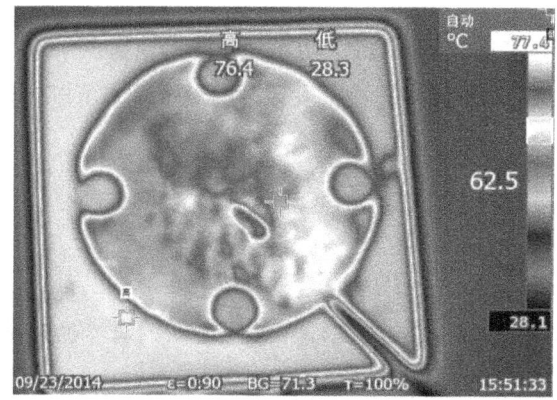

图5 热像仪测试图

Fig.5 Test pattern from thermal imager

图5所示,通过热像仪测得热源的最高温度约为78℃。

(2)翅片温度分析

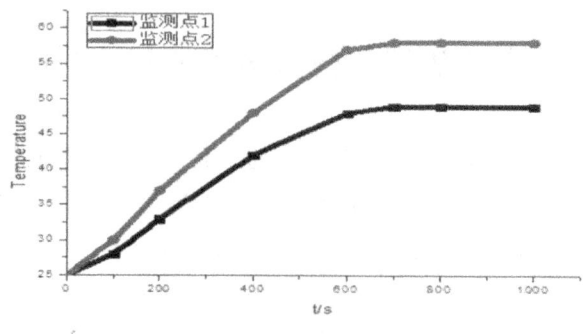

图6 翅片温度分布

Fig.6 Temperature distribution of fin

如图 6 所示,通过数据采集仪测得散热器翅片某一监测点的温度分布。监测点 1 为远离热源位置翅片上某一点的温度分布,监测点 2 为靠近热源位置同一翅片上一点的温度分布。从所测得的数据可以看出,同一翅片由于靠近热源位置的差异,温度相差 10℃ 左右,这样由于散热器的安装位置引起的。

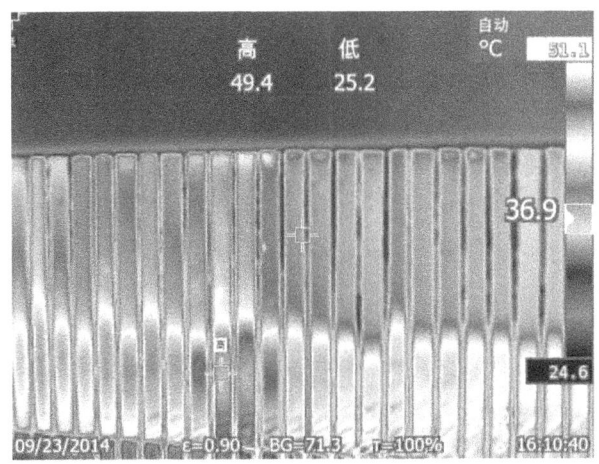

图 7　热像仪测试图

Fig.7　Test pattern from thermal imager

如图 7 所示,通过热像仪测得散热器翅片的温度分布。

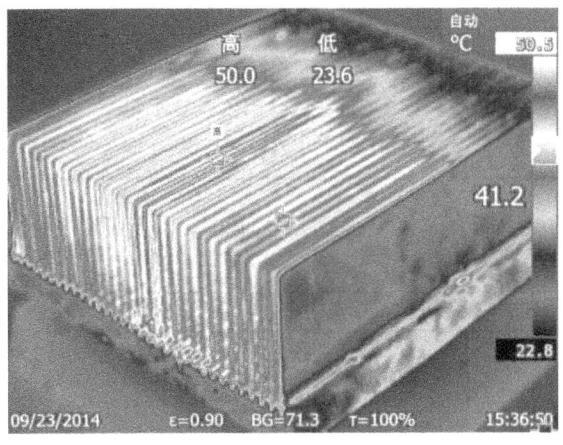

图 8　等轴侧图

Fig.8　Isometric view

如图 8 所示一定时间下,散热器的整体温度分布。

数值仿真中监测点的数值与试验中监测点的数值如表 2 所示:

表 2　模拟与试验结果温度值

Tab.2　Temperature comparison between simulation and thermal test

监测点	热沉底部	翅片 1	翅片 2
模拟结果	75.5	49.6	58.3
实验结果	78	49	58
误差	3.3%	1.3%	0.4%

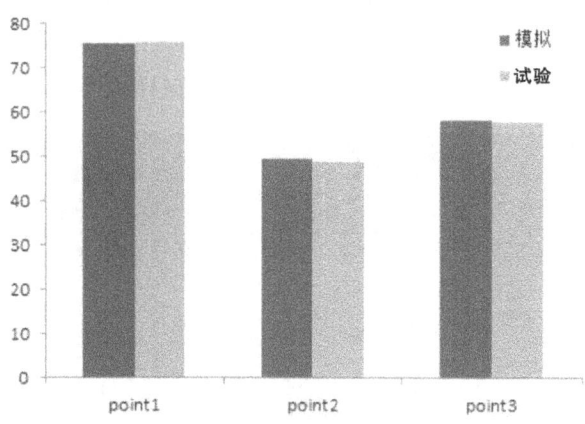

图9 数值模拟与试验测试监测点温度对比

Fig.9 Comparison of simulation and test

由图9可知,各监测点模拟结果与实验结果的温度误差在5%的范围内,说明利用有限元分析的模拟结果是可靠的。

4 最优试验方案的数值模拟

正交试验因素及水平的确定如表3所示:

表3 正交试验因素以及水平

Tab.3 Factors and levels of orthogonal design

翅片间距	翅片高度	翅片厚度	翅片长度
4	53	1.0	179
4.5	55	1.2	181
5	57	1.4	183

根据正交试验方案得出的最优组合,利用UG三维软件制图,在ANSYS中进行热分析,结果如图10所示:

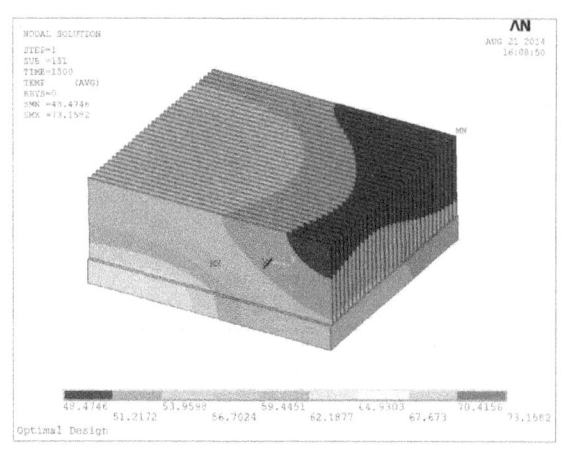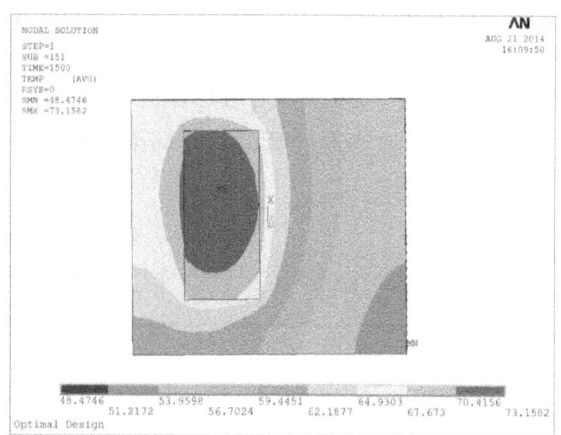

图10 最优方案热沉温度分布

Fig.10 Temperature distribution of optimal design heat sink

从图 10 可以看出,热沉的最高温度约为 73℃,最低温度为 48℃,其最高温度低于正交试验中 9 组试验,因此该试验为最优方案。

5 结论

(1)利用 ANSYS 有限元软件分析了热沉的温度场分布,从仿真结果可以看出,所设计的热沉满足系统工作的散热要求,确认了设计的正确性。

(2)利用正交试验,分析优化设计热沉的结构参数,结合数值模拟仿真结果,表明热沉翅片间距 A、翅片高度 B、翅片厚度 C、翅片长度 D 四个因素中,影响热沉性能的主次顺序为 A→B→C→D,各个因素取三个水平,分析了不同因素水平对热沉散热性能的影响,得到最优组合的方案 A1B3C2D3,即翅片间距为 4mm,翅片高度为 57mm,翅片厚度为 1.2mm,翅片长度为 183mm 的组合方案,通过数值模拟仿真得到芯片的最高温度为 73℃。

参考文献:

[1] Yang Yue–Tzu, Lin Shih–Chia, Wang Yi–Hsien, et al. Numerical simulation and optimization of impingement cooling for rotating and stationary pin‑fin heat sink [J]. International Journal of Heat and Fluid Flow, 2013, 44(0): 383–393.

[2] Ramos–Alvarado Bladimir, Li Peiwen, Liu Hong, et al. CFD study of liquid–cooled heat sinks with microchannel flow field configurations for electronics, fuel cells and concentrated solar cells [J]. Applied Thermal Engineering, 2011, 31(14‑15): 2494–2507.

[3] Costa Vítor A. F., Lopes António M. G. Improved Radial Heat Sink For Led Lamp Cooling [J]. Applied Thermal Engineering, (0).

[4] 陈明,胡安,唐勇. 绝缘栅双极型晶体管传热模型建模分析 [J]. 高电压技术,2011,(02): 453–459.

[5] 王金兰,仝良玉,刘培生.一种多芯片封装(MCP)的热仿真设计 [J]. 计算机工程与科学看,2012,04: 28–31.

[6] Lihua Liang, Bin Feng, Liping Su, et al. Investigation of Heat Dissipation for Computer CPU Radiator [J]. Applied Mechanics and Materials Vols. 2014, 635–637: 56–59.

[7] Yongzhen Liu, Zhishi Huang, Bin Feng, et al. Application of Thermal Contact Resistance in Simulation of heat dissipation by CPU Heat Sinks Based on ANSYS [J]. Advanced Materials Research Vols. 2014, 941–944: 2465–2468.

[8] 杨世铭, 陶文铨. 传热学 [M]. 高等教育出版社,2006.

二硼化镁超导线材的超声波振动拉拔加工及模拟

冯建情，王庆阳，杨　芳，熊晓梅，闫　果，张平祥

(西北有色金属研究院)

摘　要： 超声波振动拉拔是在常规拉拔过程中叠加超声波振动的新型加工工艺，属于超声塑性加工的一种。由于能够明显减小流动应力、降低拉拔力和摩擦力，使得线材拉拔时断线率降低，减少金属丝的不均匀变形，改善金属丝的性能和表面质量等特性，被广泛应用于线材拉拔，特别是难拉拔金属丝材的加工。本文以二硼化镁(MgB_2)超导线材在超声波振动拉拔过程中的应力应变行为及本构关系为基础，通过研究加工参数对 MgB_2 线材复合体微观结构、界面等不均匀性的影响，在力学分析和计算机仿真模拟基础上，获得超声波振动拉拔加工千米级多芯 MgB_2 线材的关键工艺因素，不但为高性能千米级 MgB_2 线材稳定化制备提供理论指导，还可以为其他同样采用 PIT 工艺制备的超导材料(如 Bi 系线带材、Fe 基超导线材等)加工提供理论参考。

关键词： 二硼化镁；超导材料；超声振动拉拔

The Machining and Simulation of MgB2 Superconducting Wire Assisted by Ultrasonic Vibration Drawing

FENG Jian–qing, WANG Qing–yang, YANG Fang, XIONG Xiao–mei, YAN Guo, ZHANG Ping–xiang

(Northwest Research Institute of Nonferrous Metals)

Abstract: Ultrasonic vibration drawing is a new processing technology with drawing added to the ultrasonic vibration, which belongs to the ultrasonic plastic processing. It is widely used in wire drawing, especially hard metal materials wire drawing, because which can significantly reduce the flow stress, reduce the drawing force and friction force. The breakage rate of drawing wire is reduced, which improve performance and surface quality characteristics of the wire. This qpaper is aimed to stress strain behavior and constitutive relation of magnesium diboride (MgB_2) superconducting wires drawn by the ultrasonic machining process. The effects of parameters on microstructure and surface in MgB_2 wires are studied. The key factors are obtained by ultrasonic drawing processing for kilometer level multifilament MgB_2 wires based on mechanical analysis and computer simulation. The results are not only can provide theoretical guidance for high performance kilometer MgB_2 wire, but also give theoretical reference for other superconducting materials (such as Bi2212 and Bi2223 wires, Fe based superconductor et al.).

Key words: magnesium diboride; superconducting materials; ultrasonic vibration drawing

基金项目： 国家自然科学基金(51372207)。

第一作者： 冯建情，男。超导材料。fegatin999@163.com。

自旋无能隙半导体 Mn$_2$CoAl 的第一性原理研究

胡 磊，李德贵，覃 铭

(百色学院 材料科学与工程学院)

摘 要：半金属材料在费米面附近一个自旋方向上呈现金属性，而在另一个自旋方向上具有带隙，这种性质使得我们可以获得100%自旋极化的电流，半金属材料由于在自旋电子设备中的潜在应用价值而获得越来越多的关注。最近，Heusler 合金 Mn$_2$CoAl 已经被计算是半金属铁磁体，被实验被证明是自旋无能隙材料。本文采用基于密度泛函理论的第一性原理赝势平面波方法，研究 Heusler 合金 Mn$_2$CoAl 的电子结构和磁性，首次在理论上计算了 Mn$_2$CoAl 的力学性质、热力学性质和光学性质。

关键词：电子结构；半金属铁磁性；第一性原理计算；热力学；自旋无能；隙半导体

First-principles Study on the Spin Gapless Semiconductor Mn$_2$CoAl

HU Lei, LI De-gui, QIN Ming

(School of Materials Science and Engineering, Baise University)

Abstract: Half-metallic materials present metallic behavior for one spin-direction, while they possess an energy gap around the Fermi level in the other spin direction, this nature allows us to obtain 100% spin-polarized current, they have received growing interest due to their potential value for application to spintronic devices. Recently, the Heusler alloy Mn$_2$CoAl, which was calculated to be a half-metallic ferromagnet, was experimentally proved to be a spin gapless material. This paper using the first-principles plane-wave pseudopotential method based on density functional theory, investigated the electronic structure and magnetism of the Heusler alloy Mn$_2$CoAl, what's more, for the first time, calculated the mechanical properties, thermodynamic property and optical properties of Mn$_2$CoAl in theory.

Key words: electronic structure; half-metallic ferrimagnetism; first-principle calculation; thermadynamics; spin gapless semiconductor

基金项目：广西高校科学技术研究重点项目：A356 铝合金汽车轮毂性能研究及应用(ZD2014134)；
广西高校科学技术研究一般项目：以赤泥为原料制备除氟剂及其吸附性能研究(YB2014386)；
百色学院校级项目：铸造 A356 铝合金结构和性质的第一性原理模拟计算研究(2014KB05)。
第一作者：胡 磊，男。材料的第一性原理计算。dreamhulei@163.com。
通讯作者：覃 铭，男。稀土合金的合成和性质研究、铸造 A356 铝合金汽车轮毂生产关键技术及其应用。gxbs6327mm@163.com。

装载机半轴感应淬火层研究

肖毅强[1]，蒙秋红[1]，刘升旭[1]，曾建民[2]

(1.广西柳工机械股份有限公司；2.广西大学 材料科学与工程学院)

摘　要：对一种装载机半轴用非调质钢 LGB38MnV 在不同加热温度作用后的冲击性能和微观组织进行研究，并与实际装载机半轴表面淬火层的微观组织进行比较，通过物理模拟的试验方法，确定表面淬火层对装载机半轴冲击性能的影响。结果表明：随着加热温度的从 200℃逐渐升高到 900℃，LGB38MnV 钢的冲击韧性提高，冲击吸收功从 3J 逐渐提高到 36J，其冲击断口的裂纹扩展区形貌由河流花样状形貌逐渐变化为浅小韧窝状形貌。显微组织观察发现，加热温度升高使 LGB38MnV 中的先共析渗碳体重新溶解析出，弥散分布在铁素体基体之上，当加热高于 800℃时，组织全部转变为马氏体。

关键词：半轴；非调质钢；冲击韧性；感应淬火

Research on Induction Hardening Layer of Semi-axle Steels for Shovel Loader

XIAO Yi-qiang[1], MENG Qiu-hong[1], LIU Sheng-xu[1], ZENG Jian-min[2]

(1. Guangxi LiuGong Machinery Co., Ltd；2. School of Materials Science and Engineering, Guangxi University)

Abstract: The impact property and microstructure of LGB38MnV, which was one of the microalloying forged steels of semi-axle for shovel loader, were researched under different heating temperature. Compared with the microstructure of surface quenched layer in actual semi-axle for shovel loader, physical simulation method was applied to study the relationship between surface and impact toughness. The Experimental investigation showed that the increasing of heating temperature yielded an increasing of the impact toughness. With the increasing of heating temperature from 200℃ to 900℃, the impact absorbing energy of LGB38MnV increased from 3J to 36J, the fracture appearance changed from predominant cleavage river pattern to a lot of fine and uniform ductile dimples. The results of microstructure observation showed that heating temperature increasing made the proeutectoid cementite to re-dissolved, precipitated, and dispersed in ferritic matrix. When the heating temperature was above 800oC, the microstructure of LGB38MnV steel turned to be martensite completely.

Key words: semi-axle; LGB38MnV microalloying forged steel; impact toughness; induction quenching

---

基金项目：国家自然科学基金(51064003)；广西自然科学基金(2011GXNSD018009)。
第一作者：肖毅强，男。黑色金属新型材料。13657880437@163.com。
通讯作者：曾建民，男。数值仿真。zjmg@gxu.edu.cn。

纳米石墨片微粉喷射成型工艺数值模拟

郭　辉，吴海华，熊　盼

（三峡大学 机械与动力学院）

摘　要：微粉喷射成型法具有成型效率高、对微粉落点精确可控等优点，有望成为纳米石墨片复杂元器件制备新方法。本文针对纳米石墨片微喷射成型工艺参数进行流场模拟分析，研究入口压力、喷射距离、送粉量等工艺参数对纳米石墨片微粒加速行为影响，发现纳米石墨片在基体中分布规律，并通过实例验证了微喷射成型工艺的可行性。

关键词：纳米石墨片；微粉喷射成型；数值模拟；分布规律

Nano Graphite Sheet Micro Powder Injection Molding Process Numerical Simulation

GUO Hui, WU Hai-hua, XIONG Pan

(College of Mechanical & Power Engineering, China Three Gorges University)

Abstract: Micro powder injection molding with high efficiency, and the advantages of the powder point precisely controlled is expected to become a new method for preparation of nano graphite sheet of complex components. In this paper, the nano graphite sheet of micro injection molding process process flow field simulation analysis, studied such as the inlet pressure, jet distance, send powder amount on the process parameter's influence for nano graphite particles accelerated behavior, and found the distribution of nano graphite sheet in the matrix and through the example is given to verify the feasibility of micro injection molding process.

Key words: nano graphite sheet; micro powder injection molding; numerical simulation; regularities of distribution

基金项目：三峡大学人才科研启动基金项目（KJ2012B012）；湖北省教育厅自然科学研究计划项目（B2013173）；
　　　　　华中科技大学材料成形与模具技术国家重点实验室开放基金课题（P2015-11）。

通讯作者：吴海华（1970— ），男，教授，博士。增材制造技术、石墨精深加工技术。wusuke1970@163.com。

超级电容器微电极电场强度及不均匀性模拟分析

熊　盼，郭　辉，魏正英，吴海华

（三峡大学 机械与动力学院）

摘　要： 超级电容器电极结构及其结构参数对其内部电场强度及均匀性有直接影响，从而影响电容器漏电流特性和自放电特性。本文利用有限元分析方法，对圆形、方形、条形和梳齿形二维电极及圆柱状、方柱状、条状和梳齿状三维电极的电场进行模拟分析。研究表明：对于相同电极结构，电极间的电场强度随着电极间距、电极基本结构尺寸的增大而减小；与二维电极相比，因电场叠加，三维电极的电场强度更大；在电场均匀性方面，方柱状电极和条状电极的电场均匀性较好，而梳齿状电极的电场均匀性较差，但所有结构电极的电场都为稍不均匀电场。

关键词： 超级电容器；电场模拟；电极结构

Supercapacitor Microelectrode Electric Field Strength and Inhomogeneity Simulation Analysis

XIONG Pan, GUO Hui, WEI Zheng-yin, WU Hai-hua

(College of Mechanical & Power Engineering of China Three Gorges University)

Abstract: Supercapacitor electrode structure and structural parameters have a direct impact on the internal electric field strength and uniformity, thus affecting capacitor leakage current characteristics and self-discharge characteristics. In this paper, finite element analysis was used to the 2D round, square, bars and comb electrodes and 3D cylindrical, square column, bar and electric comb electrodes. Studies have shown that: for the same kind of structure of the electrode, the electric field strength between the electrodes decreases with increasing the electrode spacing and the electrode size. Compared with planar electrodes, the electric field between the 3D electrodes is bigger due to the electric field superimposed. The field uniformity of the pillar electrode and strip electrodes are better, and the comb electrodes is worse. And the field of all electrode structure are slightly nonuniform electric field.

Keywords: supercapacitor; electric field simulation; electorde structure

0 前言

超级电容器成本低廉、储能密度高、循环寿命长、材料丰富、充放电寿命长及安全性能高，它的功

基金项目： 三峡大学人才科研启动基金项目(KJ2012B012)；湖北省教育厅自然科学研究计划项目(B2013173)。

通讯作者： 吴海华(1970—)，男，教授，博士。增材制造、石墨精深加工、快速精铸技术。wusuke1970@163.com。

率密度是传统电池的 10 倍,能量密度高出常规电容器 10～100 倍,可以瞬间提供大电流,特别适用于脉冲或瞬态大功率供电[1]。目前的电极结构大多为二维电极,如圆形电极、方形电极、梳齿形电极、条形电极,对于三维电极的研究较少。可以采取改变电极形状,制备三维结构来增大电极表面积,增加电极材料与电解液的接触面积,提升超级电容器性能。

超级电容器的性能与电极的形状、电极的大小、电极距离都有密切的关系。超级电容器的容量主要取决于电容器电极表面积活性物质及活性物质的利用率,同时电容器的大电流放电能力、内阻等也与电极的形状、电极间距有关[2]。当电极与电解液之间的接触面积较大、两电极之间的间距较小时,超级电容器的内阻就会比较小,大电流放电能力比较强;电容器的内部电场分布越均匀,电容器的漏电流特性和自放电特性就越好[3]。电容器两电极之间的间距越小,电解液中的离子传输距离就越短,离子扩散和迁移的时间就会缩短,电极材料反应活性就会增强,进而提高电容器容量及大电流充放电能力,降低电极阻抗,增强电容器的电化学性能。

超级电容器在工作时,对超级电容器两端的电极施加电压后,就会在两电极之间的介质中产生电场,两电极之间的电场强度为[4]:

$$E = \frac{U}{D} \qquad 公式\ 1$$

由公式 1 可知,当两端施加的电压一定的时候,两电极之间的间距 D 越小,电场强度就越大。当电极高度和横截面积固定的时候,电极表面积就越大,电容器性能就越好,但随着电极之间的距离变小,两电极之间产生的电场强度就会随着变大;电极的形状对电场强度也有一定的影响,如果电极的外形有尖锐的棱角,会使其附近的电场强度增大。如果超级电容器内部电场强度过大,会导致尖端放电,击穿电介质,造成电容器报废[5]。因此需要对电容器内部电场进行仿真,了解其场强分布情况,以便合理布置电极间距。

本文通过模拟二维圆形、方形、条形和梳齿形电极及三维圆柱状、方柱状、条状和梳齿状电极,研究电极结构及其结构参数对电极内部电场强度和电场均匀性的影响,为设计合理的电极提供参考。

1 电极静电场模拟

本文使用 ANSYS14.0 对电极静电场分布进行仿真分析[6]。

建立二维模型,电极结构为:圆形、条形、方形、梳齿形,选择 Plane121 单元,自由度为节点电压,设置相对介电常数。单位制使用 MKSV 制,尺寸设置如表 1 所示,单位 μm。

表 1 二维电极结构建模参数
Tab.1 planar electrode structure modeling parameters

电极结构	电极尺寸	电极高度	电解质尺寸
圆形	Φ50	100	1600×1600
方形	50×500	100	1600×1600
条形	50×50	100	1600×1600
梳齿形	50×500	100	1600×1600

进行智能网格自由划分,面加载电压为 1.2V。加载后求解,求解结果如下:

二维电极电场强度分布情况用矢量云图表示,如图1—图4。

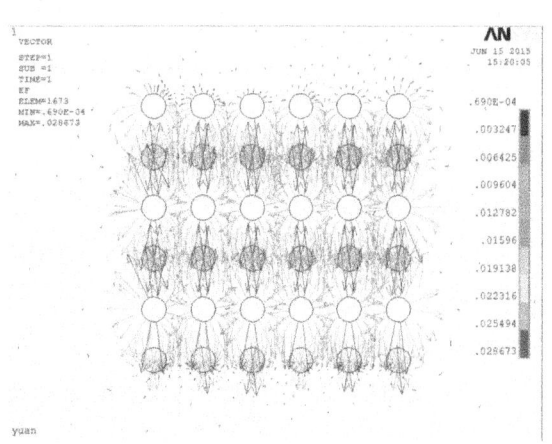

图1　圆形电极电场强度分布图

Fig.1　circular plane electric field intensity distribution

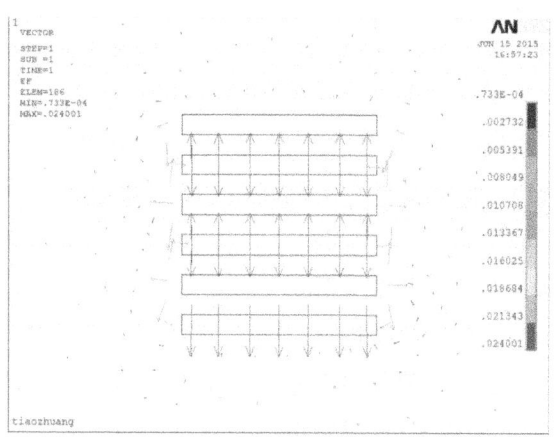

图2　条形电极电场强度分布图

Fig.2　stripe plane electric field intensity distribution

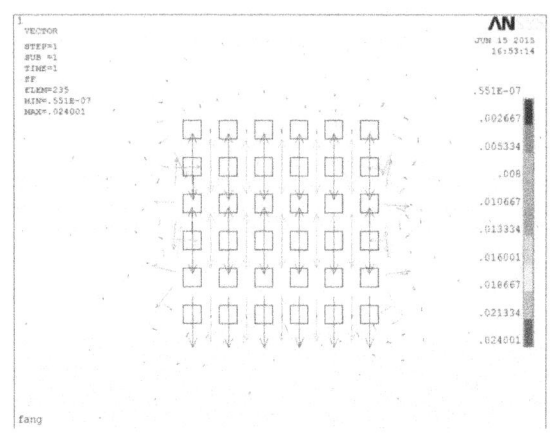

图3　方形电极电场强度分布图

Fig.3　square plane electric field intensity distribution

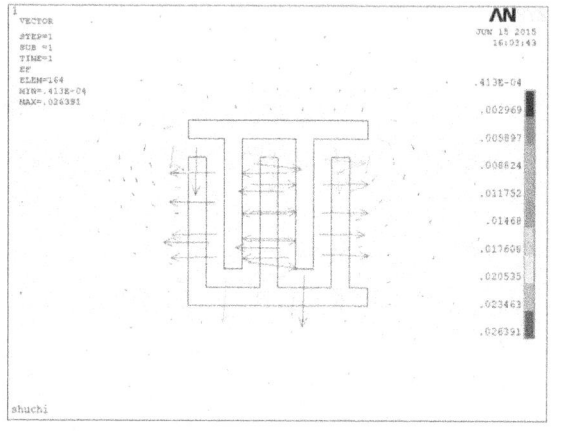

图4　梳齿形电极电场强度分布图

Fig.4　comb plane electric field intensity distribution

二维圆形、条形、方形电极的电场呈左右对称分布。二维圆形电极之间电场分布不均匀,电极距离最短处电场最大,一个二维圆形电极的电场成环形分布,电极与电极之间电场叠加。二维条形电极就是一个个的平行板电极,电极之间为均匀电场,且是最大电场。方形电极最大电场在电极之间,是均匀电场,电极周边的电场都相等。梳齿形电极最大电场也在电极之间,但是电场不是均匀分布的。

同样建立三维模型,电极结构为:圆柱状、条状、方柱状、梳齿状。选择 Solid 122 单元,自由度为电势[7],设置相对介电常数,尺寸设置如表2所示,单位为 μm。

表2　三维电极结构建模参数

Tab.2　Three-dimensional electrode structure modeling parameters

电极结构	电极尺寸	电极高度	电极间距	电解质尺寸
圆柱状	Φ50	100	50	1600×1600×100
方柱状	50×500	100	50	1600×1600×100
条状	50×50	100	50	1600×1600×100
梳齿状	50×500	100	50	1600×1600×100

三维电极电场强度分布情况用矢量云图表示,如图5—图8。

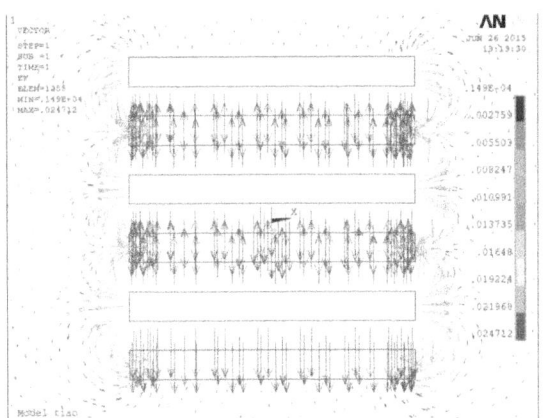

图5 圆柱状电极电场强度分布图
Fig.5 columnar electric field intensity distribution

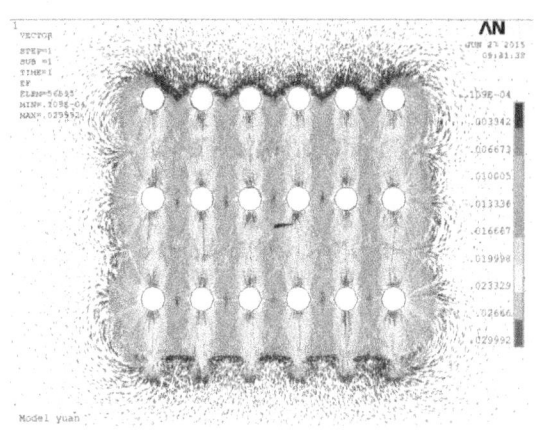

图6 条状电极电场强度分布图
Fig.6 long strips electric field intensity distribution

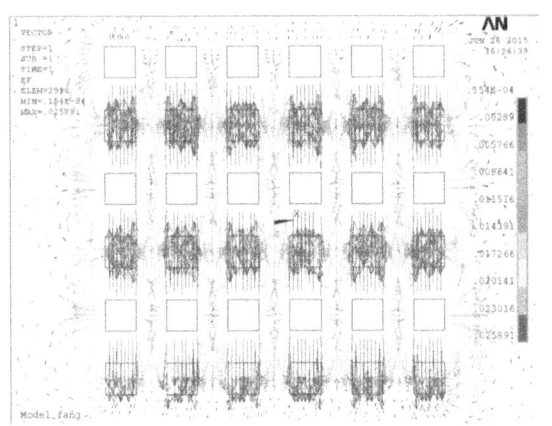

图7 方柱状电极电场强度分布图图
Fig.7 square pillar electric field intensity distribution

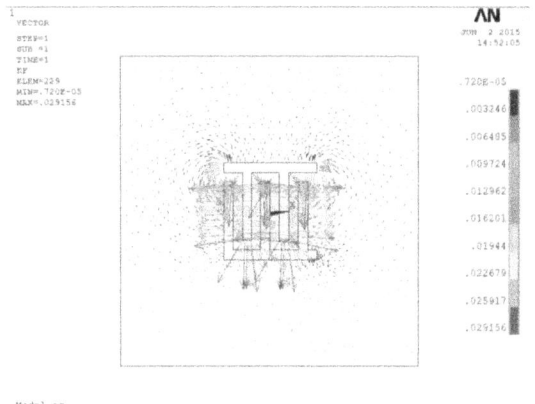

图8 梳齿状电极电场强度分布图
Fig.8 3D comb electric field intensity distribution

从图5至图8可看出,三维电极电场分布俯视图与其对应的二维电极类似,只是电场线更加密集。两电极之间最短距离处电场强度最大,电极外形越平整,电场强度分布越均匀;电极有尖角或转角如梳齿状电极,电极转角处电场强度会增大。电极结构设计时,应尽量避免电极表面出现尖角或突变。

2 电极结构参数对电场强度影响分析

电场的性能主要由能量密度[8]来衡量,电场中的能量密度公式为:

$$\omega_e = \frac{1}{2}\varepsilon E^2 \quad 公式\ 2$$

从公式2可知,电场的能量密度与电场强度的平方成正比,电场强度越强,能量密度就越大。提高能量密度是提高超级电容器性能的有效途径,而电极结构影响着电极间电场,所以研究电极结构参数对电场的影响是十分有必要的。

2.1 电极间距

按电极高度为 $100\mu m$，电极尺寸为 $50\mu m$，电极间距分别为：$10\mu m$、$30\mu m$、$50\mu m$、$70\mu m$ 进行模拟，得到不同电极间距与最大电场强度关系如下图所示。

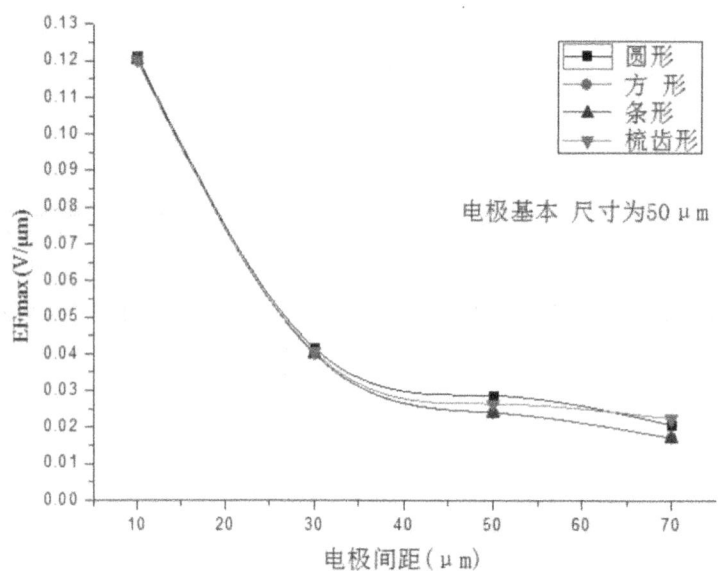

图9 二维电极不同电极间距下的最大电场强度

Fig.9 the maximum field strength of different electrode spacing of the planar electrode

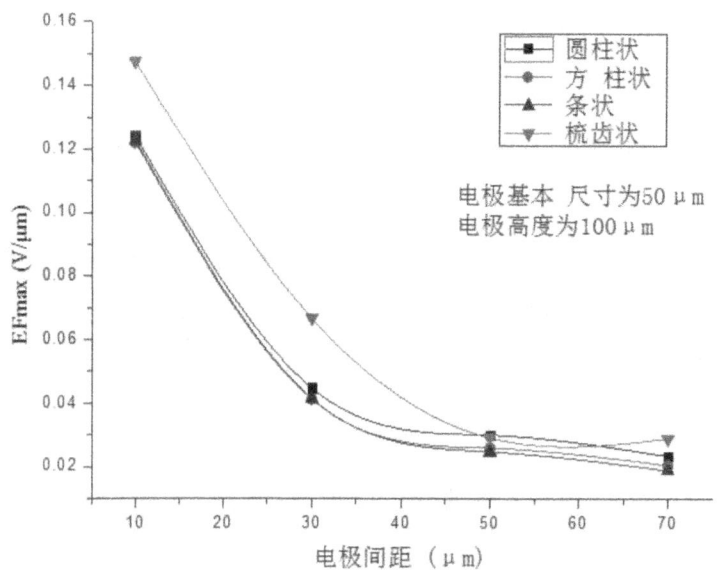

图10 三维电极不同电极间距下的最大电场强度

Fig.10 the maximum field strength of different electrode spacing of the 3D electrode

从图9和图10可知：4种电极产生的电场最大场强随着电极距离的增加而减小，电极间距对电极之间的电场有很大的影响，曲线近似为反比例曲线。二维电极不同结构之间的电场曲线非常接近，同种结构的三维电极比二维电极的电场强度要大，可能是电场叠加的结果。当两电极之间的间距越小时，电极的表面积就越大，电容器的性能越好，但同时电极两端所产生的电场强度就越大；如果电场强度超过击穿强度就会造成电容器的报废,因而在保证电极内部电场强度小于击穿强度的条件下,选取小的电极间距。

2.2 电极基本尺寸

电极高度为 100μm,电极间距为 50μm,电极宽度分别为:10μm、30μm、50μm、70μm 进行模拟,电极尺寸与最大电场的关系如下图所示。

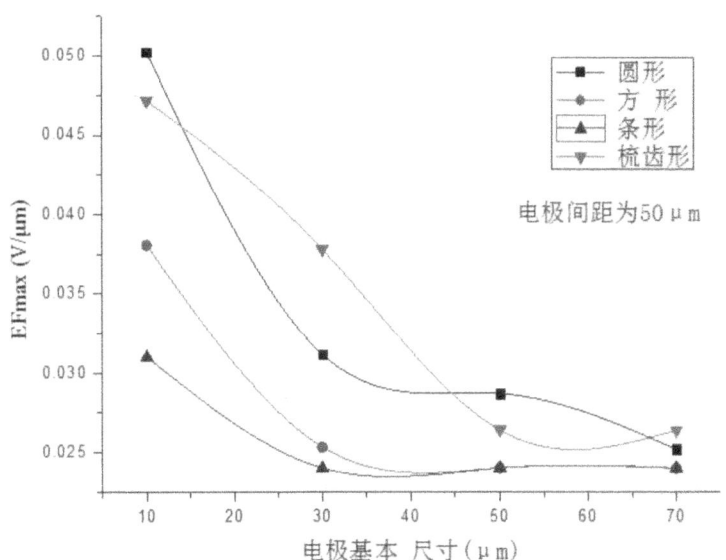

图 11　二维电极不同电极尺寸下的最大电场强度

Fig.11　the maximum field strength of different electrode size of the planar electrode

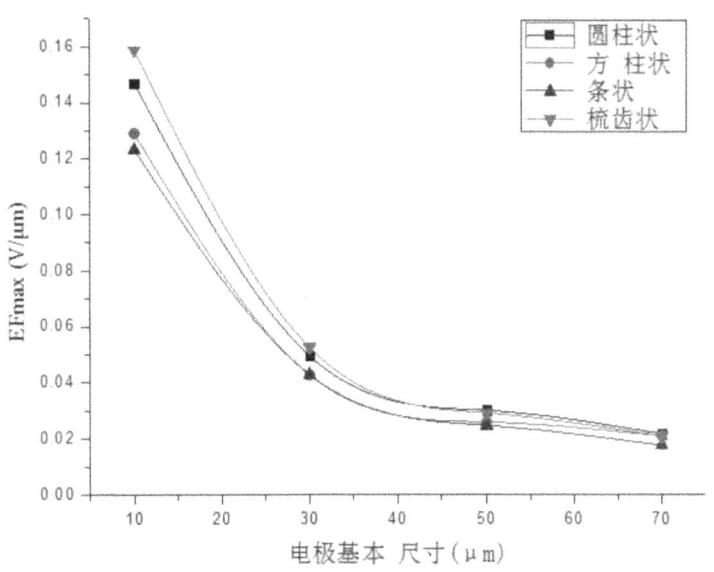

图 12　三维电极不同电极基本尺寸下的最大电场强度

Fig.12　the maximum field strength of different electrode size of the 3D electrode

从上图可看出,同种结构的三维电极电场强度比二维电极的要大。在电极基本尺寸为 10μm 时,电场强度最大;随着电极基本尺寸的增大,电场强度减小,在 10μm 到 30μm 电场的强度变化较明显,在

30μm 以后电场强度变化较小,基本持平。当电极基本尺寸较小时,电极类似一个尖针,电极内部电场强度比较大,符合日常生活中的尖端放电原理。在电极结构设计时,在保证电极不击穿的前提下。为增大表面积,因根据电极基本尺寸对应的最大电场强度,选择最小电极基本尺寸。

2.3 电极高度

以梳状电极为例,电极基本尺寸、电极间距不变,分别取电极高度为:10μm、30μm、50μm、70μm、100μm、120μm 进行模拟,得出电场的最大场强和电极高度的关系如下图所示。

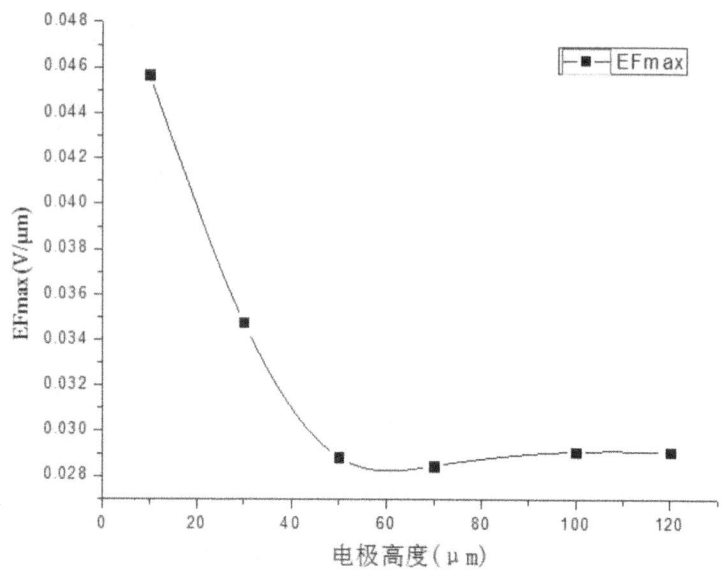

图13 最大场强和电极高度的关系

Fig.13 The relationship between the maximum field strength and the height of electrode

从图中可以看出当三维梳齿状结构高度从 10μm 增加到 50μm 时,电极产生的最大电场强度呈下降趋势;当梳齿状结构高度从 50μm 增加到 120μm 时,电极产生的最大电场强度基本趋于稳定。整个图像类似反比例函数图像。电极结构设计时,电极高度应该取大,但应该综合考虑电极间距和电极基本尺寸,满足电极的整体机械性能。

2.4 电极结构

保持电极高度为100μm,电极尺寸和电极间距取值相同,分别取 10μm、30μm、50μm、70μm,得出不同结构的最大电场值 EFmax 如图14所示。由上图可知:条状电极的电场强度最低,方柱状电极与条状电极几乎重合;三维梳齿状电极的最大电场强度最大,圆柱状电极与梳齿状电极十分相近。由此可知,电极结构对电场强度有影响。

不同电极结构的电场均匀性是不一样的,当电场稍不均匀时,若电极内部电场高于介质的击穿场强[9],会使电容器击穿报废。若电场的不均匀程度[10]非常严重,电极越容易被击穿。为了描述不同结构的电场不均匀程度,引入电场不均匀系数 f,

$$f = \frac{E_{max}}{E}$$

Emax 表示电场中最大场强,E 表示理论场强,均匀电场的电场均系数 f 为 1,f<2 表示稍不均匀电

场;f>4 属于极不均匀电场,当 f 大于 4 时,电容器更容易被击穿。不同结构的电场的均匀系数 f 如图 15 所示。

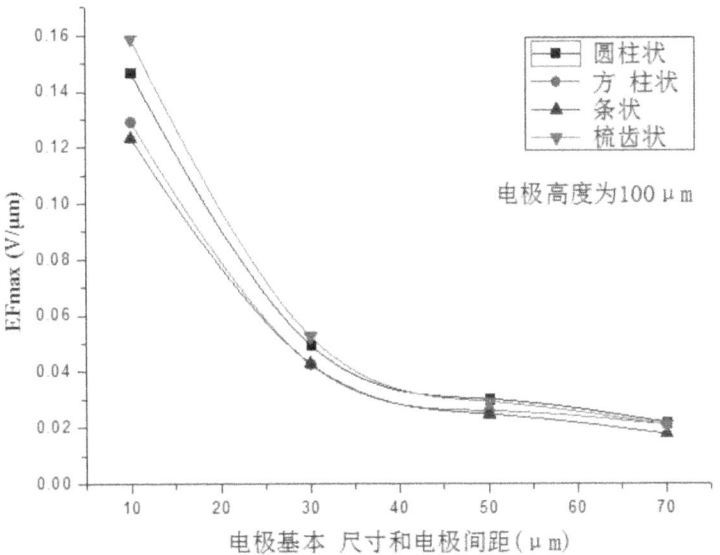

图 14 三维电极结构对电场的影响

Fig. 14 effect of 3D electrode structure for the electric field

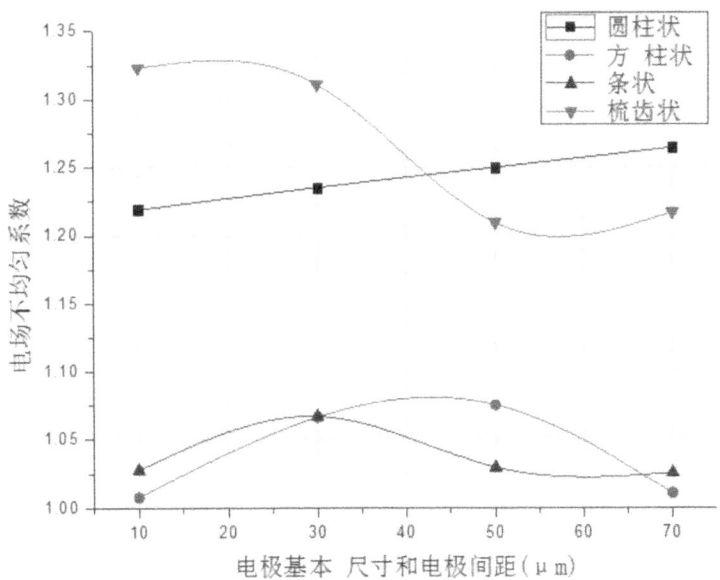

图 15 不同结构电极的电场均匀系数

Fig. 15 the electric field coefficients of the different electrodes structure

由图 15 可知:不同结构的电极产生的电场均匀度是不一样的,电极结构参数的改变也会影响电场的不均匀性。在四种电极结构中,条状电极和方柱状电极的电场不均匀系数较低,大致在 1.05 左右,接近均匀电场;三维梳齿状电极和圆柱状电极的电场不均匀系数较高,大概在 1.2~1.33。圆柱状电极的电场不均匀系数与电极基本尺寸和电极间距成正比;其他结构电极的电场不均匀系数上下波动。所以,条状和方柱状电极的电场均匀度较好,三维梳齿状电极和圆柱电极的均匀性较差。但是总体而言,所有电极的电场都是稍不均匀电场,所以应满足电极内部最大电场强度小于介质击穿场强。

3 结论

(1)同种结构的三维电极比二维电极的电场线密集,且电场强度增大,为电场叠加的结果。

(2)电场强度随着电极间距的增加而减小。

(3)电场强度随着电极基本尺寸的增大而减小,当电极基本尺寸较小时,电极内部电场强度变化率较大,当尺寸较大时,电场强度变化率较小。

(4)对于微电极,电极高度与最大电场强度成反比。

(5)条状电极和方柱状电极的电场均匀性较好,三维梳齿状电极的电场均匀性相对较差,但所有电场都是稍不均匀电场,电极内部最大电场强度小于介质击穿场强。

参考文献:

[1] Miller J R, Simon P Electrochemical capacitors for energy management [J]. Science, 2008, 321: 651—652.

[2] 文春明.基于 MEMS 技术的超级电容器三维电极制备及表征方法研究 [D]. 重庆大学光电工程学院,2012.

[3] 马西奎. 电磁场理论及应用 [M]. 西安: 西安交通大学出版社, 2000, 367—385.

[4] 专题特写: 无源电器,1994—2010 China Academic Journal Electronic Publishing House.

[5] 马西奎. 电磁场理论及应用 [M]. 西安: 西安交通大学出版社,2000,367—385.

[6] 唐兴伦、范群波、张朝晖等. ANSYS 工程应用教程——热与电磁学篇 [M]. 中国铁道出社, 北京, 2003,265–271.

[7] 文春明、温志渝等. 硅基微型超级电容器三维微电极结构制备 [J].电子元件与材料,2012. 5(5): 43.

[8] 翟登云. 高能量密度超级电容器的电极材料研究 [D]. 清华大学材料科学与工程,2011. 5.

[9] 戴玲、韩永霞.介质阻挡放电击穿场强影响因素的研究 [J]. 华中科技大学学报,2007 年 2 月,93—95.

[10] 张波,储金宇,许小红. 利用 PDE tool box 计算电场不均匀系数 [J]. 高电压技术. 2005 年 6 月,第 31 卷第 6 期.

纳米银焊膏搭接接头蠕变损伤三维测试与模拟

谭沿松，唐　珊，陈　旭*

（天津大学 化工学院）

摘　要：低温烧结后的纳米银焊膏搭接接头的焊膏层成多孔形貌，有初始裂纹。本文探究纳米银焊膏搭接接头在高温蠕变条件下的损伤演变过程，试验温度为 325℃，剪切应力为 2MPa。在蠕变进行 0h、5h、21h、27h、29.3h 时，分别利用 X 射线计算机断层扫描系统检测焊膏层的裂纹增长情况，利用 MIMICS 和 3-MATIC 软件对实验的三维立体结构进行处理，再将处理后试样焊膏层三维图像导入 ABAQUS 分析裂纹周围的应力分布。研究发现：纳米银焊膏搭接试样裂纹周围存在应力集中，进而促进了裂纹的扩展和合并。

关键词：纳米银焊膏；X 射线；蠕变损伤；裂纹

Three-dimensional Visualization and Modeling of Creep Damage in Nano-silver Sintered Lap Shear Joint

TAN Yan-song, TANG Shan, CHEN Xu*

（School of Chemical Engineering and Technology, Tianjin University）

Abstract: Initial cracks larger than 50 μm were formed during sintering of the nano-silver sintered lap shear joint. These were the passages of volatilized organic components of the nano-silver paste. In this study, evolution of creep damage of nano-silver sintered lap shear joints were studied at 325℃. During creep tests, the constant shear stress of 2MPa was subjected on nano-silver sintered lap shear joint. X-ray tomography clearly revealed the crack-growth behavior of the joint during shear creep. In addition, the 3D virtual microstructure was processed by MIMICS and 3-MATIC, then incorporated into ABAQUS to quantify the effect of cracks on shear creep behavior of nano-silver sintered lap shear joint. Stress concentration was observed at the boundary of cracks, which accelerated the growth and merging of cracks.

Key words: nano-silver paste; X-ray; creep damage; crack

基金项目：国家自然科学基金资助项目（No. 11072171, 51401145）。

第一作者：谭沿松，女。纳米银焊膏力学性能。tomorrow2012@163.com

通讯作者：陈　旭，男。材料力学性能。xchentju.edu.cn。

Y$_2$O$_3$ 含量对 Ni-20Cr-5Al 合金高温抗氧化性的影响

孙端君，赖灿伟，尚金龙，梁春园，尹继辉，宋亚茹，张修海

(广西大学 有色金属及材料加工新技术教育部重点实验室)

摘 要：采用粉末冶金技术，在 Ni-20Cr-5Al 基体粉末中，添加不同质量分数(0%~5%)Y$_2$O$_3$ 颗粒作为弥散相，制备出 ODS(oxide dispersion strengthened)镍基高温合金，并对合金进行恒温氧化实验，借助 XRD 研究了 Y$_2$O$_3$ 的含量对 ODS 镍基高温合金高温抗氧化性能的影响。结果表明：Y$_2$O$_3$ 含量过量(大于 1%时)，将导致合金的高温抗氧化性能下降；Y$_2$O$_3$ 可以促进 Cr$_2$O$_3$ 的形成；1150℃下恒温氧化超过 60h 后，Y$_2$O$_3$ 含量为 0%、1%、3%、5%的镍基合金均出现氧化膜开裂甚至剥落的现象。对添加不同质量分数 Y$_2$O$_3$ 颗粒的合金进行 XRD 物相分析，研究 Y$_2$O$_3$ 对合金高温抗氧化性能影响机制。

关键词：高温氧化；Y$_2$O$_3$ 含量；氧化膜；镍基高温合金

Effect of Y$_2$O$_3$ Content on the High Temperature Oxidation Resistance of the Ni-20Cr-5Al Alloy

SUN Duan-jun, LAI Can-wei, SHANG Jin-long, LIANG Chun-yuan, YIN Ji-hui, SONG Ya-ru, ZHANG Xiu-hai

(Key Laboratory of Nonferrous Materials and New Processing Technology of Ministry of Education, Guangxi University)

Abstract: Ni-20Cr-5Al matrix powder was added different mass fraction (0%~5%) Y$_2$O$_3$ particle as dispersion phase to create the ODS (oxide dispersion strengthened) high temperature nickel-based alloy by powder metallurgy technology. By isothermal oxidation experiment, Y$_2$O$_3$ content on the effect of high temperature oxidation resistance of ODS nickel-based alloy was studied with the help of XRD. The results showed that Y$_2$O$_3$ content is over 1% will result in a decline in the high temperature oxidation resistance of the alloy; Y$_2$O$_3$ can promote the formation of Cr$_2$O$_3$; After more than 60h under 1150℃ isothermal oxidation, Y$_2$O$_3$ content is 0%, 1%, 3%, 5% of the nickel-based alloy oxide film was observed in all the phenomenon of cracking or even flaking. The different mass fraction of Y$_2$O$_3$ particle alloy was analysed by XRD, and the mechanism of high temperature oxidation resistance of the alloy influenced by Y$_2$O$_3$ was studied.

Keyword: high temperature oxidation; Y$_2$O$_3$ content; oxide film; nickel based superalloy

基金项目：广西大学有色金属及材料加工新技术教育部重点实验室开放基金资助项目(GXKFJ09-21)和 GXKFJ09-04)；
国家自然科学基金(51371059 和 51001032)。

第一作者：孙端君(1995—)，男。本科生。高温合金研究。Canweilai@163.com。

通讯作者：张修海(1978—)，男。副教授。高温合金研究。xiuhaizhang@gxu.edu.cn。

Al-Zn-Mg 系高强铝合金喷射成型工艺

阁世景，黄　干，汤宏群*，曾建民，王荣妹，罗　曼

(广西大学 材料科学与工程学院；广西有色金属及特色材料加工省部共建国家重点实验室培育基地)

摘　要：分析了 Al-Zn-Mg 系高强铝合金喷射成型工艺的发展，总结了热处理对喷射成型的 Al-Zn-Mg 系高强铝合金性能的影响以及喷射成型工艺对 Al-Cu-Mg 系高强铝合金腐蚀性能的影响。指出均匀化处理、双级固溶处理和 T6 时效处理等热处理工艺对喷射成型的高强铝合金的抗拉强度、屈服强度和延伸率都有一定程度的提高。喷射成型工艺流程短、成本低、沉积效率高,是一种富有发展前景的近终成形快速凝固技术。

关键词：高强铝合金；喷射成型；热处理

Spray Forming for Al-Zn-Mghigh Strength Aluminum Alloy

GE Shi-jing, HUANG Gan, TANG Hong-qun, ZENG Jian-min, WANG Rong-mei, LUO Man

(School of Materials Science and Engineering, Guangxi University ; Ministry-province Jointly-constructed Cultivation Base for State Key Laboratory of Processing for Non-ferrous Metal and Featured Materials, Guangxi Zhuang Autonomous Region)

Abstract: This paper reviewed the development of high strength aluminum alloy Al-Cu-Mg by Spray Forming. Comment on effect of spray forming of Al-Zn-Mg properties .And Introduced effect of heat treatment on spray forming of Al-Zn-Mg. Tensile strength, yield strength and elongation of high strength aluminum alloy with spray forming have a certain degree of improvement by homogenization treatment, the two-stage solid solution treatment and T6 aging treatment. Spray forming has many advantages as a kind of very promising rapid-solidifying technology for near net shaping,such as shorter process, lower cost and higher deposition efficiency.

Keywords: high strength aluminum alloy; spray forming; heat-treatment

0 引言

超高强度 7xxx 系 Al-Zn-Mg 系铝合金是一种典型的可时效处理强化的铝合金,其具有密度低、强度高、热加工性能好、高的比强度和硬度、优良的焊接性能较好的耐腐蚀性能和较高的韧性等优点[1],并且通过锻造、挤压和轧制等成型工艺处理及固溶时效处理制备的 Al-Zn-Mg 系合金具有高的抗拉强度的同时又能保持较高的韧性和耐腐蚀性,而且成本低廉,因此成为结构材料开发的热点之一[2,3]。近

基金项目:广西有色金属及特色材料加工重点实验室开放基金资助项目(14-A-02-06);

广西大学自治区级"大学生创新创业训练计划"资助项目(No.2014 1059 3157)。

第一作者:阁世景,女。有色金属及特色材料加工。1157425818@qq.com。

通讯作者:汤宏群,女。有色金属及特色材料加工。hqtang@gxu.edu.cn。

几十年来,国内外学者对高强铝合金的组织及其性能等进行了大量的研究,取得了重大进展,极大地促进了该类材料在工业生产中的应用[4,5]。本文针对国内外喷射成型的 Al-Zn-Mg 系合金的发展,综述了喷射成型下热处理工艺对该系合金的影响,旨在为该系合金的发展提供指导。

1 国内外 Al-Zn-Mg 喷射成型的发展

喷射成型(Spray Forming)技术是一种新的金属成型工艺,又可称为喷射沉积(Spray Deposition)、喷射铸造(Spray Casting)及液体动态压实(LiquidDynamics Compaction)等[6]。它是 20 世纪 70 年代以来,工业发达国家在传统快速凝固／粉末冶金(RS/PM)工艺基础上发展起来的一种新的先进材料制备与成型技术。其将金属熔体雾化和雾化液滴沉积合为一体,可直接由液态金属制备具有快速凝固组织特征的大块金属坯体[7]。喷射成型工艺流程短、成本低、沉积效率高、制造系统灵活柔性,是一种极富有发展前景的近终成型快速凝固技术[8]。由于喷射成型工艺减轻了材料的偏析和氧化程度,使所得铝合金材料比其他快速凝固方法制得的铝合金材料具有更高的性能。

喷射成型于 1968 由英国 Swansea 大学的 Singer 教授首先提出并发明了著名的 Osprey 工艺。先后应用于 Al-Cu、工具钢、高温合金和高合金铸铁中,取得了良好效果的效果[9,10]。20 世纪 90 年代,英国的 Osprey Metals Ltd.和美国的宾州大学为用喷射成型成功开发含 Zn 量最高达 11.5wt.%的 7000 系超高强铝合金,在采用了 T6,T77 等热处理工艺进行处理后,其室温极限抗拉强度高达 800Mpa 以上,同时其延伸率可保持在 5%以上,大幅度超过了采用传统工艺生产的各种 7000 系铝合金[12]。到 20 世纪 90 年代末,美国、英国、日本等工业发达国家利用喷射成型技术开发出了含锌量在 8%以上(最高达 14%),抗拉强度为 760～810Mpa,延伸率为 8%～13%的新一代超高强铝合金,用于制造交通运输领域的结构件及其他高应力结构件[12,13]。喷射成型技术的出现,使得各国工业界突破传统 8wt.%含 Zn 量的限制,研制开发新一代 7000 系超高强铝合金材料变成了现实[14,15]。

我国的喷射成型技术由张永昌教授于 1987 年第一次提出[16]。自此喷射成型超高强 7000 系铝合金是国内开展喷射成型研究的热点。从 1998 年起,北京科技大学、东北轻合金有限责任公司和北京有色金属研究总院开始了有关喷射成型超高强 2000 系和 7000 系铝合金的研究工作,取得了一定的成绩,成功研制了 Zn 含量高达 10wt.%以上的喷射成型超高强铝合金[17-19]。北京有色研究院石力开教授等在"九五"期间对喷射成型制备 7000 系超高强铝合金技术进行了大量的研制和开发,成功研制了大型喷射成型设备[20]。同时北京有色金属研究总院和东北轻合金加工厂开展了仿 B96u 合金成分的超高强 7XXX 系铝合金以及具有更高锌含量的喷射成型超高强铝合金的研制开发工作,使得合金的屈服强度已分别达到 750～780MPa 和 630～650MPa,延伸率则分别达到 8%～10%和 4%～7%,接近国外 20 世纪 90 年代中期的水平[21-23]。哈尔滨工业大学李庆春教授等也加入了对喷射成型技术的研发,使得该技术应用 7xxx 系产品的制造[24]。近年来人们对喷射成型高强铝合金的研究热度不减,马瑞等[25-26]对高强铝合金喷射成型的研究表明,采用喷射成型法制备合金可克服铸造法制备合金的缺点,所得合金微观组织均匀,晶粒细小,无明显微观和宏观偏析。白朴存等[27,28]采用喷射成型技术成功制备出 Zn 含量超过 12wt.%、抗拉强度高达 800MPa 的高强铝合金,大大高于传统铸造技术获得的 Zn 含量小于 8wt.%、抗拉强度低于 700MPa 高强铝合金。王洪斌[29]通过对合金热处理的优化,使得合金同时具有 T6 的强度和 T7X 的抗应力腐蚀性能。

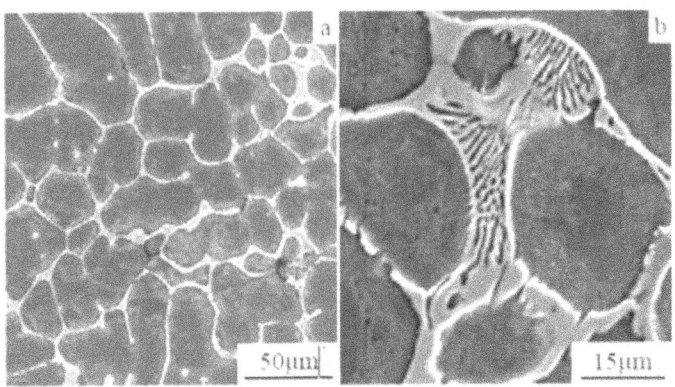

图 1 铸态 7A60 合金 SEM 微观组织
(a)铸态组织;(b)铸态组织放大图
Fig.1 SEM micrographs of cast 7A60 alloy
(a) as−cast; (b)local magnification of (a)

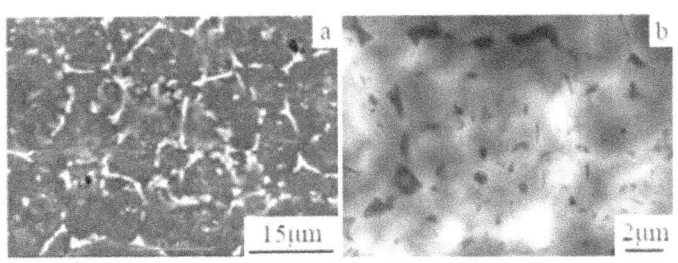

图 2 喷射成型 7A60 合金的微观组织
(a)SEM 照片;(b)TEM 照片
Fig.2 Microstructures of spray−deposited 7A60 alloy
(a)SEM micrograph;(b)TEM micrograph

图 1、图 2 是何小青等[30]研究的 7A60 合金的普通铸态和利用喷射成型技术制备的合金的相图。从图 1a[31]可以看出,铸态 7A60 合金的组织呈粗大的蔷薇型分布且很不均匀,沿晶界分布着大量的共晶组织。高倍显示晶界组织为典型的层片状共晶组织,如图 1b 所示。从图 2a 可以看出,喷射成型合金由晶粒大小在 20μm～30μm 左右的等轴晶组成,在晶内存在细小的第二相。从图 2b 可以看出,第二相不仅在晶界上存在而且在晶内也有较大数量存在。其中晶内第二相在基体中分布比较均匀并以各种形态存在,常见的有针状、条状和块状颗粒。另外可观察到有些晶界在形态上不完整。整体来看经过喷射成型的合金的组织更加细小均匀。

2 热处理对喷射成型的 Al–Zn–Mg 系高强铝合金性能的影响

2.1 均匀化处理

铸造合金在凝固时都存在枝晶偏析,Al–Zn–Mg 系高强铝合金也不例外。而晶界和晶内各组元分布不均匀,则必须通过均匀化处理消除或降低合金组织和成分的不均匀性。均匀化处理一方面能促进铝合金中低熔点可溶解共晶相完全(或接近完全)溶解,使合金铸锭化学成分分布趋于均匀,组织达到

(或接近)平衡状态,提高合金元素在基体中的固溶度;另一方面也可改善合金的塑性,提高合金的强度,最终改善合金的加工性能及使用性能。

左玉婷等[32]研究了喷射成型 Al-9.97Zn-2.65Mg-1.94Cu-0.12%Zr 合金在均匀化过程中微观组织的演变。结果表明:均匀化处理可使合金中的一次析出相明显减少,经均匀化处理的合金的晶粒尺寸没有明显长大,大多数 AlZnMgCu 四元相回溶到基体中;均匀化态组织除 Al 外,主要存在 3 种不同的相,分别为 AlZnMgCu 四元相、Al9FeNi 相以及 Al3Zr(L12)弥散粒子。

2.2 固溶处理

合金的凝固过程对粗大化合物结晶相的形成也有一定影响,均匀化处理的过程中部分的粗大不溶相不能够充分地融入基体中,这时候可通过固溶热处理将此类粗大结晶相溶入基体。

郝广瑞等[33]研究了单级固溶和双级固溶热处理工艺对喷射成型 Al-Zn-Mg-Cu 铝合金力学性能的影响。结果表明,双级固溶时效和单级固溶时效处理制度相比,前者得到的组织和力学性能较为理想;双级固溶处理综合了低温单级固溶和高温单级固溶的优点,即再结晶晶粒尺寸较小,同时回溶颗粒较多,时效后的组织也较理想。采用双级固溶处理和 T6 时效处理后,合金的抗拉强度和屈服强度都有一定程度的提高,延伸率也提高。刘敬福等[34]利用喷射成型制备高强度 Zn-35Al-3.5Mn-2.2Cu-0.1Mg 合金的固溶研究也得到了相似的结论。王洪斌等[35]对沉积态、挤压态固溶显微组织进行观察发现沉积态合金晶粒均匀细小但是合金孔隙较多;挤压态合金存在大量的第二相颗粒,为富铜相;固溶处理后,合金出现了再结晶现象,当固溶温度达到 490℃时,晶界出现熔化现象。

2.3 时效处理

合金的时效过程就是在过饱和固溶体脱溶过程,通过一系列脱溶结构(偏聚区、有序区、过渡区、平衡相)的出现、消失、形核及长大来实现。

崔华等[36—38]系统地研究了喷射成型高锌 Al-Zn-Mg-Cu 合金在不同时效处理条件下的显微组织与力学性能,发现自然时效合金在晶界处已析出强化相;单级时效合金随时效时间的延长,晶内和晶界析出相逐渐粗化;双级时效合金的析出相进一步粗化,并且晶界析出相为断续特征双级时效后,晶内析出相略有长大,此时合金的强化机制是 GP 区和 η′ 相的综合强化。与峰时效条件相比,双级时效后合金的抗拉强度和屈服强度有所降低,但合金组织中的晶界析出相完全断开,这对提高合金的抗应力腐蚀能力具有重要意义。回归再时效合金具有较细的晶内组织及类似于双级时效的晶界组织;同时发现,双级时效合金的抗拉强度比峰时效合金的强度下降了 13%左右,而回归再时效合金的强度优于峰时效合金的强度。对欠时效预处理的回归处理研究表明,欠时效预处理比 T6 峰值时效预处理更有利于在回归处理后合金晶内析出相的回溶,回归处理过程可使合金晶界析出相充分断开,再时效后晶内的析出相细小弥散,合金抗拉强度和屈服强度均高于 T6 峰值时效和常规回归再时效水平。

3 喷射成型对 Al-Cu-Mg 系高强铝合金腐蚀性能的影响

研究结果显示[39],由于喷射态合金与传统方法制备的合金在原始组织、热处理后的组织变化仍存在较大区别,如固溶度、晶粒度等,合金的抗 SCC 性能也产生较大影响,其影响规律需要深入研究。

李荣德等[40]对喷射成型的 7075 合金的研究表明,随着回归处理的进行,基体中的 GP 区和 η′ 相逐渐回溶,当回归处理超过一定时间后,合金晶粒内再次析出 η′ 相,使合金强度小幅上升;回归过程中 η′ 相先回溶再析出,因此不同回归状态下 RRA 处理后的合金晶内组织形貌差异较大。RRA 处理后晶内大量再析出的 η′ 相与晶界处粗化并断开的 η 相可以在保持合金高强度的基础上显著改善其抗 SCC 性能。

苏睿明等[41]的研究发现,经 T6 处理后,7075 铝合金晶内大量细小弥散的 η′ 相使合金的抗拉强度达到 760 MPa,但晶界处连续分布的 η 相和窄小的晶界无析出带(PFZ)使合金抗晶间腐蚀性能变差,晶间腐蚀深度达 131.4mum;经 T73 处理后晶界 η 相断开及 PFZ 大幅增宽可改善合金的耐蚀性,晶间腐蚀深度仅为 2.0mum,但晶内 η′ 相粗化及体积分数的减小使合金抗拉强度大幅下降,仅为 676 MPa;RRA 处理后合金晶内 η′ 相再次大量析出,致使抗拉强度达 758 MPa,略低于 T6 态的抗拉强度,而晶界处断续分布的 η 相和宽度略增的 PFZ 使合金抗晶间腐蚀性能也显著改善,晶间腐蚀深度为 16.8mum,与经 T73 处理后的接近。

近年来研究结果[42]还表明,随着应力载荷的提高、腐蚀时间的延长,试样的强度并未出现明显的下降,但材料的塑性明显降低。腐蚀应力越接近材料的屈服强度,塑性下降越明显,抗拉应变和延伸率均降低,在腐蚀液中,极化曲线测试也表现出同样趋势。试样表层部分区域有明显的点蚀和晶间腐蚀空洞。拉伸时,腐蚀缺口处会产生明显的应力集中成为起裂源,导致材料塑性下降,从拉伸断口分析可知,断口边缘被腐蚀区域为明显的沿晶断裂。

4 结语

Al-Cu-Mg 系高强铝合金作为一种高性能材料,被广泛应用于工业各领域,其发展水平的高低是一个国家铝工业先进程度的体现。尽管铝合金一直对航空、航天等高新技术起着十分重要的支撑作用,但随着钛合金、复合材料的快速发展,铝合金正面临前所未有的挑战。现如今人们对高强铝合金的研究已经进入了较为火热的阶段,微合金化与热处理制度已经进入了全面研究阶段,全面提升铝合金综合性能的重点越来越聚焦材料的制备工艺上。我国的 Al-Cu-Mg 系高强铝合金研究与国外相比还较落后,尤其是在基础理论、工艺技术、技术装备水平及其完善程度与国外的差距很大,这些问题亟待解决。全面提升铝合金综合性能的重点越来越聚焦在制备工艺上,未来喷射成型将是 Al-Cu-Mg 系高强铝合金强化研究的一个重要方向;喷射成型工艺参数的优化及结合材料的热处理制度将会成为高性能高强铝合金材料的重要生产方式之一。

参考文献:

[1] Tsuguo Imamura. Toward actual application of Al-Li alloy to air craft [J]. Light Metal,1986,36(11): 705-711.

[2] 陈亚莉. 波音 777 及其侯选动力装置的选材(一) [J]. 航空制造工程,1995,25(1): 32-34.

[3] 曾渝,尹志民. 超高强铝合金的研究现状及发展趋势[J]. 中南工业大学学报,2002,33(6): 592-596.

[4] Heinz A., Haszler A.. Reeent development in aluminum alloys for aerospaee applications [J]. Material Science and Engineering A, 2000, 280(1): 102-107.

[5] Lukasak D. A., Strong aluminum alloy shaves air frameweight [J]. Advanced Materials & processes, 1991, (10): 46-49.

[6] Xu Q., Lavernia E. J.. Fundamentals of the spray forming process [C]//Proceedings of the international conference on spray depositionand melt atomization. Bremen: Germany, 2000: 17–36.

[7] 李荣德,刘敬福. 喷射成形技术国内外发展与应用概况 [J]. 铸造,2009,58(8): 797–803.

[8] 王文明,潘复生,Lu Yun. 喷射成形技术的发展概况及展望 [J]. 重庆大学学报,2004,27(1): 101–106,111.

[9] 张林. 喷射成形 7475 铝合金热处理及超塑性研究 [D]. 上海交通大学硕士学位论文. 2010: 1–10.

[10] 张永昌,白丽华. 金属雾化喷射沉积工艺的研究进展 [J]. 兵器材料科学与工程,1993,16(1): 39–46.

[11] 张永安,熊柏青,石力开. 材料导报,2002,3 (16): 11–16.

[12] 张永安,韦强. 喷射成形制备高性能铝合金材料 [J]. 机械工程材料,2001,25(4): 22–25.

[13] PLIES J. B., GRANT N. J.. Structure and properties of spray formed 7150 containing Fe and Si [J]. International Journal of Powder Metallurgy, 1994, 30(3): 335–343.

[14] Singer A. R. E.. Spray rolling of metals [J]. Metallic Materials, 1970, 4(4): 246–250.

[15] Mathur P., Annavarapu S., Apclian D., et al. Process control modeling and applicantionof spraycasting [J]. The Journal of the Minerals Metals & Materials Society, 1989, 41(10): 23–24.

[16] Liu D. M., Zhao J. Z., Ye H. Q.. Modeling of the solidification of gas–atomized alloy droplets during spray forming [J]. Materials Science and Engineering: A, 2004, 372:229–234.

[17] 康福伟,孙剑飞,张国庆,李周,沈军. 喷射成形镍基高温合金热变形特性及微观组织变化 [J]. 金属学报,2007,43(10): 1053–1058.

[18] 罗光敏,樊俊飞,宋国斌,单爱党. 喷射成形高温合金 FGH4096 的热变形行为 [J]. 宝钢技术,2008(5): 57–60.

[19] 杨滨,陈美英,尧军平等. 南昌航空工业学院学报 (自然科学版),2004,18(1):1–4.

[20] Wei Q., Xiong B. Q., Zhang Y. A.. Production of high strength Al–Zn–Mg–Cu alloys by spray forming process [J]. Transactions of Nonferrous Metals Society of China, 2001, 11(2): 258–261.

[21] 张峥. 搅拌摩擦焊的特点和应用 [J]. 材料工程. 1999,2: 35–37.

[22] 史耀武,唐伟. 搅拌摩擦焊的原理与应用 [J]. 电焊机. 2000,30(1): 6–9.

[23] 顾曾迪,陈根宝,金心溥. 有色金属焊接(第 2 版) [M]. 机械工业出版社,1995:6–9,301–307.

[24] 李庆春, 沈军. 超音雾化喷射沉积技术制备高性能铝合金材料, 全国喷射成形技术学术研讨会论文集 [C]. 哈尔滨,1998: 110–116.

[25] 马瑞,周娟,肖于德. 制备工艺对高强铝合金组织及性能的影响 [J]. 热加工工艺, 2010,39(13): 38–40.

[26] 张永安, 熊柏青, 韦强等. 喷射成形制备 Al–Zn–Mg–Cu 系高强高韧铝合金的研究 [J]. 稀有金属,2002,26(6):425–428.

[27] 张永安,熊柏青,朱宝宏. Al–(10)Zn–(2.9)Mg–(1.9)Cu 超高强铝合金的喷射成形制备研究 [J]. 稀有金属,2003,27(5): 609–613.

[28] 白朴存,张秀云,陈伟. 喷射成形 Al–Zn–Mg–Cu 合金的组织与性能研究 [J]. 金属热处理,2006,31(6): 17–20.

[29] 王洪斌, 黄进峰, 杨滨等. Zn–Mg–Cu 系超高强度铝合金的研究现状与发展趋势 [J]. 材料导报;2003,17(9): 14–15.

[30] 何小青,熊柏青,孙泽明. 喷射成形 7A60 合金的微观组织与力学性能研究 [J]. 稀有金属,2007,31(2): 137–141.

[31] 何小青,熊柏青,孙泽明等. 喷射成形与铸造 7A60 合金的微观组织对比研究 [J]. 稀有金属材料与工程,2007,36(4): 656–659.

[32] 左玉婷,王锋,熊柏青等. 喷射成形 Al–9.97Zn–2.65Mg–1.94Cu–0.12%Zr 合金均匀化过程中的组织演变 [J]. 中国有色金属学报, 2010, 20(5): 820–826.

[33] 郝广瑞,王洪斌,黄进峰. 喷射成形超高强度 Al–Zn–Mg–Cu 合金的固溶处理 [J]. 北京科技大学学报, 2004,26(5): 502–506.

[34] 刘敬福, 李荣德, 白彦华. 喷射成形 Zn–Al–Mn–Cu–Mg 合金的固溶处理 [J]. 特种铸造及有色合金, 2011,31(4): 305–308,394.

[35] 王洪斌,刘慧敏,黄进峰. 热处理对喷射成形超高强 Al–Zn–Mg–Cu 系铝合金的影响 [J]. 中国有色金属学报,2004,14(3): 398–404.

[36] 崔华, 王洪斌, 程军胜等. 喷射成形高锌 Al–Zn–Mg–Cu 合金的时效行为 [J]. 北京科技大学学报, 2006,28(10): 937–940.

[37] 王锋,熊柏青,张永安等. 双级时效处理对喷射沉积 Al–Zn–Mg–Cu 合金微观组织和力学性能的影响 [J]. 中国有色金属学报,2007,17(7): 1058–1062.

[38] 曲迎东,苏睿明,唐才宇. 喷射成形 7075 铝合金欠时效回归再时效热处理 [J]. 特种铸造及有色合金, 2014,34(5): 463–466.

[39] Cai Y. H.,Liang R. G.,Su Z. P.,et al. Microstructure of spray formed Al–Zn–Mg–Cu alloy with Mnaddition [J]. Transactions of Nonferrous Metals Society of China,2011,21: 9–14.

[40] 李荣德,苏睿明,曲迎东. 喷射成形 7075 合金回归再时效处理的组织和抗应力腐蚀性能 [J]. 机械工程学报,2013,49(20): 22–29.

[41] 苏睿明,曲迎东,李荣德等. 时效处理对喷射成形 7075 合金晶间腐蚀的影响 [J]. 中国有色金属学报, 2014,24(3): 659–667.

[42] 乔及森,杨敏杰,夏浩等. 喷射成形 7055 铝合金的应力腐蚀性能研究 [J]. 材料导报,2014,28(5B): 93–96.

H70 铜合金高温变形行为研究

边 鸽[1,2]，程 明[1,2]，王瑞雪[2]

(1.广东精艺金属股份有限公司；2.中国科学院 金属研究所)

摘 要：采用 Gleeble-3800 热模拟实验机，对 H70 铜合金在热变形条件下流动应力特征进行研究。结果表明：在实验范围内，H70 铜合金热变形时可以发生明显的动态再结晶；指数形式的 Arrhenius 方程能较好地描述 H70 铜合金的高温变形行为；分区间求得热变形激活能 Q，分段建立 H70 铜合金的本构方程，可用于高温变形的数值模拟与工艺参数的确定。

关键词：H70 铜合金；流动应力；本构方程；变形激活能

Investigation on the Thermal Deformation Behavior of H70 Brass

BIAN Ge[1,2], CHENG Ming[1,2], WANG Rui-xue[2]

(1.Guangdong Jingyi Metal Co. Ltd., Foshan 528311 China;
2.Institute of Metal Research, Chinese Academy of Sciences)

Abstract: The deformation behavior of H70 brass during hot formation has been investigated by using Gleeble-3800 simulator. The results show that the flow stress feature of H70 brass alloy during high temperature plastic deformation, which can be represented by the exponent format of Arrhenius relationship. The deformation activation energy Q is different, which means the three temperature ranges correspond to various deformation mechanism. The constitutive equation is established using the method of subsection which can be used to realize numerical simulation and deformation process parameter optimization.

Key words: brass H70; flow stress; constitutive equation; activation energy

0 引言

目前物理模拟实验研究在金属材料加工中的应用更加广泛而深入[1]。通过对加工材料进行物理模拟研究，既可以掌握所加工材料在高温下的力学性能，又可以作为材料性能参数输入数值模拟的材料库中作为材料输入参数，为有限元模拟提供材料性能参数，因此进行物理模拟实验具有十分重要的应用价值。

金属热变形流动应力是材料在高温下的基本性能之一，它不仅受变形温度、变形程度、应变速率和合金化学成分的影响，也是变形体内部显微组织演变的综合反应。无论在制定合理的热加工工艺方

第一作者：边 鸽，女。材料成型及控制工程模具。biangezi@163.com。

通讯作者：程 明，男。材料加工。mcheng@imr.ac.cn。

面还是在以塑性有限元为代表的现代塑性加工力学中，其精确的流动应力数值或表达式是提高理论计算精度的关键。为此，国内外近些年来在这方面的研究十分活跃。文献[2]通过热模拟实验分析了纯铜形变中的流动应力和动态再结晶特点。文献[3]对 KFC 铜合金热压缩变形流动应力进行了分析，掌握了 KFC 铜合金流动应力特点。文献[4]对铬铜的热变形流动应力进行了一定的实验研究，但对于 H70 铜合金的流动应力和动态再结晶的研究却很少，因此需要深入的分析和研究。本文基于 Gleeble3800 热模拟试验机对 H70 铜合金进行了不同温度下的力学性能测试。

1 试样制备和试验方法

1.1 试样制备

测试的铜试样的化学成分(质量分数,%)如表 1.1 所示。试样尺寸为 Φ8×12mm，基于 Gleeble-3800 热模拟试验机对 H70 铜合金水平连铸坯料的力学性能进行测试。实验过程中对试样两端进行了必要的加工和润滑处理。

表 1 试验材料的化学成分

Tab.1 Chemical compositions (in wt.%) of H70 copper billet

Cu	Pb	Fe	Ni	Zn	杂质
68.5~71.5	0.03	0.10	0.5	余量	0.3

对于铜及铜合金等材料来说，最大的困难是 Gleeble-3800 试验机的温度测量装置热电偶无法牢固地焊接在试样表面，一旦加热受力后即将发生分离，无法完成测试工作。对于这种情况，在对 H70 铜合金材料进行物理模拟分析时，克采取一种简单而有效的措施解决。

首先在铜试样的侧面加工出直径 Φ0.6 的小孔，如图 1 所示，然后将两个热电偶分别插入孔内两侧，注意两个热电偶不要形成短路，这样既能保证热电偶插入铜合金试样中的稳固程度又可以保证物理模拟实验测试的精确度。经过实验验证该方法测出的数据稳定可靠，但该方法仅对于材质较软、塑性好的铜及铜合金试样比较合适，而对于其他塑性差硬度大的材料，其应用有一定难度，可考虑其他更有效的解决方法。

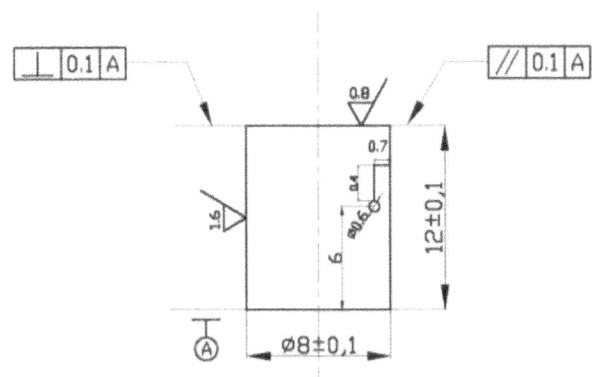

图 1 压缩式样加工图

Fig.1 The view of the compression specimens

1.2 试验方法

对 H70 铜合金进行等温热压缩实验。由于在旋轧成型过程中，坯料温度由室温升高到 700℃ 左右，这与一般热轧过程相比有不同的工艺特点，即轧制过程中温度变化大，温度梯度大，因此测试温度分别设为 20℃、100℃、200℃、300℃、400℃、500℃、600℃、700℃ 和 800℃；实验测试的应变速率为 $0.01s^{-1}, 0.1s^{-1}, 1s^{-1}$；总压缩变形量为 70%。热模拟实验的升温速度为 5℃/s，保温时间为 3min。通过对铜合金热压缩变形流动应力与变形程度、应变速率以及变形温度之间的关系分析，建立本构方程，为合理制定铜合金热变形工艺提供参考，以及为有限元数值模拟进一步分析提供准确数据或数学模型。

2 试验结果与分析

图 2、图 3 所示为 H70 铜合金在变形温度为 20℃~800℃、应变速率为 $0.01s^{-1}, 0.1s^{-1}, 1s^{-1}$ 热压缩条件下流动应力曲线。

从图中可以看出，峰值应力对应的应变随温度的升高而不断减小。流动应力 - 流动应变曲线大致可以分为三个阶段。第一阶段变形量较小时，随着应变的逐步增加，位错密度也增加，位错消失速度也随之增大。反映在流动应力 - 流动应变曲线上是随着变形量的增大，加工硬化速度减弱，但是在第一阶段总的趋向还是加工硬化超过动态软化，因此随着变形量的增加，变形应力还是不断增加。第二阶段当应变超过一定值后，应力下降，表明材料在该温度下已经发生动态再结晶，动态再结晶的发生与发展使更多的位错消失，材料的变形应力很快下降。第三阶段，应变达到一定的时候，应力与应变呈现稳态流动的特征，由于流动应力在此条件下维持一稳定值，加工硬化和动态再结晶软化达到平衡[5]。

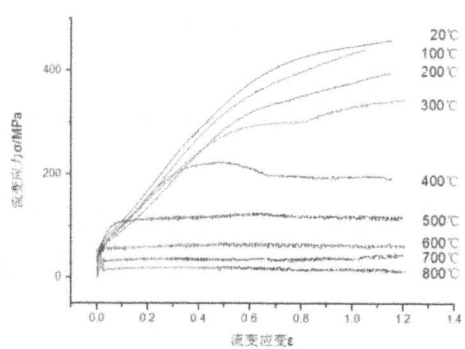

图 2 应变速率为 0.01s⁻¹ 时流动应力曲线
Fig.2 True stress−strain curves at strain rate 0.01s⁻¹

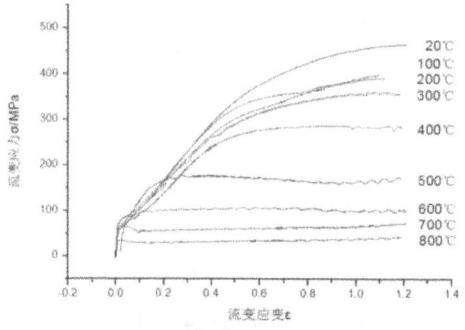

图 3 应变速率为 0.1s⁻¹ 时流动应力曲线
Fig.3 True stress−strain curves at strain rate 0.1s⁻¹

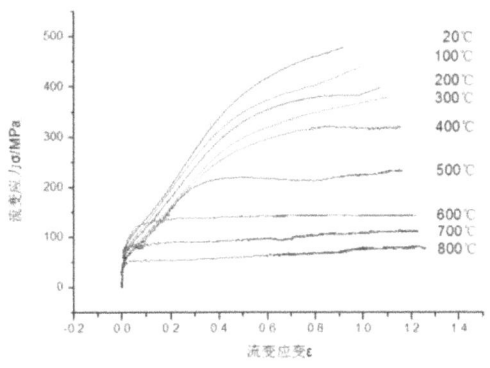

图 4 应变速率为 1s⁻¹ 时流动应力曲线

Fig.4 True stress−strain curves at strain rate 1s⁻¹

如图 5 所示,变形温度保持不变时,应变速率越低,稳态变形阶段的流动应力也越低。从图 3 可以看出,温度和应变速率是影响流动应力的重要因素。在同一温度下,材料的应力峰值随应变速率的增大而增大。一般认为,应变速率较低时材料中的储存能较高,从而有利于材料在热变形过程中发生动态再结晶[6]。在较高的应变速率下,塑性变形时单位应变的变形时间缩短,能发生运动的位错的数目增加。同时,由于动态回复、动态再结晶等提供的软化过程时间缩短,塑性变形不充分,导致流动应力增大。在同一应变速率下,温度越高,原子的热激活能的作用越大,原子振动加强,原子间的临界切应力减弱。此外,动态回复、动态再结晶引起的软化过程也随着温度的升高而增大,从而导致应力峰值的降低[7],而且随着温度升高,变形速率越小,动态再结晶的临界切应变值变小,即表示材料在高温下动态再结晶很快发生[8]。

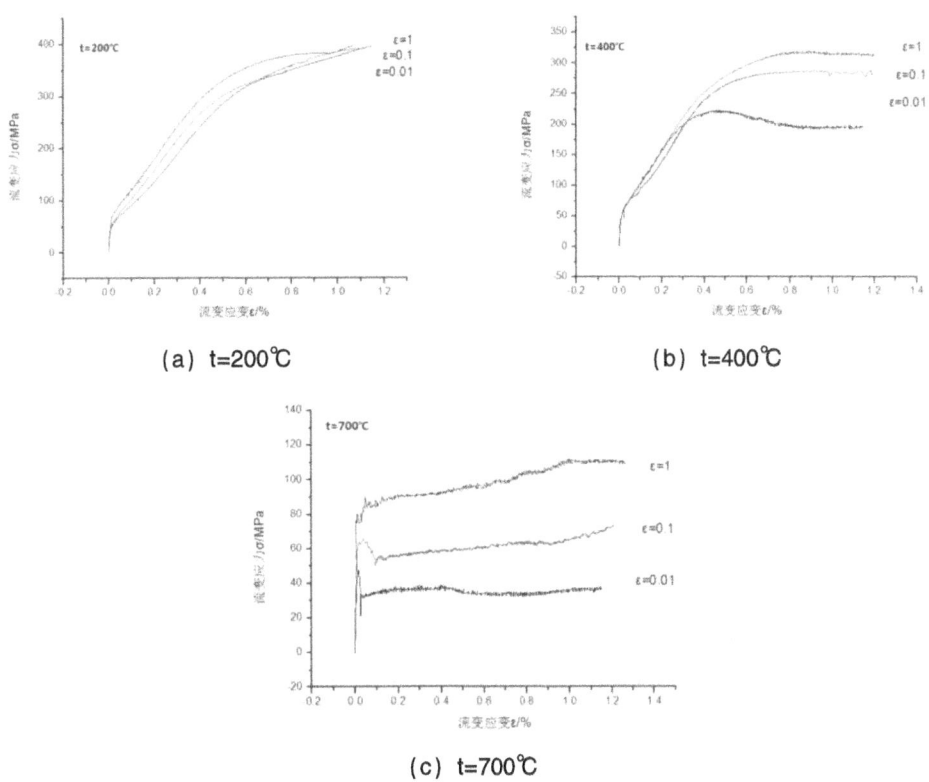

(a) t=200℃   (b) t=400℃

(c) t=700℃

图 5 不同温度下的 H70 铜合金的流动应力-流动应变曲线

Fig.5 True stress− strain curves at different temperature for H70 brass

3 流动应力本构方程建立

H70 铜合金在变形过程中对变形温度和应变速率都很敏感,研究表明[9—12],对于金属材料高温变形 σ、应变速率 ε、变形温度 T 之间的关系可表示为如下式 3.1-3.3。

$$\varepsilon = A\sigma^n \cdot \exp(-Q/RT) \qquad\qquad 3.1$$

$$\varepsilon = A\exp(\beta\sigma) \cdot \exp(-Q/RT) \qquad\qquad 3.2$$

$$\varepsilon = A[\sinh(\alpha\sigma)]^n \cdot \exp(-Q/RT) \qquad\qquad 3.3$$

其中,$\alpha = \beta/n$ \qquad\qquad 3.4

另外,根据 Zener 和 Hollomon 的研究[13],应变速率与温度之间的关系可用 Z 参数表示:

$$Z = \varepsilon \cdot \exp(-Q/RT) = A[\sinh(\alpha\sigma)]^n \qquad\qquad 3.5$$

上面各式中,α 为应力水平参数;n 为应力指数;A 为结构因子;Q 为变形激活能(J/mol),反映了材料热变形过程中的难易程度,是材料热变形过程中一个重要的力学性能参数;R 为气体常数,取 8.314J/mol;ε 为应变速率(s^{-1});T 为热力学温度(K);σ 为表示峰值应力或稳定流动应力(MPa);Z 为 Zener-Hollomon 参数,物理意义是温度补偿的变形速率因子。

在回归 H70 合金本构方程时,首先分别以 σ 和 $\ln\varepsilon$、$\ln\sigma$ 和 $\ln\varepsilon$ 为坐标作图,并用最小二乘法做线性回归,如图 3.1 所示,由于斜率差别较大,应对该八组数据分开讨论。根据式 3.7、3.8 分别求得直线 $\ln\varepsilon$-$\ln\sigma$ 与直线 $\ln\varepsilon$-σ 的斜率为 n、β 值。根据式 3.4[($\alpha_1 = \beta_1/n_1 = 0.003$,$\alpha_2 = \beta_2/n_2 = 0.003$,$\alpha_3 = \beta_3/n_3 = 0.020$。

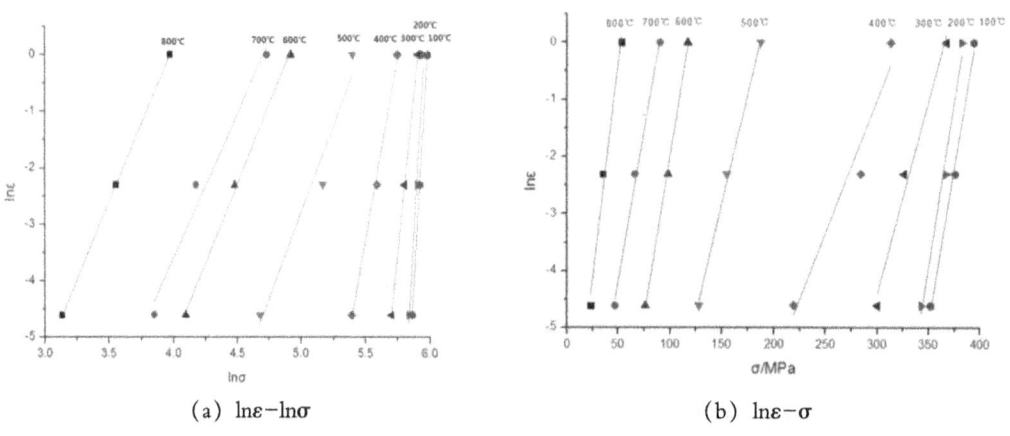

(a) $\ln\varepsilon$-$\ln\sigma$ \qquad\qquad (b) $\ln\varepsilon$-σ

图 6 不同变形温度下峰值应力 σ 与应变速率 ε 之间的关系
Fig.6 Relationship between strain rate and peak stress at different deformation temperature

在整个应力水平,对 3.3 式两边取对数,分别在一定应变速率和变形温度下对其求偏导,可以确定:

$$Q = R \left[\frac{\partial \ln[\sinh(\alpha\sigma)]}{\partial(1/T)} \right]_{\dot{\varepsilon}} \cdot \left[\frac{\partial \ln\dot{\varepsilon}}{\partial \ln[\sinh(\alpha\sigma)]} \right]_T \qquad\qquad 3.6$$

又根据 3.6 式,取不同应变速率和温度下的峰值应力、应变值,分别作出 $\ln[\sinh(\alpha\sigma)]$-1/T 和 $\ln\varepsilon$-$\ln[\sinh(\alpha\sigma)]$ 关系。易知 n 为拟合直线 $\ln\dot{\varepsilon}$-$\ln[\sinh(\alpha\sigma)]$ 的斜率。又令 K 为拟合直线 $\ln[\sinh(\alpha\sigma)]$-1/T 的斜率,则可知:

$$Q = RnK \qquad\qquad 3.7$$

分别求得不同温度区间应变速率为 0.01s^{-1}、0.1s^{-1} 和 1s^{-1} 三条拟合直线斜率平均值 K、n。将所求得的 K、n 值代入式 3.7，可求得不同温度区间 H70 铜合金变形激活能 $Q_1=168kJ/mol$，$Q_2=120kJ/mol$，$Q_3=280kJ/mol$。

对式 3.5 两边同时取对数，又联立式 3.6、3.7 可得：

$$lnZ=lnA+n \cdot ln[sinh(\alpha\sigma)] \qquad 3.8$$

取不同应变速率和温度下的峰值应力、应变值，分别将 Q、R 以及对应的温度(T)，应变速率(ε) 代入做出 lnZ-ln[sinh($\alpha\sigma$)]的关系，如图 3.2 所示。根据式 3.8 可以得到更为准确的应力指数 n 值以及拟合直线的截距 lnA 值，则 $A_1=e^{-2.65}$，$A_2=e^{19.64}$，$A_3=e^{30.17}$。

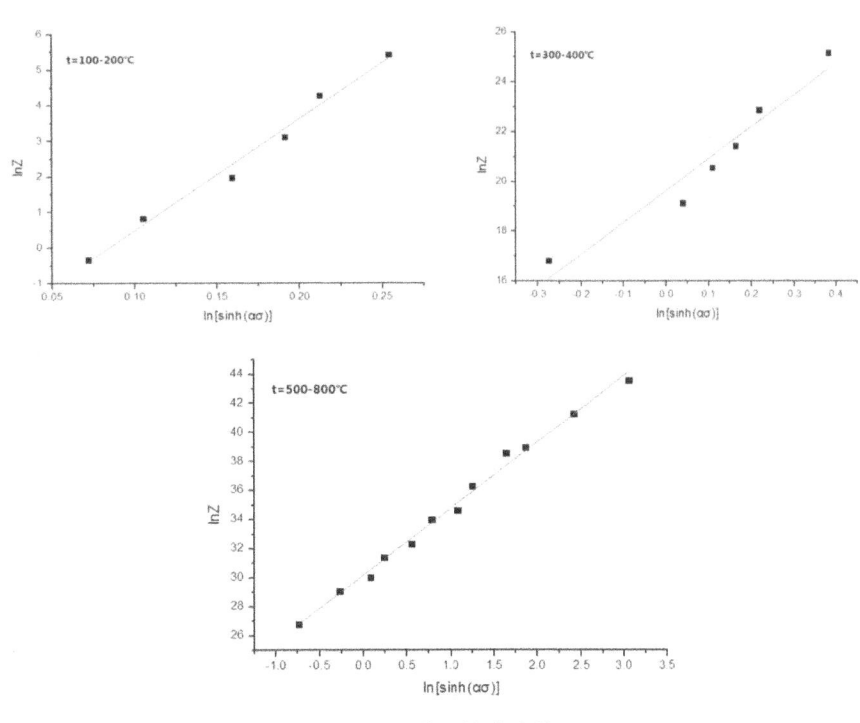

图 7 关系拟合直线

Fig.7 Fitting straight line between

图 7 为回归得到的 lnZ-ln[sinh($\alpha\sigma$)]关系图，通过回归分析可观察到二者基本符合线性关系，证明 Z 参数与 σ 之间的 Z=A [sinh ($\alpha\sigma$)]^n 关系是正确的，也证明铜合金高温变形时的流动应力方程遵从 Zener-Hollomon 参数的双曲正弦函数形式。因此通过该方法获得变形激活能结果是可靠和有根据的。

然后，将以上推导过程计算出的相关参数分别代入公式 3.3 中，即可得到：

100℃～200℃时，本构方程为：

$$\varepsilon=e^{-2.65} [sinh(0.003\sigma)]^{31.433} \cdot exp(-16827/RT) \qquad 3.9$$

300℃～400℃时，本构方程为：

$$\varepsilon=e^{19.64} [sinh(0.003\sigma)]^{12.841} \cdot exp(-119938/RT) \qquad 3.10$$

500℃～800℃时，本构方程为：

$$\varepsilon=e^{30.17} [sinh(0.020\sigma)]^{4.584} \cdot exp(-280051/RT) \qquad 3.11$$

4 金相组织观察

由图 8 所示，为室温下单相 H70 铜合金金相组织照片，其特征是：铸态 α 固溶体呈树枝状，铸态冷却较快时,α 枝晶间可能出现 β 相(用氯化铁溶液腐蚀后,枝晶主轴富铜,呈亮白色,而枝间富锌呈暗色),经变形和再结晶退火其组织为多边形 α 晶粒,有退火枝晶特征。由于各个晶粒方位不同,所以具有不同的颜色。退火处理后的铜合金能承受极大的塑性变形,可以进行深冲变形。

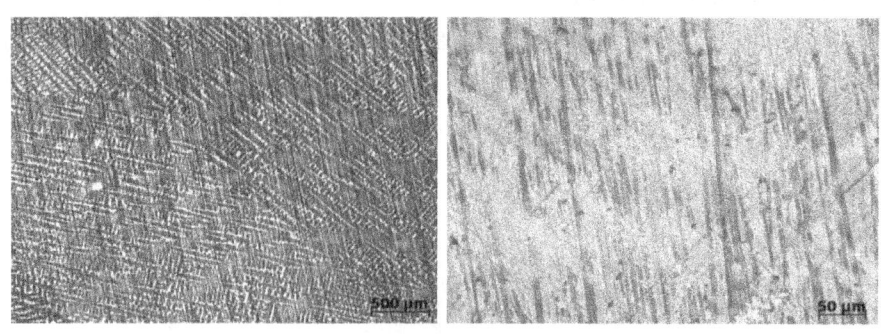

图 8 H70 铜合金原始铸态组织
Fig.8 The initial microstructure of casting brass H70

经计算,在不同的温度区间得到不同的变形激活能,这与高温变形时 α 相和 β 相的相对含量的变化有关。图 9 为不同变形温度的金相组织,可以观察到 H70 铜合金 400℃ 开始发生动态再结晶,600℃ 时动态再结晶结束。发生动态再结晶后,晶粒得到细化,可以看到明显的晶界。当温度达到 700℃ 时组织形貌明显不同,开始从 α 相中析出针状相,随着温度的继续升高,α 相含量增大并以针状均匀分布。表现在曲线上的特征为,而 100℃~400℃ 未出现明显峰值,500℃~800℃ 时出现峰值再趋于平稳,曲线趋于水平且流动应力值减小。这合理地解释了不同温度区间具有不同的变形激活能的理论。同时,也充分证明了分段建立本构方程的合理性和准确性。分段建立的材料模型可提高数值模拟精度。

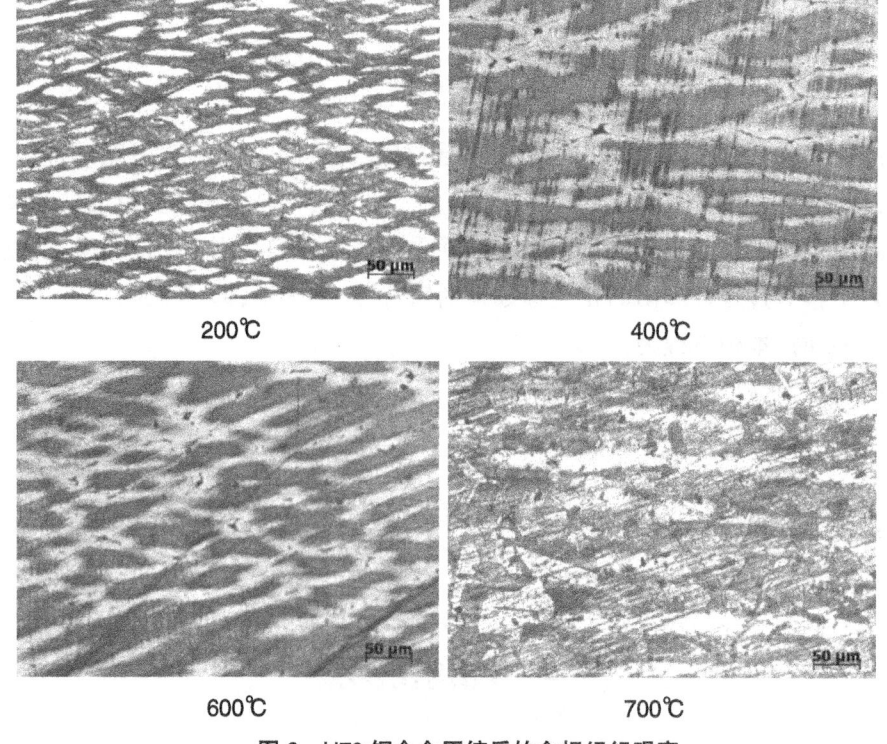

200℃   400℃

600℃   700℃

图 9 H70 铜合金压缩后的金相组织观察
Fig.9 The representative metallographic of Brass H70 after compressing

5 结论

(1)在恒定的应变速率下,流动应力随变形温度的升高而减小;而在同一应变速率下较高温度时,流动应力–流动应变曲线具有明显的峰值;温度高于 700℃时,流动应力–流动应变曲线趋于水平,变形机制为动态再结晶软化。

(2)用 Arrhenius 方程的双曲正弦形式能较好地描述 H70 铜合金合金高温变形时的流动应力行为,回归相关系数在所有温度下均超过 0.99。

(3)应力敏感指数随温度的升高而减小,在不同的温度区间得到不同的变形激活能,在 100℃～200℃时 $Q_1=168kJ/mol$,在 300℃～400℃时 $Q_2=120kJ/mol$,在 500℃～800℃时,$Q_3=280kJ/mol$。激活能的大小与 α 相和 β 相的相对含量的变化有关。

参考文献:

[1] 牛济泰. 材料和热加工领域的物理模拟技术 [M]. 北京: 国防工业出版社,1999.

[2] 侍新琳,刘铁. 纯铜形变中动态再结晶分析 [J]. 沈阳航空工业学院学报,2000(4): 19–21.

[3] 张红刚,张辉,彭大暑,林齐权,孟力平. KFC 铜合金热压缩变形流变应力 [J]. 热加工工艺,2004(1): 21–24.

[4] 樊百林,管克智,周纪华. 铬铜热变形流变应力的实验研究 [J]. 塑性工程学报. 1996(3): 3–8.

[5] 张红刚,张辉,刘婉容等. C194 铜合金热压缩变形流变应力 [J]. 湘潭大学自然科学学报,2003,25(3): 82–86.

[6] 刘万辉,鲍爱莲. 热压缩 7075 铝合金流变应力特征 [J]. 热加工工艺,2006,35(8): 28–30.

[7] 郭强,严红革,陈振华等. 铸态 AZ80 镁合金高温热变形行为研究 [J]. 塑性工程学报,2006,13(5): 16–20.

[8] 肖盼,刘天模. AZ61B 镁合金热模拟挤压变形的研究 [J]. 兵器材料科学与工程,2006,29(4): 22–25.

[9] Sellars C M,Tegart W J M. On the mechanism of hot deformation [J]. Acta. Metallurgica,1996,14(9): 1136–1138.

[10] 蔺永波,陈松明,钟掘. 42CrMo 钢的热压缩流变应力行为 [J]. 中南大学学报(自然科学版),2004,39(3):549–553.

[11] 王战峰,张辉,张昊等. 喷射沉积 5A06 铝合金热压缩变形的流变应力行为 [J]. 中国有色金属学报,2006,16(11): 1938–1944.

[12] 施发满,王泽华,林萍华等. Al–20Cu–4.5Si–3Ni–0.25RE 合金的高温流变本构方程 [J]. 热加工工艺,2008,37(20): 15–17.

[13] Zener C,Hollomon J H. Effect of strain–rate upon the plastic flow of steel [J]. J Appl. Phys., 1994,15(1): 22–27.

谈谈最小作用量原理及虚功原理

刘祖岩

(哈尔滨工业大学 金属精密热加工国家级重点实验室)

摘 要: 在对最小作用量原理起源和发展、应用和意义综述的基础上,给出了一个加深对此原理本质的理解和重要意义认识的思路,并对其与虚功原理和牛顿定律之间的关系做了一些说明。最小作用量原理是有限元方法的最根本原理, 而有限元方法是材料变形最全面实用的基础理论和技术之大成,是将材料力学、结构力学、弹性力学、塑性力学统一于材料变形力学的最有力的武器。

关键词: 最小作用量原理;虚功原理;牛顿定律;分析力学

Preparation and Mechanical Characterization of Magnetic Fluid with Core-Shell Particles

LIU Zu-yan

(National Key Laboratory for Precision Hot Processing of Metal, Harbin Institute of Technology)

Abstract: Based on the summary of origin, development, application and significance of principle of least action, a new thinking way is presented to deeply understand the essence and important function of the principle, and some explanations of the relationship among the principle of least action, principle of virtual work and Newton's law are also given out. The principle of least action is the basic principle of finite element method, and the software of finite element method is the great collection of basis theory and technics of material deformation, and the useful tool to uniform mechanics of materials, mechanics of structure, mechanics of elasticity and mechanics of plasticity into materials deformation mechanics.

Key words: principle of least action; principle of virtual work; newton's law; analytical mechanics

1 最小作用量原理的起源和发展

在物理学里,最小作用量原理是描述客观事物规律的一种原则或方法,即从一个角度比较客体一切可能的运动或经历,认为客体的实际运动或经历可以由作用量求极值得出,即作用量最小的那个经历。最小作用量原理体现在数学上,就是变分原理。有了最小作用量原理,自然就有了变分原理。

最小作用量原理从一开始朦胧、模糊的观念到定量化的具有完美数学表达式的物理学基本原理,经历了漫长的发展演变过程。其原始思想可以追溯到古希腊时期。当时的哲学家和科学家们根据哲学、神学

第一作者:刘祖岩,男。塑性加工。lzy@hit.edu.cn。

和美学的原则,认为事物总是被最简单、完美和天然的规律所支配,大自然总以最短捷的可能途径行动。

欧几里德在他的《反射光学》一书中,就把光看作沿直线传播,服从几何学规律,提出了光的反射定律。希罗也写过一本名为《反射光学》的书,指出了光在空间中两点间传播时总是沿着长度最短的路径进行,这是最小作用量原理最原始的表述。

中世纪,奥卡姆在反对经院哲学中,提出了理论思维的简单性原则和明晰性的要求,概括起来就是"如无必要,勿增实体"。他指出:"能以较少者完成的事,若以较多者去做,即是徒劳。"这就是西方哲学史上著名的"奥卡姆剃刀"原则。达·芬奇也认为自然是经济的。

最小作用量原理发展史上重要的一步当属费马原理的发现。费马曾得出这样的结论:"自然界总是通过最短的途径发生作用的。"在对光的折射定律的研究中发现,光从一种媒质进入另一种媒质的过程中,最短路径原理并不成立。然而他坚信自然界的行为总是按照简单而又经济的原则进行的。1662 年,费马提出了"最短时间原理",指出光在媒质中从一点到达另一点时,总是沿着花费时间最少的路径传播,这就是著名的"费马原理"。其用"最短时间"原理取代"最短路径"原理,是科学认识上的一大进步。费马原理作为几何光学的高度概括性原理,使此前相互独立的光的直线传播定律、反射定律、折射定律以及光路可逆性原理有了一个统一而又简洁、优美的表述[1—3]。

此后,莫培督认为,最小观念不仅适用于光的传播过程,也应普遍适用于各种物理现象。他提出"物质系统实际发生的过程应是使某个反映其经历特征的参量取最小值"。莫培督实际上是把"费马原理"发展成为了一个具有全新内容和极大适用范围的崭新原理。这一原理被称为"最小作用量原理"。

1696 年,伯努利兄弟由于对"最速降线"的研究而导致变分法的萌芽。欧拉发展了变分法,并将它用于解决抛射体的运动,也独立地得到了最小作用量原理。他还首次用变分的方式 $\delta \int v dl = 0$ 表述最小作用量原理,使其形式更加简洁,而内涵也更加深刻和具体,从而开辟了一个处理力学问题的全新途径[3]。

当最小作用量原理应用于一个机械系统时,可以得到此机械系统的运动方程。原理的研究引导出经典力学的拉格朗日表述和哈密顿表述的发展。18 世纪末,拉格朗日建立《分析力学》。雅可比称最小作用量原理为分析力学之母。19 世纪上叶,哈密顿建立哈密顿原理和哈密顿方程。这些成果标志最小作用量原理和相应的变分原理的理论体系已经发展到相当完善的地步,但其应用仍被局限于经典力学这一狭窄的领域[4]。

20 世纪初是近代物理学新发现层出不穷的时期。1900 年普朗克为解释黑体辐射引入了能量子的概念。其中普朗克常数 h 的单位是能量与时间的乘积,和作用量的单位是一致的。他断定,作为建立统一的世界物理图景之基础的最小作用量原理,是所有可逆过程的普遍原理。它虽然产生于力学,但应用的范围包括了热力学和电动力学。1923 年,德布罗意对光现象和力学现象做了类比,发现反映这两个不同领域运动规律的原理具有完全相似的数学表达式。而光学的发展已证明了原来被认为纯粹是波动的光也具有粒子性,即光具有波粒二象性。他由此推测原来被认为纯粹是粒子的电子也可能具有波动性,从而大胆提出了"物质波"的假设。这个假设于 1928 年被电子衍射实验所证实。薛定谔发展了这一思想,认为经典力学和几何光学的一些规律具有完全相似的数学形式,而量子力学又和物理光学类似,因此也必然存在一个物质波的波动方程和光的波动方程相类似,由此得出了薛定谔的波动方程,这是不同于海森伯矩阵力学的另一种形式。薛定谔证明了它和矩阵力学的等价性[1—3]。

进入 20 世纪后,爱因斯坦摒弃了绝对时空观念、引入相对性原理,建立了狭义相对论,随后向引力场中发展导致了广义相对论。爱因斯坦与希尔伯特分别找到了相应的作用量的表示式,从而由最小作用量原理建立了引力场方程[4]。1948 年,费曼也是从最小作用量原理出发,创造了和薛定谔、海森伯方法并列的一种表述量子力学的等价方法—路径积分方法[1—3]。

在现代物理学里,最小作用量原理非常重要,在相对论、量子力学、量子场论里都有广泛的用途。

最小作用量的费马表述

几何光学中,费马原理可用下式表示,其中 n 为介质折射率, s 为路径元。

$$\delta \int_A^B n(x, y, z)ds = 0$$

最小作用量的莫培督表述

莫培督发表的最小作用量原理阐明,对于所有的自然现象,作用量趋向于最小值。他定义一个运动中的物体的作用量是物体质量,移动速度与移动距离的乘积: $A = mvs$ 。

最小作用量的拉格朗日和欧拉表述

拉格朗日第一个对最小作用量原理作了正确的表述:

$$\delta A = \delta \int Mvds = 0$$

欧拉 - 拉格朗日最小作用量原理如下: p_i 广义动量, q_i 广义坐标, T 动能。

$$\delta A = \delta \int \sum_i p_i dq_i = 0 \quad 或 \quad \delta A = \delta \int_{t_i}^{t_f} 2Tdt = 0$$

最小作用量的哈密顿表述

哈密顿把最小作用量原理发展到了它的巅峰。他把力学中粒子的运动轨迹与几何光学中光线轨迹进行类比,把力学中的最小作用量原理与光学中的费马原理进行类比,利用广义坐标 q_i、广义动量 p_i 和拉格朗日函数 L,定义了一个新的哈密顿作用量函数 S:

$$S = \int_{t_1}^{t_2} L(q_i, p_i, t)dt \qquad (1)$$

其中: $L=T-V$ 为系统的动能和势能之差。从而他得出了哈密顿原理:在相同的时间和相同的约束条件下,完整有势系统由某一初位形转移到另一已知位形的一切可能运动中,真实运动的哈密顿作用函数具有极值,其数学表达式为:

$$\delta S = 0 \qquad (2)$$

哈密顿原理阐明,一个物理系统的拉格朗日函数,所构成的泛函的变分问题解答,可以表达这物理系统的动力行为。这泛函称为作用量。

哈密顿原理以更概括的概念和更少的必要公设出发,运用数学手段把已有的力学知识组织成更严密、更系统和更抽象的逻辑演绎体系,不仅给出了解决一切力学问题的统一的观点和方法,而且为包容更广泛的经验事实创立了条件,从而成为新的科学研究的起点。虽然哈密顿原理本来是用来表述经典力学的,但这原理也可以应用于经典场,像电磁场或引力场,可以延伸至量子场论等。

最小作用量原理可用示意图 4-1 来进一步说明。随时间变化的系统的作用量在任意瞬间,都是系统可几状态的函数。而在系统可几状态中,只有一个真实的状态,就在这个真实状态中,作用量取最小值。因为可几状态是状态函数,作用量对状态函数取极值,这就是数学上的变分。所以说最小作用量原理在数学上就是变分原理。

系统作用量函数可由图1中的近似于船壳型的曲面来表示，就是可几作用量曲面。而真实状态由一平面代表。两者相交的曲线 ABCD 就是最可几作用量曲线。其中 BC 段示意系统处于静止状态。

最小作用量原理中的作用量或许比能量概念更为重要。作用量的量纲是[能量·时间]，是能量与时间的统一，能量仅反映了作用量中的一部分联系。

在图1中。一个坐标轴是时间，而另一个坐标轴是作用量，还包含有时间，这是这个示意图的一个特点，也是理解作用量与时间关系的难点。

2 最小作用量原理的应用和意义

在经典力学中，由哈密顿作用量函数，即公式(1)和(2)，可得拉格朗日方程：

$$\frac{d}{dt}(\frac{\partial L}{\partial q_i'}) - \frac{\partial L}{\partial q_i} = f_i$$

对于保守系统，$f_i = 0$，可得牛顿第二定律。对于非保守系统，f_i为广义力，它包括两部分：一部分是由系统的耗散引起的，另一部分是非保守力作功引起的。

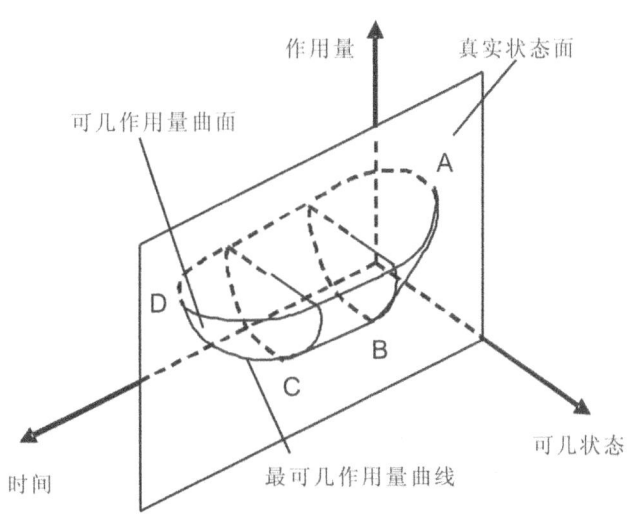

图1　最小作用量原理三维船壳型示意图

在流体力学中，对于惯性参照系下的粘性流体，选取适当的拉格朗日函数，利用变分计算，可得粘性流体的力学方程。可以证明，最小作用量原理普适于地球上实际发生的任意时限与任意体积的流体运动，以及各种流体动力学的原始方程模式的精确评估，对于流体力学模式的改进具有重要的实际意义[2,5]。

在电磁学理论中，通过选择适当的拉格朗日函数和作用量，并利用最小作用量原理，就得到麦克斯韦方程组，从而建立起了整个电磁学理论[6-7]。

在热力学理论发展过程中，亥姆霍兹把最小作用量原理推广应用于热力学领域，此后普朗克在这一方向上进一步研究。结果表明，从最小作用量原理可以建立热力学的基本方程、基本定律及热力学量的变换关系[2]。文献[8-9]初步探讨了最小作用量原理与熵增原理以及相对论热力学之间的关系。

现代物理学的发展，一是向小的方面深钻，一是向大的方面拓展，它们在宇宙的极早期又统一起来了。最小作用量原理对这两个方面的探索都起着至关重要的作用，选取不同的作用量，就等于建立了一种物理理论。这说明物质世界的统一性以及最小作用量原理在物理学中有至高无上的地位。在研究新的物理场时，所能依据的就是最小作用量原理，并由它导出场方程和守恒定律，它已成为粒子物理学、规范场论、现代宇宙学等物理理论的基本柱石[2]。

从最小作用量原理在物理学中的地位而论，没有哪一个定律或定理能有如此的魅力，始终吸引着众多的哲学家和科学家们；也没有哪一个规律能像它一样，把经典物理与近代物理，甚至把物理学与数学如此紧密地结合起来。最小作用量原理不仅反映了自然界的简单、和谐、对称与美，也反映了人们对自然规律的普遍性与简单性的追求[2-3]。朗道和粟费席兹合著的理论物理学教程系列丛书可称为是以最小作用量原理统率物理学的典范[6]。

3 虚功原理

虚功原理是分析静力学的重要原理，是拉格朗日于 1788 年确立的。其内容为：一个静止的质点系，如果约束是理想双面定常约束，则系统继续保持静止的条件是所有作用于该系统的主动力对作用点的虚位移所作的功的和为零。

实际的结构在荷载作用下，要产生内力和变形，结构的变形引起结构的位移，而虚位移指的是结构的附加的满足约束条件及连续条件的无限小可能位移。所谓虚位移的"虚"字表示它与真实的受力结构的变形而产生的真实位移无关。对于虚位移要求是微小位移，即要求在产生虚位移过程中不改变原受力平衡体的力的作用方向与大小，亦即受力平衡体平衡状态不因产生虚位移而改变。真实力在虚位移上做的功称为虚功。

虚功原理，拉格朗日方程，哈密顿原理是分析力学的主要内容。分析力学不同于在牛顿定律基础上建立起来的矢量力学，它的特点是以确定位形空间的广义坐标代替矢径，以能量和功的描述代替力矢量的分析，然后利用微积分和变分的数学分析方法，表述出力学统一的原理和公式。

分析力学或者说理论力学包括三个部分，矢量力学，分析静力学和分析动力学。矢量力学的基础是牛顿三个定律，它确定了质量、力、位移、速度、加速度之间的关系；分析静力学的基础是虚功原理，它确定了静止状态下系统能量状态及其变化关系；而分析动力学的基础是最小作用量原理，它确定了系统真实状态和其它可几状态下作用量之间的关系。应用达朗贝尔原理可以将动力学问题转化为静力学问题来处理。

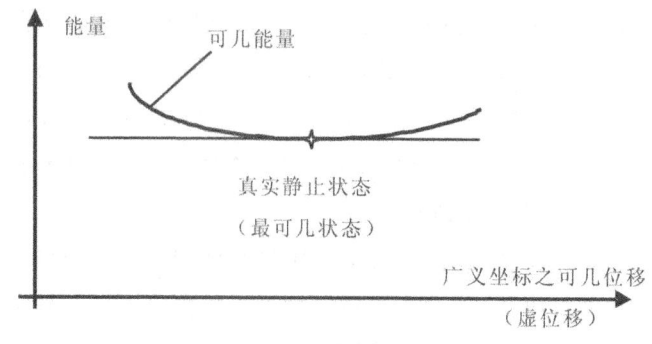

图 2 虚功原理示意图

可以这样理解，系统处于静止状态是系统处于运动状态的特例，系统处于运动状态时遵循最小作用量原理,系统处于静止状态时既遵循最小作用量原理,也遵循虚功原理,所以虚功原理是最小作用量原理的一个特例,或者说是一个蜕变。图 2 是虚功原理的示意图。真实静止状态系统的能量处于最低点,此时系统任何微小的虚功变化都是在水平的切线上的,即为零。和图 1 相比较,可见动态系统的作用量蜕变成静态系统的能量,动态系统的可几状态蜕变成静态系统的可几广义位移或虚位移。图 1 中 BC 段表示着系统处于静止状态。图 1 是个三维图,图 2 是个二维图,但是这个二维图绝不是三维图中固定时间维而简化得来的二维图。只不过两者有些相似。

最小作用量原理作为公理是不能被证明的,虚功原理可由最小作用量原理推得,同样,牛顿定律可由最小作用量原理或虚功原理推得,这样它们都是等价的。但它们的应用范围有所不同。最小作用量原理适用于任何可逆状态的物理系统(当然除了和不可逆过程的熵增原理的关系不清楚外),虚功原理适用于处于静止状态的物理系统,而牛顿定律只适用于机械系统。按层次分类,显然最小作用量原理处于高的层次,虚功原理次之,而牛顿定律处于更低一点的层次。最小作用量原理要用三维图来描述,虚功原理要用二维图解释,牛顿定律只需要用一维矢量就可以说明。由此可见,用虚功原理去证明最小作用量原理,用牛顿定律去证明虚功原理等等就像用牛顿定律去证明能量守恒定律一样是不合适的。另外强调一点,有限元法或有限元软件的基本原理是变分原理,也就是最小作用量原理。

现有弹性力学和塑性力学体系的基点或出发点是"列方程,推公式,解析解",并认为其解就是精确解、真实解。在这样的基点上,就会区分弹性力学和塑性力学,就产生了弹性力学的若干解析法和塑性力学的解析法。这些方法除了能解决几个极其简单的问题之外,任何实际问题都解决不了,其致命弱点就是推广不开,不能推广。

有限元法的基点或出发点是离散化,最小作用量原理或能量极值原理,线性方程组,数值解。这应该成为新的弹性力学和塑性力学体系的基点或出发点。

有限元先离散后综合的方法,是认识复杂事物的有效手段。离散可以使问题的复杂性、非线性转化为简单性、线性,之后的综合又使得有限个离散的碎片还原为一个整体。综合过程中一定要遵循一个原理,这个原理可以是物质不灭定律、能量守恒原理或者最小作用量原理等。

有限元法的技术载体是有限元软件。有限元软件的成功应用标志着有限元法的正确、成熟和实用,也标志着将材料力学、结构力学、弹性力学、塑性力学统一于一身的材料变形力学是一个完善实用的科学体系!

参考文献:

[1] 熊斌,吴国林. 最小作用量原理及其意义[J]. 四川师范大学学报(自然科学版),1993,16(6): 64–69.

[2] 许良. 最小作用量原理[J]. 自然辩证法研究,1992,8(12): 15–21.

[3] 塔拉. 最小作用量原理与简单性原则[J]. 内蒙古大学学报(人文社会科学版),2003,35(1): 116–120.

[4] 李秀芬,梁廷高. 对最小作用量原理的一些思考[J]. 四川师范大学学报(自然科学版),1998,21(2): 227–230.

[5] 邬鸿勋,龚晓任,龚平. 流体力学的 Lagrange 函数和最小作用量原理[J]. 海军工程学院学报,1991,4: 79.

[6] 朗道,栗弗席兹 著,鲁欣 等译,场论[M]. 高等教育出版社,2012: 27–60.

[7] 谢拥军,梁昌洪. 电磁场中的广义最小作用量原理及其应用[J]. 科学通报,1997,42(2): 211–214.

[8] 戴陵江,蔡绍洪. 极值原理统一的一种可能形式[J]. 贵州师范大学学报(自然科学版),1998,16(3): 64–70.

[9] 李复龄. 相对论热力学的进展[J]. 物理学进展,1989,8(3): 362–384.

压力容器中心对称J型坡口接头的残余应力计算与测试

雷永平[1]，谢濡泽[1]，林　健[1]，符寒光[1]，吴中伟[1]，徐　晓[2]

(1.北京工业大学 材料科学与工程学院；2.深圳中广核工程设计有限公司)

摘　要：以压力容器中心对称位置的J型坡口焊接试件为对象,采用全释放切割法对焊接接头的表面残余应力进行测试。结果表明:中心对称J型坡口焊接试件在焊缝较近处的应力为较大的拉应力,而随着距焊缝距离的增加,残余应力的逐渐减小,在距离焊缝较远处则为较小的压应力。试图用固有应变法测量中心对称J型坡口焊接试件的内部残余应力，获得了T和L型样上固有应变分量引起的残余应力,为进一步计算整个结构内部的残余应力提供了数据。利用ABAQUS软件对中心对称J型坡口焊接试件的温度场和残余应力进行数值模拟,将模拟结果与实验测量结果进行对比,模型的计算结果与实验测试结果吻合良好。

关键词：J型接头；残余应力；测试；数值分析

Calculation and Measurement of the Welded Residual Stress for a Center Symmetric J Type Groove Joints in Pressure Vessel Structure

LEI Yong-ping[1], XIE Ru-ze[1], LIN Jian[1], FU Han-guang[1], WU Zhong-wei[1],XU Xiao[2]

(1.College of Materials Science and Engineering, Beijing University of Technology; 2. China Nuclear Power Design Company)

Abstract: A cording to the J groove welded joints at the center symmetrical position of a pressure vessel, the surface residual stress of welded joints has been measured by section cutting method. The results show that the residual tensile stress is obtained near the welded specimen center, and the values is larger. With increasing distance from the welded center, the residual stress decreases gradually. In order to get the inner residual stress distribution, the inherent strain distribution in a T and L type specimens have been measured by using the inherent strain method. It is providing the data support for the further calculation of the residual stress in the whole structure. The temperature field and residual stress have been calculated using ABAQUS software. The simulation results and experimental results were compared. The results show that the model calculation results and test results are in good agreement.

Keywords: J type welding joint; residual stress; measurement; numerical analysis

基金项目:国家自然科学基金资助项目(51275006)。

第一作者:雷永平,男。焊接结构分析及可靠性。yplei@bjut.edu.cn。

感应熔化过程温度和流动耦合行为数值模拟

汪 洪，周建新，殷亚军，吴 凯

(华中科技大学 材料成形与模具技术国家重点实验室)

摘 要：熔炼是铸造过程中非常重要的环节，其质量决定着后续工艺的性能。感应电炉因其自身优势被广泛应用，因此对其熔炼过程进行定量化研究具有重要意义。本文在对感应电炉物理模型加以简化的基础上，进行温度场和流动场的耦合数值模拟研究，建立描述其物理过程的数学模型。采用有限差分法求解电磁场和温度场，采用有限体积法求解流动场。分析了感应熔炼过程金属熔体传热与流动行为特点，可视化描述熔炼过程的温度与流动分布，探讨了工艺参数的影响机制。

关键词：感应熔化；温度场；流动场；数值模拟

Numerical Simulation of Coupled Temperature and Fluid Flow Behavior during Induction Melting Process

WANG Hong, ZHOU Jian-xin, YIN Ya-jun, WU Kai

(State Key Laboratory of Materials Processing and Die & Mould Technology, Huazhong University of Science and Technology)

Abstract: Melting plays a critical role in casting production and determines the follow-up casting process performance. Since the induction furnace is more and more widely applied because of its advantages, it's of great significance to do the quantitative research through the whole process. In this paper, numerical simulation research of coupled temperature and fluid field is proposed based on the simplified physical model, and the mathematical governing equations describing the process is established. Finite difference method is used to solve the electromagnetic and temperature field, and finite volume method is used to solve the fluid field. Heat transfer and fluid behavior of metallic melting during the induction melting process are analyzed, the visual results describing the temperature and fluid distribution are given, and the influence mechanism of process parameters are discussed.

Key words: induction melting; temperature field; fluid field; numerical simulation

基金项目：国家数控重大专项(2012ZX04012–011)，材料成形与模具技术国家重点实验室 2015 年重点自主研究项目(数字化智能化铸造技术及应用)。

第一作者：汪 洪，男。材料成形过程模拟仿真及数字化技术。micwanghong@hust.edu.cn。

通讯作者：周建新，男。材料成形过程模拟仿真及数字化技术。zhoujianxin@hust.edu.cn。

0 引言

熔炼是铸造过程中极其重要的环节,其熔炼质量很可能决定后续铸件产品的性能。感应电炉因其自身的优越性,被应用在更多的场合,其熔炼的基本原理是电磁感应原理和电流的热效应原理,产生的感生电流在存在一定电阻的金属炉料内部流动,电磁能转换为热能将固态金属炉料熔化。虽然感应加热技术已经得到广泛的重视和应用,但是鉴于理论基础和计算工具的限制,目前感应加热设备的设计仍主要依靠经验与简化计算,而且熔炼过程因炉体结构为封闭式结构,内部金属熔体的传热、流动与传输行为无法直接观察,所以感应加热过程依旧很难精确预测和控制,计算机数值模拟技术为瞬态观察感应电炉熔炼过程金属熔体传热和流动行为提供了非常有力的手段。

从 20 世纪 70 年代以来,感应电炉被更加广泛的应用到铸造熔炼过程中去。在感应电炉熔炼数值分析方面,国内外学者在不同感应电炉用于不同的物理过程中的电磁场分析、磁热流耦合模拟等方面也做了一些研究。进入 21 世纪后,相关的研究越来越多。M. Pal 对钢液的感应搅拌过程进行建模和数值模拟研究,构建描述液态钢液感应搅拌的三维物理模型,提出磁热流耦合数学模型,采用有限元法进行数值计算,预测液态钢液流速和磁通密度,分析电流密度对流速的影响[1,2]。Jiin-Yuh Jang 对空心圆柱体电磁感应加热过程进行磁热分析[3]。A. Bermudez 对工业硅感应炉提纯进行数值模拟研究,将对称的感应炉模型简化为二维模型,提出了一种有限元求解方法[4,5]。A. Umbrasko 对冷坩埚感应炉的熔化过程进行数值模拟,采用大涡模拟方法建模,结果表明高度直径比越大,熔化效率越高[6]。Matej Kranjc 对感应加热过程进行数值分析,以圆柱形钢材为研究对象,对热物理现象进行磁热耦合建模,并使用有限元求解,热探测相机结果被用来证明数值模拟结果的正确性[7]。浙江工业大学的吴金富利用 ANSYS 软件对圆坯件感应加热过程中不同工艺参数条件下的涡流场和温度场进行了数值模拟,并就感应加热在透热和淬火方面对实际工况进行了模拟[8]。重庆大学的朱寿礼对中频无芯感应熔炼炉熔化过程和中断后的温度特性进行数值模拟研究,建立该物理过程的数学模型,计算炉内金属炉料在不同时刻的温度分布和熔化相界面的移动特性,讨论内热源与集肤效应对数值结果的影响,并对熔化过程中断后的降温和凝固过程进行数值计算,研究熔化中断后的相界面变化、温度分布与变化规律[9,10]。大连理工大学的薛冠霞对感应凝壳熔炼过程耦合温度场和流动场进行数值模拟,基于 C 语言建立 ANSYS 与 Fluent 软件数据接口,实现电磁场、温度场与流动耦合数值计算,研究了感应线圈加载电流、频率、坩埚缝数、线圈位置和炉料种类对于感应凝壳熔炼过程的悬浮驼峰形状、温度场和流动场影响规律[11,12]。

本文在简化的感应电炉熔炼物理模型基础上,构建三维数值计算模型,以 Fe-1.5%C 合金为研究对象,对感应电炉熔炼过程多物理场进行耦合数值计算,讨论多物理场耦合数值计算结果,分析电磁参数对电磁场、温度场和流动场的影响规律。

1 物理建模

感应电炉熔炼的基本原理是电磁感应原理和电流的热效应原理。坩埚外侧的螺旋形水冷线圈接入交变电流,通过电磁感应现象,在感应线圈周围将产生交变的磁场;该磁场的磁力线将部分穿透坩埚内的金属炉料,因为交变磁场的磁通量变化产生感应电动势,金属炉料内部将会产生感生电流;感生电流在存在一定电阻的固态金属炉料内部流动,必然会产生一定的热能,该能量将其加热并熔化成液态金属。简化后感应电炉熔炼过程三维物理模型如图1所示。

将金属炉料简化为圆柱体形状,并将感应线圈考虑为无螺距结构,且该螺旋线圈轮廓为正方形,其尺寸为10cm×10cm。金属炉料尺寸为120cm×250cm,炉体与炉盖的厚度均为24cm。包含外围空气层,整个模型外形尺寸为240cm×240cm×350cm。模型的网格剖分使用的华铸CAE软件进行剖分,网格剖分步长为20mm×20mm×20mm,总网格数为2.52×10⁶,网格剖分结果的二维截面图如图2所示。

本章数值计算中,将合金成分简化为Fe-1.5%C合金,所用到的固液相合金、炉体、炉盖和空气的热物性参数见表1,相关热物性参数见表2。

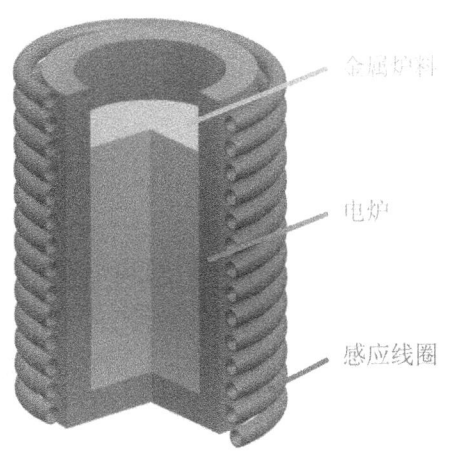

图1 三维数值算例模型

Fig.1 Three-dimensional numerical model

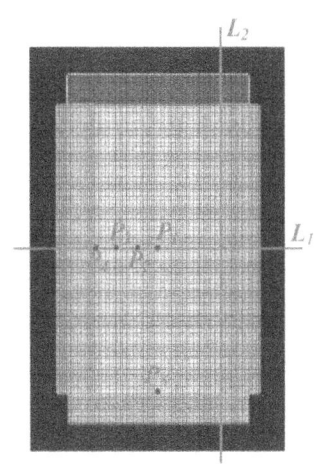

图2 网格剖分结果

Fig.2 Mesh subdivision result

表1 金属炉料、炉体结构和空气热物性参数

Tab.1 Thermal physical parameters of metal, furnace and air

材料	ρ (kg/m3)	λ (W/m·℃)	CP (J/kg·℃)	初始温度 (℃)
金属炉料(固相)	7750	470	20.8	30
金属炉料(液相)	7270	686.7	30.2	30
炉体	1900	1093	1.98	30
炉盖	2100	1130	0.93	30
空气	1.21	1005	0.0379	30

表2 合金热物性参数

Tab.2 Thermal physical parameters of alloy

参数	值	参数	值
纯熔点 (℃)	1536.34	相变潜热 (J/kg)	2.7×10^5
固相线温度 (℃)	−78	运动粘度 (Pa·s)	6.0×10^{-4}
液相线温度 (℃)	−17.1	电导率 (S/m)	9.93×10^6

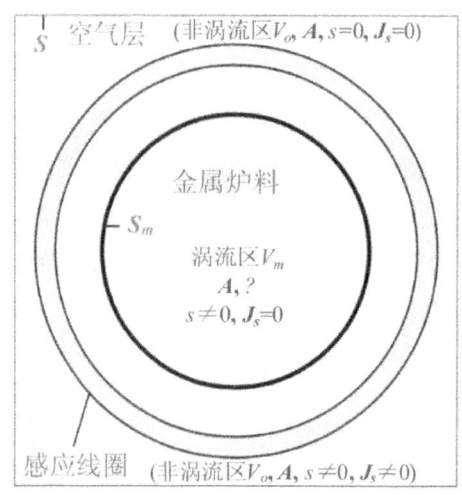

图3 电磁场求解区域

Fig.3 Solving areas of electromagnetic field

三维螺旋电磁场求解区域定义为如图2所示。将求解全域定义为 Vw。金属炉料区域为涡流区,定义为 Vm,其余区域为线圈、其他结构和空气,为非涡流区,定义为 Vo,并将源电流 Js 引入线圈所在区域。涡流区含有导电媒介,不含有源电流,故 $\sigma \neq 0$,Js=0;线圈所在区域内含有源电流和导电媒介,故 $\sigma \neq 0$,Js$\neq 0$;其余区域不含有源电流和导电媒介,故 $\sigma = 0$,Js=0。引入复矢量的" $A, \varphi - A$ "法,将三维电磁场控制方程描述为

$$\text{涡流区} \quad \begin{cases} \nabla \times \left(\dfrac{1}{\mu} \nabla \times A \right) - \nabla \left(\dfrac{1}{\mu} \nabla \cdot A \right) + \sigma \dfrac{\partial A}{\partial t} + \sigma \nabla \varphi = \mathbf{0} \\ \\ -\nabla \cdot \left(\sigma \dfrac{\partial A}{\partial t} + \sigma \nabla \varphi \right) = 0 \end{cases} \quad (1)$$

$$\text{非涡流区} \quad \nabla \times \left(\frac{1}{\mu} \nabla \times A \right) - \nabla \left(\frac{1}{\mu} \nabla \cdot A \right) = \boldsymbol{J}_s$$

采用三维非线性瞬态传热方程来描述感应电炉熔炼过程传热行为,传热行为主要为:金属炉料内部的热传导,金属炉料与接触介质的热传导和热辐射。根据傅里叶传热定律和能量守恒,考虑液态金属流动对传热行为的影响,将对流项加入到传热方程中,得到涡流场中温度场耦合非线性瞬态传热方程为

$$\rho C_p \frac{\partial T}{\partial t} + \boldsymbol{v} \cdot \nabla T = \nabla \cdot (\lambda \nabla T) + \rho L \frac{\partial f_s}{\partial t} \qquad (2)$$

在感应电炉熔炼过程中,固液相的转变不是瞬间完成的。在固液相线之间,存在糊状区,凝固的过程实际上可以认为是糊状区的推进过程。在构建动量守恒方程时,将糊状区域内两相作用力、液相粘性力和热溶质浮力考虑进去,源项中除了重力以外,考虑电磁场产生的洛伦兹力,于是有

$$\frac{\partial (\rho \boldsymbol{v})}{\partial t} + \rho \boldsymbol{v} \cdot \nabla \boldsymbol{v} = \nabla \cdot (\mu_l \nabla \boldsymbol{v}) - \frac{\mu_l}{K} \boldsymbol{v} - \nabla p + \rho_b \boldsymbol{g} + \boldsymbol{J}_e \times \boldsymbol{B} \qquad (3)$$

2 结果与讨论

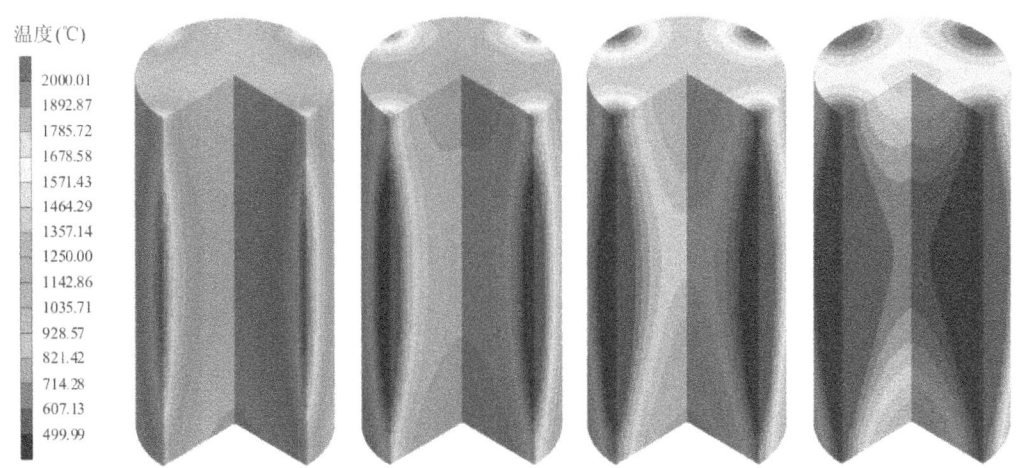

图 4 f=1000Hz, Js=1000A 条件下不同时刻温度分布
Fig.4 Temperature distribution of melt under f=1000Hz, Js=1000A

图 4 为 f=1000Hz, Js=1000A 条件下感应电炉熔化过程不同时刻金属熔体温度分布。从温度分布云图来看,金属炉料从最外层中心开始熔化,然后沿着半径方向向内、沿着高度方向向上和向下逐步熔化。

在金属炉料高度一半的最外层,电流与磁场强度最大,此处应是最先被加热的位置,并且升温速率最快。根据磁场的边缘效应可知,金属炉料最上部和最下部的中心位置磁场强度最小,此处应是最后被加热的位置,并且升温速率最慢。

为更好的分析感应电炉熔炼过程的升温特性,绘制图 2 中的 P1、P2、P3、P4 和 P5 点的升温曲线。P1 是整个金属炉料最先被加热且升温速率最快的位置,P5 是整个金属炉料最后被加热且升温速率最慢的位置。

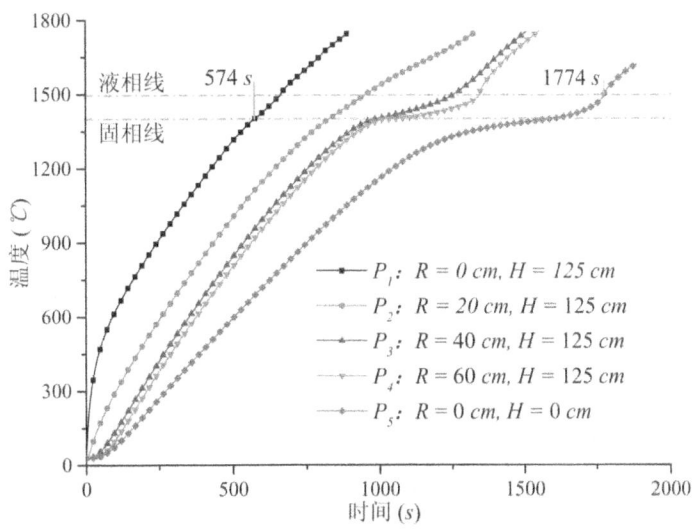

图 5 f=1000Hz, Js=1000A 条件下取样线 L1 上不同点升温曲线
Fig.5 Heating curves of different points on L1 under f=1000Hz, Js=1000A

图 5 为 f=1000Hz, Js=1000A 条件下取样线 L1 上不同点升温曲线。从曲线的结果得知:随着感应电炉熔炼过程的进行,P1 点所处的位置首先被加热,并且此处的升温速率最快;P5 点所处的位置最后被加热,并且此处的升温速率最慢。P1 点所处的位置在第 574s 开始熔化,达到固相线温度,然后从此处开始,固相液混合的糊状区开始向金属炉料中心处、最顶部和最底部推进,并于第 1774s 时 P5 点所处的位置达到液相线温度,固态金属炉料被完全熔化成液态合金液。

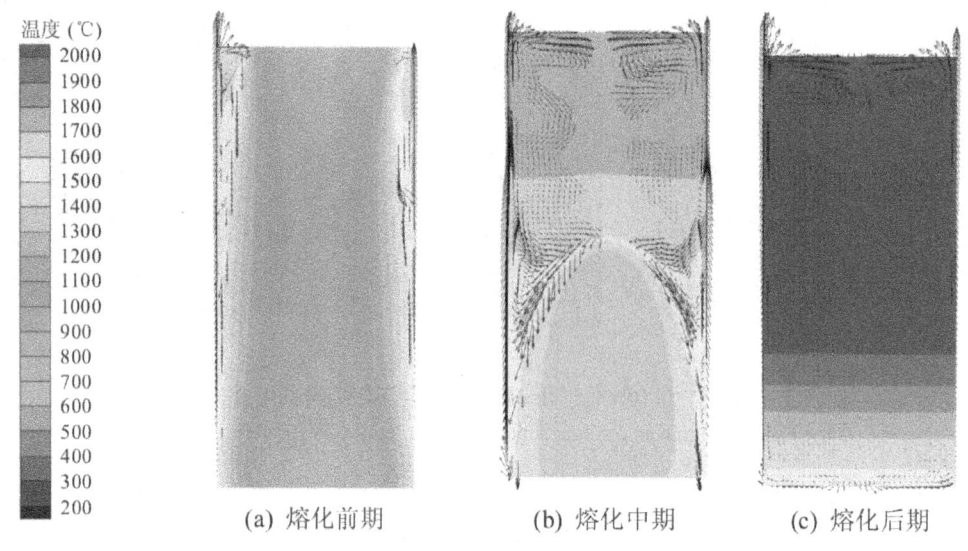

图 6 不同时刻金属炉料熔化温度分布速度矢量图

Fig.6 Temperature distribution and velocity vectors of melt at different time

图 6 为不同时刻金属炉料熔化温度分布与速度矢量图。重力与电磁力的相互作用,改变了液相合金液的流动行为,从而改变了金属熔体的传热行为。在熔化前期,固态金属炉料依然从最外层开始熔化,这是由电磁场的内热源项定的。当初生的液相合金液形成时,重力与电磁力相互作用,牵引液相合金液开始流动。向上的电磁力要明显大于向下的重力,且液相合金液的密度要小于固相金属炉料。熔化的液相合金液向上流动,并已熔化的液相合金液上部形成漩涡流,固液相界面前沿随着漩涡流逐步向下推进,在流动的过程中,液相合金液的温度逐渐趋于一致,并最终完全熔化形成体系温度较为均匀的液相合金液。

3 总结

论文构建了感应电炉熔炼过程的三维数值计算模型,详细阐释了模型的网格与计算参数,探讨了磁热耦合行为中金属炉料传热行为机理,分析了流动对磁热耦合金属炉料传热行为的影响。

参考文献:

[1] Pal M. Modeling of Induction Stirred Ladles [D]. Stockholm: Royal Institute of Technology, 2012.

[2] Pal M, J?nsson P. Multiphysics modeling of an induction-stirred ladle in two and three dimensions [J]. International Journal for Numerical Methods in Fluids, 2012, 70(11): 1378-1392.

[3] Jang J, Chiu Y. Numerical and experimental thermal analysis for a metallic hollow cylinder subjected

to step-wise electro-magnetic induction heating [J]. Applied Thermal Engineering, 2007, 27 (11-12): 1883-1894.

[4] Bermudez A, Gomez D, Muniz M C, et al. Numerical simulation of induction furnaces for silicon purification [M]//Bonilla L L, Moscoso M, Platero G, et al. MATHEMATICS IN INDUSTRY. 2008: 48.

[5] Bermúdez A, Gómez D, Mu?iz M C, et al. Numerical simulation of a thermo-electromagneto-hydrodynamic problem in an induction heating furnace [J]. Applied Numerical Mathematics, 2009, 59 (9): 2082-2104.

[6] Umbrasko A, Baake E, Nacke B, et al. Numerical studies of the melting process in the induction furnace with cold crucible [J]. COMPEL - The international journal for computation and mathematics in electrical and electronic engineering, 2008, 27(2):359-368.

[7] Kranjc M, Zupanic A, Miklavcic D, et al. Numerical analysis and thermographic investigation of induction heating [J]. International Journal of Heat and Mass Transfer, 2010, 53(17-18): 3585-3591.

[8] 吴金富. 基于 ANSYS 的感应加热数值模拟分析 [D]. 杭州: 浙江工业大学,2004.

[9] 朱寿礼,廖强,李曾敏,等. 中频无芯感应熔炼炉熔化过程温度特性的数值模拟 [J]. 热科学与技术,2004 (01): 38-42.

[10] 朱寿礼,廖强,李曾敏,等. 中频无芯感应熔炼炉熔炼过程中断后的温度特性数值模拟[J]. 热科学与技术,2005(01): 29-33.

[11] 薛冠霞. 感应凝壳熔炼过程温度场及流场耦合数值模拟 [D]. 大连理工大学,2007.

[12] Xue G,Wang T, Su Y, et al. Numerical Simulation of Thermal and Flow Fields in Induction Skull Melting Process [J]. RARE METAL MATERIALS AND ENGINEERING, 2009, 38(5): 761-765.

螺旋电磁场下铸钢锭凝固过程宏观偏析数值模拟

汪 洪，周建新，殷亚军，吴 凯

(华中科技大学 材料成形与模具技术国家重点实验室)

摘 要：宏观偏析是铸钢锭凝固过程中常出现的缺陷之一，导致铸钢锭的微观组织和力学性能产生差异，因此寻求抑制宏观偏析产生的方法一直都被学者们关注。本文提出了通过螺旋电磁场来抑制宏观偏析产生的方法，构建该过程在电磁场条件下耦合温度场、流动场和传质场数学模型，采用有限差分法求解电磁场、温度场及有限体积法求解流动场、传质场。数值模拟结果表明：螺旋电磁场在高度方向上产生的洛伦兹力会对自然对流进行抑制，当磁感应强度达到 0.5T 时，宏观偏析现象抑制效果非常明显。

关键词：螺旋电磁场；铸钢锭；宏观偏析；数值模拟

Numerical Simulation of Macrosegregation during Solidification of Cast Steel Ingot under Spiral Electromagnetic Field

WANG Hong, ZHOU Jian-xin, YIN Ya-jun, WU Kai

(State Key Laboratory of Materials Processing and Die & Mould Technology, Huazhong University of Science and Technology)

Abstract: Macrosegregation is one of common defects of cast steel ingot during solidification process, and may result differences in microstructure and mechanical properties. Seeking a method to reduce the macrosegregation has always been a challenge by scholars. In this paper, a method of spiral electromagnetic condition has been proposed, and the mathematical model of coupled temperature, fluid and mass field under electromagnetic field is established. Finite difference method is used to solve the electromagnetic and temperature field, and finite volume method is used to solve the fluid and mass field. The simulation results demonstrate that the Lorentz force generated in the height direction of spiral electromagnetic could reduce the natural convection, and an obvious reduce of macrosegregation phenomenon occurs while the magnetic induction density reached 05 T.

Key words: spiral electromagnetic field; cast steel ingot; macrosegregation; numerical simulation

基金项目：国家数控重大专项(2012ZX04012-011)，材料成形与模具技术国家重点实验室 2015 年重点自主研究项目(数字化智能化铸造技术及应用)。

第一作者：汪 洪，男。材料成形过程模拟仿真及数字化技术。micwanghong@hust.edu.cn。

通讯作者：周建新，男。材料成形过程模拟仿真及数字化技术。zhoujianxin@hust.edu.cn。

热浸镀铝层的阻尼机制研究

王友彬，曾建民

（西北工业大学 凝固技术国家重点实验室）

摘　要：采用热浸镀方法在 Q235 钢板表面制备镀铝层，通过扫描电镜(SEM)和电子背散射衍射技术(EBSD)观察镀层的微观形貌，利用动态热机械分析仪(DMA)研究镀件和基体的阻尼性能。结果表明：镀铝层/基体的锯齿状界面使镀层与基体的接触面增加，镀铝层的引入产生的热应力使界面处的位错密度增加。镀件的阻尼性能要大于基体材料，对阻尼性能的提高起主要作用的是位错阻尼机制和界面滑移机制。

关键词：镀铝层；界面；阻尼

The Study on Damping Behavior of the Hot-Dip Al Coating

WANG You-bin, ZENG Jian-min

（State Key Laboratory of Solidification Processing, Northwestern Polytechnical University）

Abstract: The Al coatings were prepared on Q235 steel plates by hot dipping, and the microstructure of the Al coatings were characterized through scanning electron microscopy （SEM）and electron backscatter diffraction(EBSD). The results show that the serration-like interface between the coating and substrate could improve the area of the contact surface. The increasing of the dislocation is due to the thermal stress induced by the Al coating. The damping could be significantly improve by the dipped Al coating, which is due to the mechanism of dislocation and interface slip.

Keywords: Al coating; interface; damping

基金项目：广西有色金属及特色材料加工重点实验室开放基金(GXKFJ14-03)；国家"973"计划前期研究专项项目(2012CB722804)。

第一作者：王友彬，男。钢材热浸镀层的耐蚀性及力学性能。wangyoubin114@126.com。

通讯作者：曾建民，男。铝合金铸造及加工工艺、金属表面处理。zjmg@gxu.edu.cn。

不同聚合物熔体壁面滑移的试验研究

孙秀伟 [1,2]

(1.唐山学院 机电工程系；2.河北省高校机电液一体化应用技术研发中心)

摘 要：采用双料筒毛细管流变仪,选用直径 D 分别为 0.5、1.0、1.5mm 的口模,研究 PP、HDPE、PS 和 PMMA 四种聚合物熔体的壁面滑移现象,分析壁面滑移的存在对熔体流动的影响以及壁面滑移速度与剪切应力、温度之间的关系。研究表明：当剪切应力达到一定值时,四种聚合物熔体在流动过程中均产生壁面滑移现象；壁面滑移的存在直接影响聚合物熔体的流动,而且对于不同熔体产生的影响不同,对 PP 和 HDPE 熔体流动具有减阻效果,对 PS 和 PMMA 熔体流动则具有增阻效果。口模直径越小,效果越明显；壁滑移速度随剪切应力增加而明显提高,二者之间呈指数关系；随温度升高,发生壁面滑移的临界剪切应力降低。

关键词：聚合物熔体；毛细管流变仪；壁面滑移；剪切应力

Experimental Studies on the Wall Slip of Different Polymer Melts

SUN Xiu-wei[1,2]

(1.Department of Mechanical and Electrical Engineering, Tangshan College;

2.Research Centre for Electro-hydraulic Integration Technology of the university in Hebei)

Abstract: The wall slip of four polymer melts(PS,PMMA,PP and HDPE) is researched by using a double capillary rheometer under die diameter from 1.5mm to 0.5mm. The effects of wall slip on the flow of melts are analyzed. The effects of shear stress and temperature on the wall slip velocity are discussed. The results show that when shear stress reaches a certain value, wall slip will start for four polymer melts. The wall slip will reduce flow friction for PP and HDPE, while will increase flow friction for PS and PMMA, and the smaller the die diameter is, the more obvious effect is. The slip velocity increases as an exponential function with the increase of shear stress. The critical shear stress of wall slip decreases with the rise of temperature.

Key Words: polymer melt; capillary rheometer; wall slip; shear stress

基金项目:河北省重点发展学科建设模具 CAD/CAM/DNC 一体化加工技术。

作者简介:孙秀伟(1979—),女。模具设计与模具 CAE 分析。tsssxw@163.com。

热镀锌锌铁化合物形成过程模拟

罗　立[1]，曾建民[1]，王友彬[1]，张　锐[2]

（1.广西大学 材料科学与工程学院；2.广西南宁凯源铁塔公司）

摘　要：锌渣是热浸镀生产过程中存在于镀液中的固态锌铁合金化合物，对镀件表面质量、镀层的抗腐蚀性能以及热镀锌的生产效率会产生有害影响。本工作以锌渣作为研究对象，在缩小比例的镀池中对锌渣的形成过程进行了物理模拟。运用扫描电子显微镜(SEM)、能谱分析仪(EDS)等技术手段，研究了铝元素掺入锌液后对锌渣形成的影响。研究结果表明：铝对热镀锌过程的冶金反应具有促进作用，增加了镀锌渣中的铁含量，并大大减少了锌渣含量；同时，加入铝后，镀锌渣中的悬浮颗粒尺寸减小，便于采用物理方法清除。

关键词：热镀锌；锌渣；锌铁化合物

Simulation of Zn-Fe Compound Formation for Hot Galvanizing

LUO Li[1], ZENG Jian-min[1], WANG You-bing[1], ZHANG Rui[2]

(1.School of Materials Science and Engineering, Guangxi University ; 2. Guangxi Nanning Kaiyuan Steel Tower Company)

Abstract: The Zn dross is the solid Zn-Fe alloy compound in Zn bath for hot-dip galvanizing and has a negative effect on the coating surface quality, corrosion performance and production efficiency. The aim of the present work is trying to make a physical simulation for the formation of zinc dross in a reduced scale plating bath. The effects of Al addition on the formation of Zn dross have been studied by the scanning electron microscope (SEM) and energy dispersive system (EDS). The results show that Al addition could promote the Fe-Zn reaction during the hot-dip galvanizing, which could decrease the quantity of the dross by increasing the Fe content in the dross. It is also found that Al addition could decrease the size of the suspended Zn dross particles, which makes the dross easy to clear by physical methods.

Key words: hot-dip galvanizing; zinc dross; Zn-Fe alloy compound

基金项目:南宁市科学研究与开发项目(20145197),广西自然科学基金(2011GXNSD018009)。

第一作者:罗　立,男。物理模拟。15078805034@163.com。

通讯作者:曾建民,男。物理模拟。zjmg@gxu.edu.cn。

碳纤维树脂基复合材料纤维体积分数
对固化过程影响的模拟研究

孙　飞[1]，廖敦明[1]，周建新[1]，董长春[1]，朱大雷[2]

(1.华中科技大学 材料加工工程系,材料成形及模具技术国家重点实验室；2.北京卫星制造厂)

摘　要： 本文建立了反映树脂基复合材料固化过程的热传递和固化动力学双向耦合模型，使用有限差分法，自主开发了固化度场和温度场的模拟程序。使用该程序，模拟 AS4/3501-6 平板件的固化过程，得到了固化度场和温度场的变化，结果发现该复合材料在固化过程有明显的两个阶段，每个阶段均有一个放热高峰，特别是在第二阶段，放热现象更加明显。随后，改变纤维的体积分数，结果显示随着纤维体积分数的增加，系统的最高温度逐渐降低，两者呈现反比例的关系。

关键词： 树脂；纤维；复合材料；固化；数值模拟

Numerical Study on the Influence of Fiber Volume Fraction of Resin Matrix Composites during the Exothermic Curing Process

SUN Fei[1], LIAO Dun-ming[1], ZHOU Jian-xin[1], DONG Chang-chun[1], ZHU Da-lei[2]

(1. State Key Laboratory of Materials Processing and Die & Mould Technology, Department of Material Processing Engineering, Huazhong University of Science and Technology; 2. Beijing Satellite Manufacturing Factory)

Abstract: In this study, a two-way coupling model of heat transfer and curing kinetics reflecting the resin matrix composites in the curing process has been established. Based on the model, a simulation program has been developed independently by using the finite difference method, which can analyze the degree of curing and the temperature fields. By using this program, some numerical simulations for a flat plate of AS4/3501-6 system are conducted to obtain the temperature and degree of cure fields. The results illustrate that there are two stages of the AS4/3501-6 composites in the curing process. Each stage has a peak temperature. Especially at the second stage, the peak temperature is more obvious. Then the fiber volume fraction of the composites is changed. The result shows that with the increase of fiber volume fraction, the max temperature inside the flat plate is gradually reduced, which indicates a proportion relationship.

Key words: resin; fiber; composites; cure; numerical simulation

基金项目: 教育部新世纪优秀人才支持计划(NCET-13-0229)；中央高校基本科研业务费资助(2012TS010)。
第一作者: 孙 飞,男,博士生。复合材料模拟仿真、虚拟制造。 *sunfei@hust.edu.cn*。
通讯作者: 廖敦明,男,教授,博导。材料加工 CAD/CAE。 liaodunming@hust.edu.cn。

基于有限元法的定向凝固过程中高温合金叶片温度场数值模拟

曹　流，廖敦明，陈　涛，孙　飞，周建新，庞盛永

(华中科技大学　材料成形及模具技术国家重点实验室)

摘　要：由于定向凝固过程数值模拟中需要处理随型型壳和辐射换热，采用了改进型 Monte Carlo 射线追踪法计算辐射换热，详细推导了定向凝固过程温度场的有限元(FEM)数学模型，并阐述了智能区分边界和局部矩阵等关键技术，自主开发了基于有限元法的定向凝固过程温度场数值模拟程序。模拟计算高温合金叶片在高速凝固法(HRS)过程中不同抽拉速度下的温度场，模拟结果与实际过程相吻合，验证了所推导的定向凝固过程温度场有限元数学模型的正确性和可靠性。

关键词：定向凝固；数值模拟；有限元法；射线追踪；局部矩阵；高温合金叶片

Numerical Simulation of Temperature Field of Superalloy Turbine Blade during Directional Solidification Based on Finite Element Method

CAO Liu, LIAO Dun-ming, CHEN Tao, Sun Fei, ZHOU Jian-xin, PANG Sheng-yong

(State Key Laboratory of Materials Processing and Die & Mould Technology University of Sicence and Technology)

Abstract: Considering that the irregular shell and radiation heat transfer are needed to be handled in numerical simulation of directional solidification (DS), an improved Monte Carlo ray tracing approach has been used to calculate radiation heat transfer, and the mathematical model of temperature field during DS based on finite element method (FEM)has been derived in detail. The key technologies, such as distinguishing boundaries automatically and partial matrix, have also been stated. A temperature-field numerical simulative program of DS based on FEM has been presented. The temperature fields of a superalloy turbine blade during high rate solidification (HRS)with different draw speeds have been calculated, which accord with real process, so the correctness and reliability of the mathematical model of temperature field during DS based on FEM have been verified.

Key words: directional solidification; numerical simulation; finite element method; ray tracing; partial matrix; superalloy turbine blade

基金项目：教育部新世纪优秀人才支持计划(NCET-13-0229,NCET-09-0396)；国家数控重大专项(2012ZX04010-031,2012ZX0412-011)；
　　　　　国家高技术研究发展计划("863"计划)(2013031003)。
第一作者：曹　流(1991—)，男，博士生。铸造过程有限元法数值模拟。caoliu@hust.edu.cn。
通讯作者：廖敦明(1973—)，男，教授，博导。铸造 CAD/CAE 研究。liaodunming@hust.edu.cn。

AP1000 核电主管道淬火过程温度场有限元模拟

彭新元 [1,2,3]，周贤良 [1,2,3]，华小珍 [2,3]，李 超 [2]，唐龙书 [2]

(1.南京航空航天大学 材料科学与技术学院；2.南昌航空大学 材料科学与工程学院；
3. 江西省金属材料微结构调控重点实验室)

摘 要：通过 ANSYS 有限元软件建立 AP1000 核电主管道三维有限元模型，并对其淬火过程进行有限元模拟和分析,研究了主管道表面和直管心部在淬火过程的温度变化规律。结果表明,主管道表面和直管心部温度均随淬火时间的延长而降低,主管道表面温度降低非常快,淬火 60s 时可以冷却到 260℃,淬火结束时(600s 后)已冷却至室温;而主管道中心部位及接管嘴部位温度降低较慢,淬火 180s 之后温度分别为 600℃和 700℃以上,淬火 300s 左右才都能冷到 427℃以下。

关键词：AP1000 主管道；316LN 不锈钢；淬火；温度场

中图分类号：TG156.3   文献标志码：A

Finite Element Simulation of Temperature Field during Quenching Process in Main Pipeline of AP1000 Nuclear Power

Peng Xin-yuan[1,2,3], ZHOU Xian-liang[1,2,3], HUA Xiao-zhen[2,3], Li Chao[2], TANG Long-shu[2]

(1. School of Material Science and Technology, Nanjing University of Aeronautics and Astronautics;
2. School of Material Science and Engineering, Nanchang Hangkong University;
3. Key Laboratory for Microstructural Control of Metallic Materials of Jiangxi Province)

Abstract: The three-dimensional finite element model of AP1000 main pipeline was established by ANSYS finite element analysis software. The quenching process of AP1000 nuclear power main piping was also simulated by ANSYS. The temperature distribution of AP1000 nuclear power main pipeline surface and center in quenching process was discussed. The results show that the temperatures of main pipeline surface and center are decrease with increasing quenching time. The temperature of surface decreases quickly and can cooling to 260℃ after quenching 60s. The temperature cooling to room temperature at the end of quenching (t=600s). While the temperatures of main pipeline center and filler neck decrease slowly, can reach 600℃ and 700℃ respectively after quenching 180s. Both of them drops to below 427℃.

Keywords: AP1000 main piping; 316LN stainless steel; quenching; temperature field

第一作者:彭新元,男。材料加工工程。pxy728@126.com。

金属熔滴在基板上碰撞过程的数值分析模型的建立

殷凤良，朱　胜，刘宏伟，曹　勇

(装甲兵工程学院 装备再制造技术国防科技重点实验室)

摘　要：熔滴在基板上的沉积成形过程，直接决定了 3D 打印增材再制造的精度及质量。在 FLOW 3D 流体分析软件平台上，建立熔滴在水平及倾斜基板上沉积成形过程的三维数值分析模型，分析熔滴界面的动态变化及热量在熔滴与基板之间的传递情况，模型考虑表面张力作用及凝固相变问题。研究网格尺寸及临界固相分数等模型参数对计算结果的影响，确定合适的网格尺寸与熔滴尺寸、碰撞速度等参数的关系，研究将基板处理为固体及流体对计算结果的影响。利用所建立的模型，计算分析不同条件下熔滴自由界面及温度的动态变化过程，以及熔滴凝固后的最终形态。

关键词：金属熔滴；3D 打印；数值分析模型；增材再制造

A Numerical Analysis Model for Molten Droplet Impinging Onto the Substrate

YIN Feng-liang, ZHU Sheng, Liu Hong-wei, CAO Yong

(National Defense Key Laboratory for Remanufacturing Technology, Academy of Armored Force Engineering)

Abstract: The impinging process of a molten droplet onto the substrate is the basic for the droplet deposition-based 3D printing forming remanufacturing technology. Based on FLOW 3D platform, a three-dimensional numerical simulation model was built to analysis the molten impinging process onto flat and inclined substrate, as well as heat transferring between the droplet and substrate. Surface tension force and solidification phenomenon were taken into account in the model. The influences of model parameters on the calculated results were studied, such as mesh size and critical solid friction etc. The relations between proper mesh size and droplet initial size and velocity were analyzed. The calculated results were compared when the substrate was treated as solid state and fluid state respectively. Finally, the model was employed to calculate the droplet impinging process with different conditions.

Key words: metal molten droplet; 3D printing; numerical analysis model; additive material remanufacturing

基金项目：国家自然科学基金资助项目(51205408,51375493)。

第一作者：殷凤良，男。焊接及快速成形过程数值模拟研究。yinshr@sohu.com。

铸造高硼低合金高速钢相图计算与分析

杨勇维[1]，符寒光[1]，雷永平[1]，王开明[1]，朱礼龙[2]，江　亮[2]

(1.北京工业大学 材料科学与工程学院；2.中南大学 粉末冶金研究院)

摘　要：采用 Thermo-Calc 热力学软件计算高硼低合金高速钢平衡凝固相图，绘制 Fe-B 伪二元系的垂直截面图，得到平衡凝固过程相转变温度与共晶成分。制备含碳量为 0.4wt.% 和不同硼含量(1.0wt.%、2.0wt.%、3.0wt.%)合金试样，用 X 射线衍射仪(XRD)测试凝固试样的相结构；用差热分析法(DSC)测定相变温度，用光学显微镜和扫描电镜观察试样显微组织结构。结果表明：计算相图的结果与实验数据符合度均较好，证明了计算相图的准确性，为相关领域研究的工程师和研究人员提供了一个可行的方案。

关键词：高硼低合金高速钢；Thermo-Calc 软件；相图计算；XRD；DSC

Phase Diagram Calculation and Analysis of the Cast High Boron Low Alloy High Speed Steel

YANG Yong-wei[1], FU Han-guang[1], LEI Yong-ping[1], WANG Kai-ming[1], ZHU Li-long[2], JIANG Liang[2]

(1.College of Material Science and Technology, Beijing University of Technology;
2. Powder Metallurgy Research Institute, Central South University)

Abstract: The equilibrium solidified phase diagrams of high boron low alloy high speed steel (HBLAHSS) have been calculated and the vertical section pseudo-binary phase diagrams have been drawn by the Thermo-Calc software. Specimens are prepared with different content of boron (1.0wt.%, 2.0wt.%, 3.0wt.%). The phase transformation temperatures were investigated by using differential scanning calorimetry (DSC) and the type of phases were determined by using X-ray diffraction (XRD). The microstructures in different specimens were observed by using optical microscope (OM) and scanning electron microscope (SEM). The calculation results obtained from Thermo-Calc is agreed with the ones from experiments. The work provides a practical method for engineers and researchers in related areas.

Key words: high boron low alloy high speed steel; Thermo-Calc software; phase diagram calculation; XRD; DSC

基金项目：国家自然科学基金资助项目(51475005)。

第一作者：杨勇维，男。轧辊材料。yangyw2@emails.bjut.edu.cn。

通讯作者：符寒光，男。新型金属材料开发与应用。hgfu@bjut.edu.cn。

等离子 –MIG 复合焊接电弧的数值模拟

朴圣君，金 成

（大连交通大学 材料科学与工程学院）

摘 要：建立了等离子-MIG 复合焊接电弧的二维有限元数学模型，根据等离子体与 MIG 电弧体相互作用原理及控制方程，考虑等离子气、保护气的作用，利用 FLUENT 分析软件对控制方程与相关源项进行 UDF 二次开发，获得了稳态复合电弧的温度场、磁场、速度场、电磁作用力以及阳极工件表面的电流密度等数据。结果表明：等离子-MIG 复合焊接可以进行良好的电弧复合，两电极的间距、等离子气体流量、保护气体流量对复合电弧形态及温度场有显著的影响。

关键词：等离子-MIG 复合焊；电弧形态；数值模拟；温度场

Numerical Simulation of Plasma–MIG Hybrid Welding Arc

PIAO Sheng–jun, JIN Cheng

（School of Materials Science and Engineering, Dalian Jiaotong University）

Abstract: A two-dimensional finite element numerical model of plasma-MIG hybrid welding was established, based on plasma and MIG arc interaction mechanism and govern equations. Considering the effects of plasma gas and shielding gas, the steady state temperature field, magnetic field, velocity field, electromagnetic force and current density of the anode surface were achieved by UDF secondary development of the govern equations and source terms in FLUENT. Results show that plasma and MIG acr can be well coupled. The distance between the two electrodes, the plasma gas flow rate, and shielding gas flow rate can significantly influence the hybrid arc shape and temperature distribution.

Key words: plasma-MIG hybrid welding; arc shape; numerical simulation; temperature field

基金项目：国家自然科学基金项目(51105049)。
第一作者：朴圣君，男。焊接过程数值模拟。82228610@qq.com。
通讯作者：金 成，男。焊接过程数值模拟。jincheng@126.com。

基于元模型方法的6013铝合金热变形流变行为建模

肖　罡[1,2]，杨钦文[2]，何　欢[3]，李落星[1,2]，曾建民[3]

(1. 湖南大学 汽车车身先进设计制造国家重点实验室；2. 湖南大学 机械与运载工程学院；

3. 广西大学 有色金属及材料加工新技术教育部重点实验室)

摘　要：为探寻对金属材料热变形流变行为建模的高效方法，运用Gleeble-3500热模拟机对6013铝合金进行热压缩试验，基于试验数据和三类元模型方法(克里金、径向基函数与多元多项式)，分别构建该合金的热变形流变行为模型，并就各方法建模的预测能力进行了分析与对比。结果表明：三类元模型方法均可建立较高精度的预测模型，以描述各热力学参数之间高度非线性的复杂关系。在模型预测精度和泛化能力的对比分析中，克里金方法较径向基函数和多元多项式均具有显著优势，是元模型方法中可适用于金属热变形行为建模的一种较为准确有效的方法。

关键词：6013铝合金；热变形；流变行为；元模型方法

Modeling the Flow Behavior of 6013 Aluminum Alloy during Hot Deformation Based on Metamodeling Method

XIAO Gang[1,2], YANG Qin-wen[2], HE Huan[3], LI Luo-xing[1,2], ZENG Jian-min[3]

(1. State Key Laboratory of Advanced Design and Manufacturing for Vehicle Body, Hunan University;

2. College of Mechanical and Vehicle Engineering, Hunan University; 3. Key Laboratory of Nonferrous Materials and New Processing Technology of Ministry of Education of China, Guangxi University)

Abstract: This work aims to search for an effective approach to model elevated temperature flow behaviors of materials. Hot compression tests of 6013 aluminum alloy were conducted using Gleeble-3500 thermal simulator. Based on the experimental data and three types of metamodeling method (Kriging, Radial Basis Function and Multivariate Polynomial), the models for the alloy were built and subsequently compared. The results show that the models constructed from these metamodeling methods are competent to describe the complicated nonlinear relationship among thermo dynamical parameters because of their high prediction accuracy. While evaluating the modeling accuracy and generalization capability, the performance of Kriging method is much better than the other two. It is indicated that Kriging method is the most appropriate way for modeling hot deformation behavior among the metamodeling methods.

Key words: 6013 aluminum alloy; hot deformation; flow behavior; metamodeling method

基金项目：国家自然科学基金面上基金资助项目(No.51475156)；国家重点科技项目(No.2014ZX04002071)；广西有色金属及特色材料加工重点实验室开放基金资助项目(No.GXKFJ14-08)。

第一作者：肖　罡，男。铝合金成型技术。xg_hnu@163.com。

通讯作者：李落星，男。轻合金变形行为及成型技术。luoxing_li@yahoo.com。

Fe-Mn-Al-C 低密度钢热变形与组织性能的研究

章小峰 [1,2]，杨 浩 [1]，冷德平 [1]，张 龙 [1]，黄贞益 [1]，陈 光 [2]

(1.安徽工业大学 冶金工程学院；2.南京理工大学 材料科学与工程学院)

摘 要：高 Al 量 Fe-Mn-Al-C 系低密度钢是目前轻量化结构钢研究领域中的热点，具有低密度、高强度和高塑性等优异性能。借助 Gleeble3500 热模拟实验机，本文针对含 Al 量分别为 8%、10% 和 12% 的三种 Fe-Mn-Al-C 合金在不同变形温度下进行单道次压缩实验，研究该类低密度钢高温下塑性变形行为。通过金相显微镜观察热模拟压缩后试样组织形貌，分析其微观结构的变化，分析了 950℃、1050℃、1100℃ 及 1150℃ 变形温度及 $0.01-1.0S^{-1}$ 变形速率时的低密度钢流变应力行为。结果表明，高应变速率及低变形温度提高了合金的流变应力。

关键词：低密度钢；物理模拟；热变形行为；组织

Hot Deformation and Structure Property of Fe-Mn-Al-C Low Density Steels

ZHANG Xiao-feng[1,2], YANG Hao[1], LENG De-ping[1], ZHANG Long[1],
HUANG Zhen-yi[1], CHEN Guang[2]

(1.School of Metallurgical Engineering, Anhui University of Technology;
2.School of Materials Science and Engineering, Nanjing University of Science and Technology)

Abstract：Fe-Mn-Al-C low-density steel with high Al content is a hotspot in the lightweight structure steels research field now, which has low density, high strength, high ductility and other excellent performance. This paper studied plastic deformation of Fe-Mn-Al-C steel with 8%,10% and 12% Al content under different high temperatures single compression test with Gleeble3500 thermal simulation machine. The article also observes sample microstructure changes after thermal simulation compression by metallographic microscope. According the thermal simulation compression experiment, the influences of deformation temperature under 950℃、1050℃、1100℃ and 1150℃ and deformation rate with $0.01-1.0S^{-1}$ on low density steel flow stress were analyzed. The results show that high strain rate and low deformation temperature will increase the flow stress of the steels.

Key Words: low density steels; physical simulation; hot deformation behavior; microstructure

基金项目：中国博士后科学基金资助项目(2014M561648)。
第一作者：章小峰，男。金属材料组织与性能控制。egzxf@ahut.edu.cn。

7N01 铝合金等温多道次热压缩变形行为研究

何宇棋，张　婷，符跃春

(广西大学 有色金属及材料加工新技术教育部重点实验室，材料科学与工程学院)

摘　要：采用 Gleeble-3500 热模拟机对 7N01 铝合金进行等温多道次热压缩试验以模拟其热变形过程，实验中变形温度为 300℃和 400℃，应变速率为 $0.01s^{-1}$ 和 $1s^{-1}$，道次真应变为 0.2，道次间保温时间为 10s 和 100s。实验结果表明：7N01 铝合金等温多道次热压缩过程呈现明显的静态软化特性。多道次变形后的流变应力明显小于单道次，且随着变形温度的升高、应变速率的降低和道次间保温时间的延长，流变应力降低。等温多道次变形的静态软化率随着道次的增加而增加。经多道次热变形后，7N01 铝合金主要呈典型的回复组织并出现了再结晶现象。

关键词：7N01 铝合金；等温多道次热压缩；软化；再结晶

Isothermal Multi-pass Hot Compression Deformation Behaviors of 7N01 Aluminum Alloy

HE Yu-qi, ZHANG Ting, FU Yue-chun

(Key Laboratory of New Processing Technology for Materials and Nonferrous Metal, Ministry of Education, School of Materials Science and Engineering, Guangxi University)

Abstract: The isothermal multi-pass hot compression tests of 7N01 aluminum alloy were performed on Gleeble-3500 thermal simulation machine to simulate the hot rolling process. Deformations were carried out at the temperature of 300 and 400℃, the strain rate of 0.01 and $1s^{-1}$, the true strain of 0.2 at each pass, and the holding time between passes of 10 and 100s. The results show that 7N01 aluminum alloy presents obvious static softening characteristics during isothermal multi-pass deformation process. The flow stress of multi-pass is smaller than that of single-pass, and the flow stress decreases with increasing deformation temperature, decreasing strain rate and increasing holding time. The static softening rate increases with the increasing of deformation pass. 7N01 aluminum alloy shows a typical recovery and recrystallization microstructure after isothermal multi-pass hot deformations.

Key words: 7N01 aluminum alloy; isothermal multi-pass hot compression; softening; recrystallization

---

第一作者：何宇棋，女。铝合金热处理。qiyuhesunny@foxmail.com。

0 引言

7N01 铝合金是由日本开发的 Al-Zn-Mg 系铝合金,具有比强度高、耐蚀性好及热加工性能优良等优异性能,是轨道交通车辆及大型汽车用高精度型材的主要合金[1]。目前我国高铁、动车快速发展,迫切需要制定轨道交通车辆用大型铝合金材料的挤压、轧制和锻造等热加工工艺,而金属的高温塑性变形行为规律是制定热加工工艺的理论依据[2]。尽管国内外学者对多种铝合金的高温塑性变形行为进行了较有成效的研究[3-6],但对 7N01 铝合金热加工性能的研究却并不充分。本文通过等温多道次热模拟压缩实验,研究 7N01 铝合金在热压缩过程中的变形行为以及微观结构变化规律,分析 7N01 铝合金的软化机制,为合理制定 7N01 铝合金的热加工工艺提供理论和实验依据。

1 实验内容

本研究采用南南铝加工有限公司提供的 7N01 铝合金铸锭,其名义化学成分(wt.%)为 4.5Zn,1.4Mg,0.1Cu,0.5Mn,0.2Cr,0.1Fe,0.05Si,余量为 Al。铸锭在 470℃ 均匀化退火 24h,出炉水冷后加工成 Φ10mm×15mm 的圆柱试样。5 道次热压缩实验在 Gleeble-3500 热模拟机上进行,热模拟机采用电阻加热,升温速率为 2.5℃·s^{-1}。试样在热压缩前保温时间 3min,热变形温度(T)为 300℃ 和 400℃;应变速率($\dot{\varepsilon}$)为 0.01 s^{-1} 和 1.0 s^{-1},道次的应变量(ε)均为 0.2,道次间保温时间(tp)为 10s 和 100s。热压缩结束后立即对试样进行水淬,将平行于压轴方向的中部纵截面进行研磨、抛光后用 Keller 试剂腐蚀,在日立 SU8020 扫描电镜上进行显微组织观察。

2 实验结果与讨论

2.1 等温多道次热变形真应力–应变曲线

7N01 铝合金等温多道次热压缩变形的真应力 - 应变曲线如图 1 所示。可以看到,在变形初期,单道次和多道次流变应力均迅速增加至一明显的峰值,且随着变形温度的升高和应变速率的降低,流变应力峰值降低。在应变速率为 0.01 s^{-1} 时,单道次和多道次流变应力在达到峰值后发生明显软化,随后保持动态平衡,而在应变速率为 1.0 s^{-1} 时流变应力在达到峰值后基本保持不变。由于道次间的静态软化,多道次流变应力明显低于单道次流变应力,且道次间保温时间为 100s 时的软化大于 10s 时的,表明静态软化对多道次流变应力的影响较大[7,8]。

从图 1 中还可以看出,在多道次变形过程中,后续道次的起始流变应力具有一定的规律性。为了更好地显示 7N01 铝合金多道次流变应力的演变及静态软化特征,将合金多道次变形的起始应力(σ_S)、屈服应力(σ_Y,0.2%)及峰值应力(σ_P)的变化规律总结如图 2 所示。可以看出,起始应力(σ_S)随着道次的增加基本保持恒定或稍有增加,在相同条件下道次间保温时间为 100s 时的 σ_S 均小于 10 s 时的,表明保温时间的延长将降低起始应力。屈服应力(σ_Y)在第二道次时明显增加,但随着道次的继续增加又开始降低,但均大于第一道次时的值。多道次峰值应力(σ_P)明显小于单道次峰值应力,表明多道次过程中发生静态软化。在低应变速率(0.01 s^{-1})时,随着道次的增加软化基本保持恒定,但在高应变速率(1.0 s^{-1})时,随着道次的增加软化逐渐减小,且随着道次间保温时间的增加软化也增大。

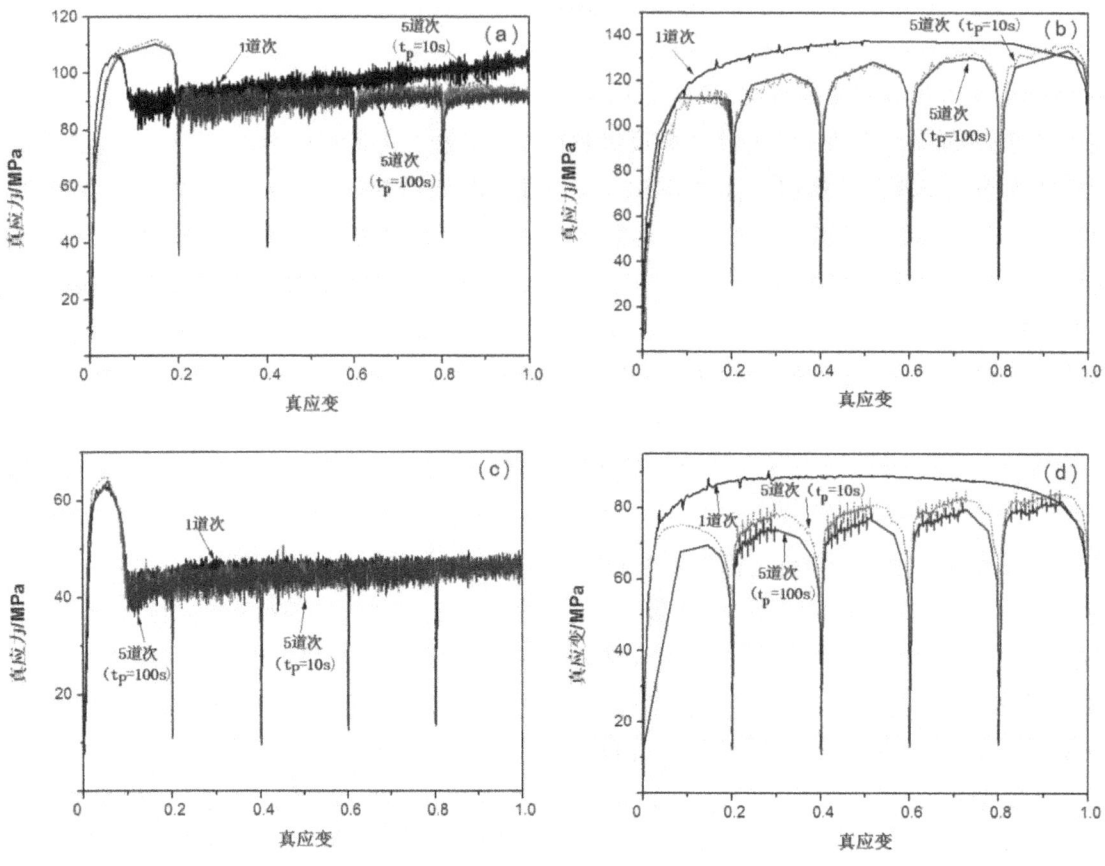

图1 7N01合金热压缩变形的真应力-真应变曲线

Fig.1 True stress-true strain curves of 7N01 alloy by hot compression

(a) T=300℃,$\dot{\varepsilon}$=0.01 s^{-1};(b) T=300℃,$\dot{\varepsilon}$=1.0 s^{-1}; (c) T=400℃,$\dot{\varepsilon}$=0.01 s^{-1};(d) T=400℃,$\dot{\varepsilon}$=1.0 s^{-1}

2.2 等温多道次热变形静态软化率

在多道次热变形的保温间隙,由热变形产生的位错、亚结构等将发生静态回复、静态再结晶、晶粒长大及静态析出等现象,从而引起合金的静态软化。静态软化使合金的显微组织和性能发生变化,并最终反映在热变形应力中。通常等温多道次热变形过程的静态软化率(R_{FS}^n)[9]可用式(1)来表示:

$$R_{FS}^n = \frac{(\sigma_{1,\,n} - \sigma_{2,\,n+1})}{(\sigma_{1,\,n} - \sigma_0)} \times 100\% \qquad (1)$$

其中,$\sigma_{2,n+1}$ 和 σ_0 分别代表第 n+1 道次和第一道次热变形的屈服应力;$\sigma_{1,n}$ 是第 n 道次变形结束时对应的瞬时变形应力值。

将图1与图2中的相关应力值代入多道次静态软化率公式中,即可获得静态软化率的计算结果如图3所示。可以看出,静态软化率随着变形道次(应变量)的增加而增加,只有在 T=400℃、$\dot{\varepsilon}$=0.01 s^{-1}、t_P=100s 时静态软化率较小且变化幅度不大。在变形温度为300℃时,$\dot{\varepsilon}$=1.0 s^{-1}、t_P=100s 条件下的静态软化率可达到70%以上,明显高于其他条件下的静态软化率,表明在变形温度为300℃时,低应变速率时道次间保温时间对合金静态软化率的影响不大,而高应变速率时随着道次间保温时间的延长,静态软化率增大。在变形温度为400℃时,高应变速率($\dot{\varepsilon}$=1.0 s^{-1})下的静态软化率明显高于低应变速率($\dot{\varepsilon}$

=0.01s⁻¹)下的静态软化率。同时还可看到,在低应变速率时,道次间保温时间为 100s 时的静态软化率低于 15%,表明在此条件下道次间的静态软化变化不大[10]。

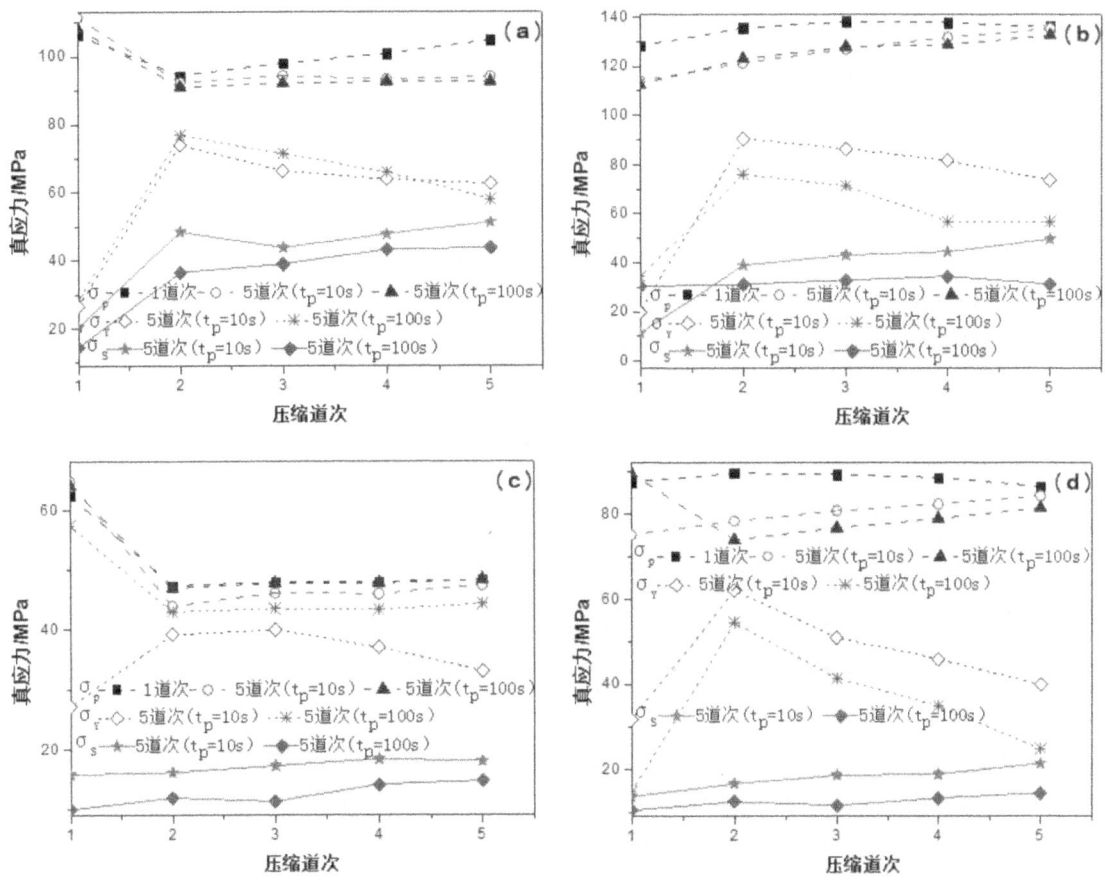

图 2 7N01 合金热压缩变形后的起始应力、屈服应力及峰值应力

Fig.2 Relationships of initial stress, yield stress and peak stress between different passes

(a) $T=300℃$, $\dot{\varepsilon}=0.01$ s⁻¹; (b) $T=300℃$, $\dot{\varepsilon}=1.0$ s⁻¹; (c) $T=400℃$, $\dot{\varepsilon}=0.01$ s⁻¹; (d) $T=400℃$, $\dot{\varepsilon}=1.0$ s⁻¹

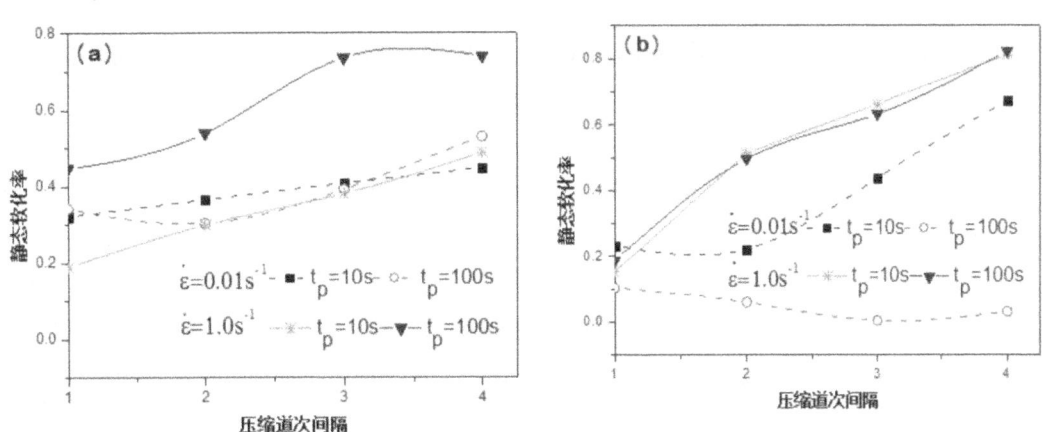

图 3 不同热变形条件下软化率 FS 与压缩道次间隔的关系

Fig.3 Relationship between softening fraction(FS) and time passes

(a) 300℃, (b) 400℃

2.3 等温多道次热变形显微组织

图 4 为等温多道次热变形后典型的 7N01 铝合金微观组织。可以看到，热变形后晶粒明显被拉长，呈典型回复特征。在变形温度为 300℃ 时,应变速率的增加和道次间保温时间的延长促进合金发生静态回复,使得晶粒变得粗大,软化程度增加。随着变形温度升高到 400℃,晶界的移动能力提高,晶粒发生一定程度的再结晶,可以看见合金中有少量黑色的再结晶晶粒,这使得合金的流变应力明显低于 300℃ 时的应力。同时,道次间保温时间的延长将降低晶粒内的位错密度[11],进一步提高合金的软化程度。

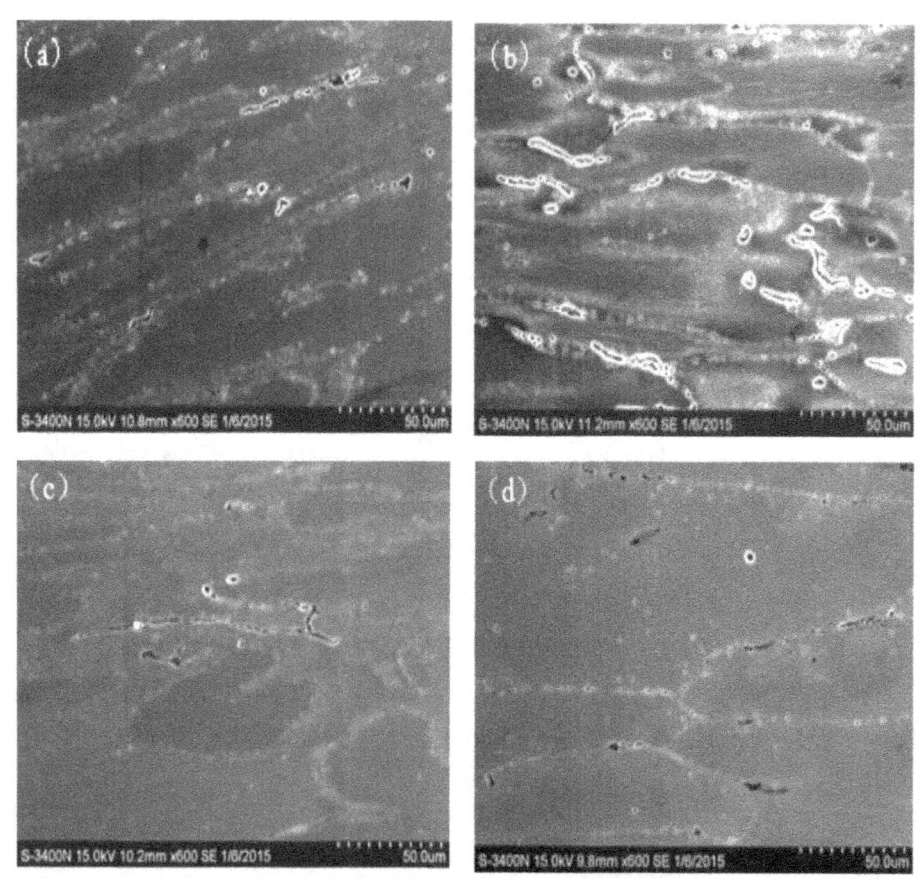

图 4　7N01 合金不同变形条件下的微观组织

Fig.4　SEM microstructures of 7N01 alloy after deformation under different Conditions

(a) T=300℃, $\dot{\varepsilon}$=0.01 s^{-1}, t_P=10s;　(b) T=300℃, $\dot{\varepsilon}$=1.0 s^{-1}, t_P=100s;

(c) T=400℃, $\dot{\varepsilon}$=0.01 s^{-1}, t_P=10s;　(d) T=400℃, $\dot{\varepsilon}$=0.01 s^{-1}, t_P=100s

3 结论

(1)7N01 铝合金等温多道次热变形过程由于道次间的静态软化导致流变应力随着道次（累积应变)的增加出现明显的静态软化特性。相同条件下,合金在道次间保温时间为 10s 时的流变应力均大于 100s 时的流变应力。

(2)7N01 铝合金的静态软化率随着变形道次(累积应变)的增加而增加,只有在 T=400℃ 、$\dot{\varepsilon}$=0.01s^{-1} 、

t_p=100s 时静态软化率较小且变化幅度不大。

（3）7N01 铝合金等温多道次热变形后主要呈典型的回复组织并出现了再结晶现象。

参考文献：

[1] McQueen H J, Blum W. Dynamie recovery: sufficient mechanism in the hot deformation of Al （B99.99）[J]. MaterialsScience and Engineering A, 2000, 290: 95.

[2] M. A. Jabbari Taleghani, E. M. Ruiz Navas, M. Salehi, et al. Hot deformation behaviour and flow stress prediction of 7075aluminium alloy powder compacts during compression at elevated temperatures [J]. Materials Science and EngineeringA , 2012, 534: 624-631.

[3] Xiong Yimin, Zhang Yong an, et al. Hot deformation behavior for Al － Gu-Mg-Ag-Zr aluminum alloy [J]. ChineseJournal of Rare Metals, 2009, 33(1): 6.

[4] 寇琳媛,金能萍,张辉,等. 7150 铝合金高温热压缩变形流变应力行为 [J]. 中国有色金属学报,2010,20 (01): 43-48.

[5] D. Feng, X. M. Zhang, S. D. Liu, et al. Constitutive equation and hot deformation behavior of homogenized Al－7.68Zn－2.12Mg－1.98Cu－0.12Zr alloy during compression at elevated temperature [J]. Materials Science and Engineering A. 2014, 608: 63-72.

[6] Sakai T, Minra H. Continuous dynamic recrystallization during the transient severe deformation of aluminum alloy 7475 [J]. Acta Materialia. 2009, 57: 153.

[7] 李俊鹏,沈健等. 温度对 7075 铝合金热变形显微组织演化的影响 [J]. 中国有色金属学报,2008,18 (11): 1951.

[8] Wan Min, Ghang Weihong, et al. A novel cutting force modelling method for cylindrical end mill [J]. Applied Mathematical Modelling, 2010, 34(3): 823-836.

[9] McQueen H J,Spigarclli S, Kassncr M E, et al. Hot Deformation and Processing of Al Alloys [M]. Boca Raton, FI.,USA: CRC Press, 2011:125-226.

[10] X. Y. Liu, Q. L. Pan, Y. B. He, et al. Flow behavior and microstructural evolution of Al-Cu-Mg-Ag alloy during hot compression deformation [J]. Materials Science and Engineering A. 2009, 500: 150-154.

[11] 翁舒楚. 7150 铝合金热变形过程中动态组织演变规律研究 [D]. 长沙: 湖南大学, 2012.

铝合金焊接接头细观多晶模型及拉伸织构演化数值模拟

丛述玲，金 成

（大连交通大学 材料科学与工程学院）

摘 要：基于voronoi算法对铝合金焊接接头进行细观组织建模，构建了等轴晶、柱状晶区结合区域的多晶模型。结合接头初始晶粒取向，并将其赋予到多晶模型中。运用晶体塑性理论，编写了用户自定义材料子程序(UMAT)，模拟了铝合金焊接接头在单向拉伸应变为10%时的织构演化，得到接头塑性变形过程中晶粒细观形变机制及应力应变分布。数值模拟结果显示，接头细观最大应力出现在柱状晶区、等轴晶结合区域。

关键词：铝合金焊接接头；多晶模型；晶体塑性有限元；细观应力应变

Mesoscopic Polycrystalline Model of Aluminum Alloy Welded Joints and Numerical Simulation of Texture Evolution in Tension

CONG Shu-ling, JIN Cheng

（School of Materials Science and Engineering, Dalian Jiaotong University,）

Abstract: Based on Voronoi algorithm, a mesoscopic structure model of aluminum alloy welded joint was built, which consisted of the polycrystalline binding region of equiaxial and columnar grain zone. According to initial grain orientation data were determined, and were assigned to the polycrystalline model. Using crystal plasticity theory, a user-defined material subroutine (UMAT) was developed. Under simulated uniaxial tensile strain of 10%, its texture evolution was obtained. The meso grain deformation mechanism and stress and strain distribution in plastic deformation process were also studied. Results show that the numerical for a mesoscopic joint, the maximum stress was appeared between the columnar and equiaxed grain binding region.

Key words: aluminum alloy welded joints; polycrystalline model; crystal plasticity finite element method; meso stress and strain

基金项目：国家自然科学基金项目(51105049)。

第一作者：丛述玲，女。焊接过程数值模拟。545219327@qq.com。

通讯作者：金 成，男。焊接过程数值模拟。jincheng@126.com。

基于两种热力学数据获取方法的 Al-4Cu-6Si 合金凝固路径计算

赵光伟，叶喜葱，黄才华

（三峡大学 机械与动力学院）

摘　要：基于两种热力学数据获取方法：直接耦合相图计算软件 Thermo-Calc 法与调用拟合函数法，对 Al-4Cu-6Si（wt.%，下同）三元共晶合金的凝固路径进行数值计算。计算结果表明，Al-4Cu-6Si 合金的凝固路径分为三个阶段：$L+\alpha \rightarrow L+\alpha+Si \rightarrow L+\alpha+\theta+Si$。耦合法计算得到的初生相、两相共晶、三相共晶的含量分别为：50.813%、37.234%、11.953%，而拟合函数法的计算结果分别为：50.809%、38.000%、11.191%。对比基于两种热力学数据获取方法的计算结果发现，二者计算精度比较接近，但拟合函数法的计算效率更高，CPU 运行时间仅为耦合法的 1.21%。

关键词：凝固路径；Al-Cu-Si；三元共晶；Thermo-Calc；计算效率

Computations of Solidification Paths of Ternary Al-4Cu-6Si Eutectic Alloys with Two Different Thermodynamic Data-acquisition Methods

ZHAO Guang-wei, YE Xi-cong, HUANG Cai-hua

(College of Mechanical and Power Engineering, China Three Gorges University)

Abstract: The solidification path of Al-4Cu-6Si alloy was calculated based on two different thermodynamic data-acquisition methods: direct coupling with the CALPHAD software Thermo-Calc, and using regression functions. The results show that the solidification path of Al-4Cu-6Si is $L+\alpha \rightarrow L+\alpha+Si \rightarrow L+\alpha+\theta+Si$. The phase content of primary phase, binary eutectic, ternary eutectic calculated by "coupling" method are 50.813%, 37.234%, 11.953%, respectively, and the calculation results by using regression functions method are: 50.809%, 38.000%, 11.191%. Comparing the calculation results from the two kinds of thermodynamic data acquisition calculation method, it is found that the accuracy of the method is relatively close, but thermodynamic data-acquisition method of using regression function is much more efficient than direct coupling with TQ6-Interface of Thermo-Calc. The CPU time of using regression function method is 1.21% of "coupling" method.

Keywords: solidification path; Al-Cu-Si; ternary eutectic alloy; thermo-calc; computational efficiency

基金项目：湖北宜昌市自然基础科学研究与应用专项项目(A14-302-a16)；三峡大学科学基金(KJ2011B032)。

第一作者：赵光伟，男。多元多相合金凝固模拟与实验。zgwhit@ctgu.edu.cn。

激光熔覆原位合成 Mo2NiB2 的热力学分析

胡肇炜，李文戈

(1,上海海事大学 商船学院)

摘　要：Mo-Ni-B 系三元硼化物金属基陶瓷具有优良的耐磨、耐蚀、耐高温的性能,具有广阔的应用前景。本文计算了 Mo-Ni-B 三元系中各种化合物不同温度下标准生成的自由焓。计算合成 Mo2NiB2 反应在不同开始温度下的绝热温度及反应产物的熔化比。证明采用 Mo、Ni、B 三种元素粉末原位合成 Mo2NiB2 在热力学上是可行的。试验表明采用激光熔覆原位合成技术可获得 Mo2NiB2。

关键词：激光熔覆；原位合成；热力学

Thermodynamic Analysis of the Mo2NiB2 Produced by Laser In-situ Cladding

HU Zhao-wei, LI Wen-ge

(1,Merchant Marine College, Shanghai Maritime University)

Abstract: Mo-Ni-B ternary boride cermet is wear-resisting, corrosion-resistant and heat-resisting, it can be applied widely in industries. The standard formation free enthalpy was calculated at different temperatures of compounds in Mo-Ni-B ternary system. The adiabatic temperature and the molten ratios of different compounds in Mo-Ni-B ternary system were calculated. The calculated results show that it was feasible to produce Mo2NiB2 by laser in-situ cladding from Mo、Ni、B elemental powders. Mo2NiB2 ternary boride was produced by laser in-situ cladding in experiment.

Key words: laser cladding; situ synthesis; thermodynamics

基金项目:国家自然科学基金(51172140, 51202143), 上海市教委科研创新项目(11ZZ141)。

第一作者:胡肇炜,男。材料加工。huzw0731@sina.com。

通讯作者:李文戈,男。材料加工。liwenge66@163.com。

基于数值模拟的锻造成形工艺优化方法研究

覃荣武，廖小平，王承辉，曾建民，夏　薇

（广西大学　广西有色金属及特色材料加工重点实验室）

摘　要：针对汽车发动机曲轴锻造成形的工艺优化与质量控制需求，从工艺参数敏感性分析入手，采用增强型平移传播Latin超立方体试验设计方法对始锻温度、模具预热温度、摩擦系数进行抽样。同时，以锻件形变均匀性、损伤值及模具磨损为优化指标，通过数值模拟仿真获取实验样本数据，构建高斯代理模型，并利用遗传算法对该优化问题模型进行求解，从而快速获取锻造成形最优工艺参数组合。结果表明提出的优化方法能有效提高锻件质量和锻造工艺的稳定性。

关键词：锻造工艺优化；实验设计；数值模拟；高斯代理模型

Research of Forging Process Optimization Method Based on Numerical Simulation

QIN Rong-wu, LIAO Xiao-ping, WANG Cheng-hui, ZENG Jian-min, XIA Wei

（Guangxi University, State Key Laboratory of Processing for Non-ferrous Metal and Featured Materials of Guangxi）

Abstract: According to the process optimization and quality control requirements of a car engine crankshaft forging forming process and starting point to the analysis of parameter sensitivity, the enhanced translational propagation Latin hypercube design was used to obtain samples of the initial forging temperature, mold preheating temperature, friction coefficient. At the same time, the experimental data are obtained through numerical simulation with the objective of improving the forging deformation uniformity and reducing damage value and mold wearing, then constructing the Gaussian surrogate model, and genetic algorithm is used to solve this optimization problem to obtain fast forging optimal combination of process parameters. The results show that the proposed optimization method can effectively improve the forging quality and process stability.

Key words: forging process optimization; experiment design; numerical simulation; gaussian surrogate model

第一作者：覃荣武，男。现代模具设计方法与制造技术。1119170756@qq.com。
通讯作者：夏　薇，女。现代模具设计方法与制造技术。xiawei@gxu.edu.cn。

800MPa 级微合金化 C-Mn 高强钢激光
焊接接头的组织及疲劳性能

王晓南[1]，孙　茜[1,2]，王　卫[3]，邸洪双[2]，陈长军[4]，吴宝强[2]，朱国辉[3]

(1.苏州大学 沙钢钢铁学院；2.东北大学 轧制技术及连轧自动化国家重点实验；
3.安徽工业大学 材料科学与工程学院；4.苏州大学 机电工程学院激光加工中心)

摘　要：研究了 4.5mm 厚抗拉强度 800MPa 级 Nb-Ti-Mo 微合金化 C-Mn 钢激光焊接接头的显微组织，并对比分析了焊接接头焊缝区与母材的疲劳性能。研究结果表明：激光焊接焊缝区显微组织为板条马氏体，板条束内存在一定量的尺寸在 100nm 左右的微合金碳氮化物；母材的显微组织主要为贝氏体铁素体和铁素体，在晶界上弥散分布一定量的碳化物，晶粒内部存在大量尺寸在 <10nm 微合金碳化物。焊缝区和热影响区的硬度明显高于母材。当应力比 R=0.1，循环基数为 10^7 时，母材的条件疲劳强度显著高于焊缝。母材和焊接接头焊缝区的疲劳裂纹源分别为驻留滑移带与基材界面、焊接气孔。疲劳裂纹在母材显微组织中扩展路径倾向于铁素体与碳化物界面，而疲劳裂纹在焊缝区中切断马氏体板条束和原始奥氏体晶界扩展，裂纹扩展无明显方向性。

关键词：微合金钢；激光焊接；显微组织；疲劳断裂；板条马氏体

Microstructure and Fatigue Properties of 800MPa Grade Microalloyed C-Mn Steel Welded Joint Using Fiber Laser

WANG Xiao-nan[1*], SUN Qian[2], WANG Wei[3], DI Hong-shuang[2],

CHEN Chang-jun[4], WU Bao-qiang[2], ZHU Guo-hui[3]

(1.Shagang School of Iron and Steel, Soochow University; 2.The State Key Lab of Rolling & Automation,
Northeastern University; 3.School of Metallurgy and Engineering, Anhui University of Technology;
4.Laser Processing Research Center, School of Mechanical and Electrical Engineering, Soochow University)

Abstract: Microstructure of tensile strength 800MPa grade Nb-Ti-Mo microalloyed C-Mn Steel welded joints were studied, and fatigue properties of base metal (BM) and welded seam (WS) were comparative analysis. The result showed that microstructure of WS was lath matensite, and there were amount of microalloyed carbonitride precipitate in the matensite lath. The microstructure of the BM was mainly fine grained ferrite and bainitic ferrite (a lot of <10nm microalloyed carbides precipitate in the ferrite), and certain amount carbides were on the ferrite grain boundaries. Micro-hard-

基金项目：国家自然科学基金资助项目(No.51305285)、江苏省基础研究计划(自然科学基金)(No.BK20130315)、
江苏省光子制造科学与技术重点实验室开放基金资助项目(No.GZ201304)。
第一作者：王晓南，(1984—)，男，博士，副教授。高性能钢铁材料及激光焊接。 wxn@suda.edu.cn。

ness of the WS and heat affected zone was significantly higher than that of the BM. When the stress ratio R=0.1, circulation base of 10^7, the fatigue strength of the BM was significantly higher than that of the WS. Fatigue crack sources of the BM and WS were interface of persistent slip bands and substrate, weld porosity, respectively. In the BM, fatigue crack propagation paths were along interface of the ferrite and carbide. However, in the WS, fatigue crack cut lath martensite and original austenite grain boundaries, no significant crack propagation direction.

Key words: micro-alloy steel; laser welding; microstructure; fatigue fracture; lath martensite

焊接工艺对 800MPa 级微合金 C-Mn 钢焊接接头组织性能的影响

王 卫[1,2]，王晓南[1]，王海生[1,2]，章顺虎[1]，朱国辉[1]，陈长军[3]，张 敏[3]

(1.苏州大学 沙钢钢铁学院；2.安徽工业大学 冶金工程学院；3.苏州大学 激光加工中心)

摘 要：研究了激光焊接和气体保护焊接对抗拉强度 800 MPa 级 C-Mn 微合金钢焊接接头组织和性能的影响。通过显微组织观察及硬度、拉伸实验研究不同焊接工艺下焊接接头组织与力学性能，结果表明：激光焊接条件下焊缝区和粗晶区的显微组织均为板条马氏体。气体保护焊接条件下焊缝区的显微组织为先共析铁素体和针状铁素体混合组织，其粗晶区显微组织为粒状贝氏体。激光焊接和气体保护焊接细晶区的显微组织主要为铁素体和 M-A 组元，在细晶区中含有少量的珠光体。气体保护焊粗晶区原始奥氏体晶粒尺寸为激光焊接条件下的 4 倍左右。激光焊接接头热影响区内硬度显著高于气体保护焊，两者接头峰值硬度均在粗晶区，其中气体保护焊在热影响出现明显的软化区。激光焊接接头拉伸断裂位置出现在母材区，而气体保护焊接头拉伸断裂位置出现在焊缝区。两者断口显微形貌主要是韧窝，为典型的韧性断裂。

关键词：激光焊接；气体保护焊接；焊接工艺；显微组织；力学性能

Effect of Welding Process on the Microstructure and Properties of Joint of 800MPa Grade Microalloyed C-Mn Steel

WANG Wei[1,2], WANG Xiao-nan[1], WANG Hai-sheng[1,2], ZHANG Shun-hu[1], ZHU Guo-hui[2], CHEN Chang-jun[3], ZHANG Ming[3]

(1. Shagang School of Iron and Steel, Soochow University; 2.Metallurgical Engineering School, Anhui University of Technology; 3.Laser Processing Research Center, Soochow University, Jiangsu)

Abstract: The effect of laser welding and arc welding on the microstructure and properties of tensile strength800 MPa grade microalloyed C-Mn steel welded joints was studied. Study on the microstructure and mechanical properties by the microstructure observation, tensile and hardness test. The results indicated that in laser welding both weld metal and coarse-grain heat affected zone (CG-HAZ) were lath martensite. In arc welding weld metal consist prodominantly of acicular ferrite (AF) and proeutectoid ferrite (PF).The microstructure of CGHAZ was granular bainite. Microstructure of fine-grain heat affected zone (FGHAZ) mainly consists of fine grain ferrite and M-A constituent in addition to all FGHAZ contained a few pearlite. The original austenite grain size in CGHAZ of arc

基金项目：国家自然科学基金资助项目(No.51305285)、江苏省基础研究计划(自然科学基金)(No.BK20130315)、江苏省光子制造科学与技术重点实验室开放基金资助项目(No.GZ201304)。

第一作者：王 卫，男。钢铁的焊接。wangweiahut@163.com。

通讯作者：王晓南(1984—)，男，博士，副教授。先进钢铁材料开发。wxn@suda.edu.cn。

welding was 4 times greater than it of laser welding. The HAZ hardness of laser welding joint was significantly higher than it of the arc welding. Both joints peak hardness were at CGHAZ, and arc welding appeared obvious softened zone. The laser welding joint tensile fracture occurred in the base metal while arc welding joint tensile fracture occurred in weld zone. Both fracture of microscopic morphology mainly was dimple, its typical ductile fracture.

Key words: laser welding; arc welding; welding process; microstructure; mechanical properties

抗拉强度 700MPa 级微合金钢激光焊接接头组织性能

王海生[1]，王晓南[*2]，王　卫[1]，朱国辉[1]，陈长军[3]，张　敏[3]

(1.苏州大学 沙钢钢铁学院；2.安徽工业大学 冶金工程学院；3.苏州大学 机电工程学院激光加工中心)

摘　要：对 4.6mm 厚 Nb-Ti 微合金高强钢激光焊接的焊接性进行研究。通过光学显微镜(OM)、扫描电镜(SEM)、显微硬度测量、拉伸实验和冲击实验等手段进行分析。实验结果表明：焊缝组织为板条马氏体以及少量的贝氏体及和铁素体；热影响区粗晶区几乎全部为马氏体组织，随着热输入增大局部出现铁素体和贝氏体，并且粗晶区晶粒的尺寸随着焊接热输入的增大而增大。焊缝和热影响区硬度大于母材硬度，而且，随着焊接热输入增大焊缝的硬度减小，其主要原因是由于焊缝中有铁素体和贝氏体生成。焊缝及热影响区的强度均大于母材，因此，拉伸断裂位置在母材区。焊接接头的冲击吸收功大于母材，两者的断裂方式为韧性断裂。

关键词：激光焊接；微观组织；力学性能；微合金钢

Microstructure and Properties of Laser Welding Joint of Microalloyed Steel With Tensile Strength 700MPa

WANG Hai-sheng[1], WANG Xiao-nan[*2], WANG Wei[1],ZHU Guo-hui[1], CHEN Chang-jun[3], ZHANG Min[3]

(1.Shagang School of Iron and Steel, Soochow University; 2.School of Metallurgy and Engineering, Anhui University of Technology; 3.Laser Processing Research Center, Soochow University)

Abstract: Research on the welding of 4.6mm Nb-Ti micro alloyed high strength steel with laser welding. We studied the microstructure and properties with optical microscopy (OM), scanning electron microcopy (SEM), tensile test, hardness test, impact experiment. The study shows that the microstructure of weld is maintly lath martensite with a small amount of bainite and ferrite. The CG-HAZ is almost of martensite, with the increase of heat input, a little mount of ferrite and bainite appeared. And the grain size of CG-HAZ is increasing with the increase of heat input. The hardness of weld and HAZ are greater than base material, and the hardness decreased with the heat input increase. The strength of WS and HAZ is higher than BM, so tensile fracture position is in base material. The impact energy of WS is higher than BM.

Key words: laser welding; microstructure; mechanical properties; microalloyed steel

---

基金项目：国家自然科学基金资助项目(51305285)、江苏省基础研究计划(自然科学基金)(BK20130315)、
江苏省光子制造科学与技术重点实验室开放基金资助项目(GZ201304)。
作者简介：王海生(1989—)，男，硕士，学生。高强钢激光焊接。511213174@qq.com。
导师简介：王晓南(1984—)，男，博士，副教授。先进钢铁材料开发及焊接性能。wxn@suda.edu.cn。

汽车用双相钢 DP590 激光焊接接头组织及性能研究

孙 茜[1]，邸洪双[2]，王晓南[2*]，杨兆华[3]，张 敏[3]，陈长军[3]

（1.东北大学 轧制技术及连轧自动化国家重点实验；2.苏州大学 沙钢钢铁学院；3.苏州大学 机电工程学院激光加工中心）

摘 要：研究了 1.0mm 厚汽车用双相钢 DP590 激光拼焊焊接接头的显微组织、显微硬度、力学性能及成形性能变化规律。研究表明：焊缝区的显微组织为板条马氏体，热影响区的显微组织为铁素体和板条马氏体，母材区显微组织为铁素体和 M/A 岛；焊缝区、热影响区、母材的平均硬度分别为 340HV、275HV 和 205HV；单向拉伸实验表明拉伸失效位置出现母材区域，与母材相比，焊接接头的均匀延伸率有所降低；杯突实验表明，裂纹在焊缝中心处形成并垂直于焊缝向热影响区和母材扩展，焊缝区的马氏体板条束在断裂前发生了明显的微观塑性变形。焊缝区的杯突值可达到母材的 81.9%，表明焊缝区具有良好的成形性能且可满足实际生产要求。

关键词：双相钢；激光焊接；显微组织；力学性能；成形性能

Study on Microstructure and Properties of Automotive Duplex Steel Sheet Welded Joint Using Pulsed Lasers

SUN Qian[1], DI Hong–shuang[1], WANG Xiao–nan[2*], YANG Zhao–hua[3], ZHANG Min[3], CHEN Chang–jun[3]

(1.The State Key Lab of Rolling & Automation, Northeastern University; 2.Shagang School of Iron and Steel, Soochow University; 3.Laser Processing Research Center, School of Mechanical and Electrical Engineering, Soochow University)

Abstract: Microstructure, micro-hardness, mechanical properties and formability of automotive duplex steel sheet welded joint using laser welding were studied. The result shows that microstructure of fusion zone (FZ) and heat affected zone (HAZ) was lath martensite, ferrite and martensite, respectively, and microstructure of base material (BM) was ferrite and M/A island. The average micro-hardness of FZ, HAZ and BM was 340HV, 275HV and 205HV, respectively; the softening zone was not found in HAZ. Tensile testing showed that tensile failure position was at base material. Comparing with the base material, the uniform elongation of welded joints decreased. Erichsen test showed that cracks formed at the center of the fusion zone, and crack expansion in the heat affected zone and the base material. Crack propagation direction was perpendicular to the fusion zone. Lath martensite beam in fusion zone generated significantly microscopic plastic deformation before fracturing. Erichsen value of the weld zone could reach 81.9% of the base material. Fusion zone showed good formability and could meet the requirements for production.

Key words: Duplex steel; laser welding; microstructure; mechanical properties; formability

基金项目：国家自然科学基金资助项目(No.51305285)、江苏省基础研究计划(自然科学基金)(No.BK20130315)、江苏省光子制造科学与技术重点实验室开放基金资助项目(No.GZ201304)。

第一作者：孙 茜,(1986年—)，女,博士研究生。冷轧汽车用钢组织性能控制。sacsq@qq.com.。

通讯作者：王晓南,(1984年—)，男,博士,副教授。高性能钢铁材料及激光焊接。wxn@suda.edu.cn。

焊缝熔池凝固组织中枝晶形貌的元胞自动机法模拟

张　敏，李露露，薛　覃，李继红

(西安理工大学 材料科学与工程学院)

摘　要：改进已有元胞自动机(CA)模型，考虑了界面能各向异性、界面扰动、溶质扩散、成分过冷、曲率过冷等因素，对焊接熔池凝固过程中等轴晶的生长进行了模拟，并讨论了不同过冷度和扰动振幅对枝晶生长形貌及溶质偏析的影响。模拟结果表明，改进后的模型成功地模拟出枝晶生长的竞争机制和溶质的偏析规律，与理论分析取得很好的一致性。

关键词：凝固；元胞自动机；枝晶形貌；溶质偏析

Simulation of Dendrite Morphology in Welding Pool with Cellular Automaton Method

ZHANG Min, LI Lu-lu, XUE Qin, LI Ji-hong

(School of Material Science and Engineering, Xi'an University of Technology)

Abstract: A modified cellular automaton model was developed, which considered factors like the anisotropy of interfacial energy, the constituent undercooling and the curvature undercooling. The growth of equiaxial crystal during the solidification of the weld molten pool was simulated, and the effect of different undercooling and the disturbance amplitude on growth morphology of dendrite and the solute segregation was discussed. The results show that the modified model successfully simulated the competition mechanism of dendrite growth and the rule of solute segregation, which is consistent with theoretic results.

Key words: solidification; cellular automaton; dendrite morphology; segregation

0 引言

　　焊接接头的微观组织直接影响着焊接构件的使用性能。焊缝熔池金属的凝固是一个不平衡的连续冷却过程，具有高温、动态等特点。采用理论分析方法进行研究存在着巨大的数学困难，而传统的实验方法不但耗时、耗资，且无法揭示枝晶组织形成的物理机理。随着焊接领域计算科学的发展和凝固理论的完善，数值模拟方法在解决焊接问题中发挥着重要的作用，成为继理论分析和实验方法之后的

基金项目：国家自然科学基金资助项目(51274162)；国家高技术研究发展计划(863 计划)(2013AA031303)；
陕西省自然科学基金资助项目(2012JM6003)；陕西省教育厅产业化培育项目(No.2012.JC16)。
第一作者：张　敏，男。新型焊接材料、焊接结构断裂强度和焊接工程结构。zhmmn@xaut.edu.cn。

另一种分析焊接过程的有效手段。准确地预测焊接接头的微观组织,进而有效地控制焊缝微观组织的形成过程,得到具有优良力学性能的焊接构件,是焊接工作者所追求的目标。

微观组织的模拟方法主要有确定性法,蒙特卡罗法,元胞自动机法和相场法[1-3],近几年来,元胞自动机法由于物理基础明确、计算量可控等优点,在凝固微观组织的模拟中得到了广泛的应用。目前学者所进行的研究中,绝大部分工作都是铸造凝固组织方面的模拟,焊接领域的研究较晚且发展相对滞后。在国外,Pavlyk 和 Dilthey[4]在 CA 法的基础上模拟了特定条件下钨极氩弧焊焊接熔池定向凝固过程中柱状晶形貌,模拟结果经实验验证较为合理。Yin H.和 Felicelli S.D.[5]将有限元法和元胞自动机技术结合在一起,模拟了激光焊焊接过程中柱状树枝晶的二维生长形貌,并讨论了不同焊接工艺参数对枝晶形貌的影响。在国内,华中科技大学黄安国[6]运用金属凝固相关理论,做出一些合理的假设或近似,建立了较为准确的预测焊缝凝固组织的 CA 模型,模拟结果再现了焊缝金属中微观组织的凝固过程。哈尔滨工业大学的马瑞、占小红等[7-8]将元胞自动机法和有限差分法结合在一起,实现了 Ni-Gr 合金焊接凝固过程熔池中结晶以及溶质场分布的模拟。北京工业大学赵玉珍[9]利用温度场计算软件计算了焊接温度场并耦合进晶界演化模型中,得到了氩弧焊熔池凝固过程柱状晶随时间变化的形貌。

焊接接头微观组织的研究集中于柱状晶组织,等轴晶形貌的研究尚不多见。本文作者在前期研究的基础上[10],改进了元胞自动机模型,考虑了液相中溶质的扩散以及过冷度等问题,模拟了焊接熔池中单个等轴晶的演变过程,再现了溶质浓度的分布规律,并讨论了不同焊接参数对等轴晶生长形貌的影响。

1 元胞自动机基本原理

元胞自动机法的基本思想是一个胞元或是系统的基元,依据与其相邻其他胞元的情况,按照事先设定的演变规则来决定自己的状态,从而通过定义局部简单的规则来描述系统整体复杂的演变规律。在计算前,将模拟区域划分为适量等大均匀的网格,每个网格需存储不同的状态值或变量值来标识,计算过程中,假定一个目标单元,并根据预先定义的转变规则和邻胞状态来控制目标单元下一时刻的状态。金属材料的微观组织模拟中,通常采用简单的正方形网格体系,模拟计算凝固过程的枝晶生长时,每个网格被赋予三种状态(液态、固态和界面),并按照凝固基本理论控制网格的演变规律。

2 模型和算法

为简化问题,模型中不考虑焊接热过程及枝晶生长过程中释放的潜热,对整个焊接熔池做等温处理,且认为界面处始终处于平衡状态。根据热力学条件,液态金属的形核和生长必须在过冷的条件下进行。由于凝固过程中的热扩散比溶质扩散高 3~4 个数量级,本模型忽略枝晶尖端的热扩散。假设枝晶尖端生长过程无热过冷,只考虑曲率过冷 ΔT_r 和成分过冷 ΔT_c。则总的过冷度 ΔT 为:

$$\Delta T = \Delta T_r + \Delta T_c \tag{1}$$

液/固界面是曲面,则会由于界面张力效应造成的附加压力而破坏原有的平衡,这时界面只有通过改变液/固两相的平衡温度来获得一个新的过冷度 ΔT_r。本文所建模型考虑界面能各向异性计算曲率过冷:

$$\Delta T_r = K\overline{\Gamma}\left\{1 - 15\varepsilon\cos\left[4(\theta - \theta_0)\right]\right\} \tag{2}$$

式中，$\bar{\Gamma}$ 为平均 Gibbs-Thomson 系数，ε 为界面能各向异性强度系数，θ 为液/固界面法线与水平方向的夹角，θ_0 为晶体择优生长方向，K 为液/固界面的曲率，取决于界面的固相分数 f_s 梯度，可由下式确定：

$$K = \left[2\frac{\partial f_s}{\partial x}\frac{\partial f_s}{\partial y}\frac{\partial^2 f_s}{\partial x \partial y} - \frac{\partial^2 f_s}{\partial x^2}\left(\frac{\partial f_s}{\partial y}\right)^2 - \frac{\partial^2 f_s}{\partial y^2}\left(\frac{\partial f_s}{\partial x}\right)^2 \right] \cdot \left[\left(\frac{\partial f_s}{\partial x}\right)^2 + \left(\frac{\partial f_s}{\partial y}\right)^2 \right]^{-3/2} \tag{3}$$

合金凝固过程中的溶质再分配使得液/固界面液相一侧及其前方的熔体形成了一个溶质扩散场，同时也相应的改变了其液相线温度，这种由溶质再分配导致界面前方熔体成分及其凝固温度发生变化而引起的过冷成为成分过冷。

$$\Delta T_c = m_l \left(w_0 - w_l^* \right) \tag{4}$$

式中，m_l 为液相线斜率，w_0 为合金原始成分，w_l^* 为界面液相成分。

在给定总过冷度后，界面液相成分可由下式得到：

$$w_l^* = w_0 - m_l^{-1} \cdot \left(\Delta T - \Delta T_r \right) \tag{5}$$

固溶体合金 w_0 在凝固时，由于合金系的平衡分配系数 $K_0 < 1$，表明已凝的固相从母液中带走较少的溶质原子，多余的溶质原子都排入液/固界面前的液相中。为了减小误差，排出的这部分固相增量直接由溶质扩散求解。则界面胞元向周围液相邻胞排出的多余溶质可表示为：

$$\Delta C = \Delta x^{-1} D_l \Delta t \cdot \sum_{nb} \left(w_l^* - w_{nb} \right) \tag{6}$$

式中，Δx 为空间步长，D_l 为液相扩散系数，Δt 为时间步长，w_{nb} 为液相邻胞成分，下标 nb 代表界面胞元的液相邻胞。

界面胞排出的溶质意味着其固相分数的增加 Δfs，又根据 MSC 理论[11]可知，分枝的形成与界面扰动有关。在数值模拟中，通过在固相分数增加过程中施加一个扰动函数来控制枝晶分枝，则二者关系满足：

$$\Delta fs = \Delta C \left[w_l^* \cdot \left(1 - K_0 \right) \right]^{-1} \left[1 + A\left(1 + 2rand\left(\right) \right) \right] \tag{7}$$

式中，A 为扰动因子，$rand(\ )$ 为能够在 $[0,1]$ 间产生的一个随机数。

最后，界面元胞排出的溶质导致枝晶周围液相溶质浓度升高，液相元胞间出现较大的浓度梯度。因此还需对液相胞进行溶质扩散计算，采用的控制方程为标准的二维非稳态扩散方程：

$$\frac{\partial w_l}{\partial t} = D_l \cdot \left(\frac{\partial^2 w_l}{\partial x^2} + \frac{\partial^2 w_l}{\partial y^2} \right) \tag{8}$$

式中，D_l 为液相溶质扩散系数。采用显式有限差分法对溶质的扩散进行数值计算，对位于计算区域边界的网格采用零通量的边界条件进行处理。

3 模拟结果与讨论

为模拟单个等轴晶生长，将二维计算区域划分为 400×400 个均匀的正方形网格，网格尺寸为 $5 \times 10^{-7} m$，以碳含量为 0.4% 的铁碳合金为模拟对象，并假定整个模拟区域温度均匀恒定。模拟初始时刻，在计算域核心胞元处放置一个成分为 $K_0 \cdot w_0$ 的晶核，设定其他胞元均为成分为 w_0 的液相。

3.1 等轴晶生长模拟

图 1 为模拟的不同时刻等轴晶生长形貌。由图可以看出,在生长初期,等轴晶沿着 8 个方向快速生长,其中平行于主轴方向的一次枝晶较其余方向生长更快,这是因为在二维空间中,择优生长方向为 <10>,枝晶沿着择优方向生长速度最快。同时,一次枝晶臂上伴有少量细小的二次枝晶。随着时间推移,择优方向的一次枝晶继续长大,而其余方向一次枝晶生长受抑制,其原因在于迅速生长的二次枝晶,将溶质排向周围,使枝晶附近液相溶质浓度升高,从而抑制其生长。在枝晶生长中后期,如图 1(c) 所示,伴随着一、二次枝晶的迅速生长,二次枝晶臂上长出三次枝晶,需要注意的是,由于一、二次枝晶生长过程排出的溶质富集在内部间隙中,形成的高溶质浓度不利于三次枝晶的萌生,而二次枝晶尖端附近富集的溶质会很快扩散到周围液相中,因此,三次枝晶主要集中于二次枝晶尖端附近。

 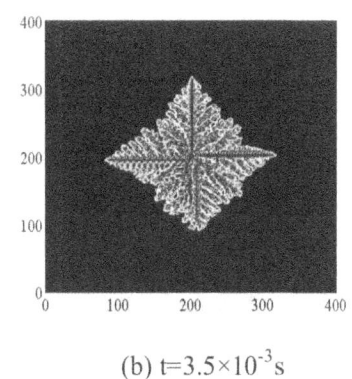 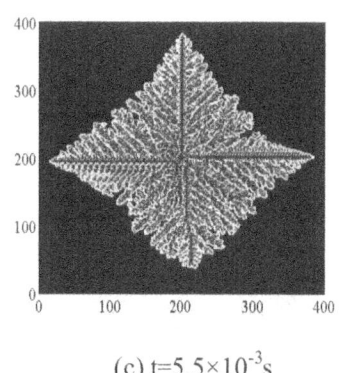

(a) t=1.5×10⁻³s (b) t=3.5×10⁻³s (c) t=5.5×10⁻³s

图 1　不同时刻等轴晶生长形貌

3.2 不同过冷度下等轴晶生长模拟

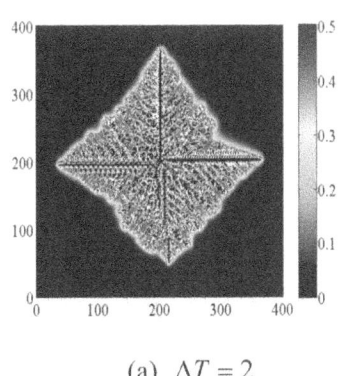 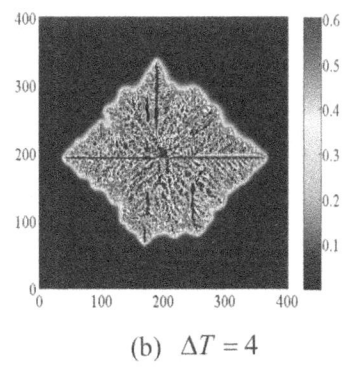 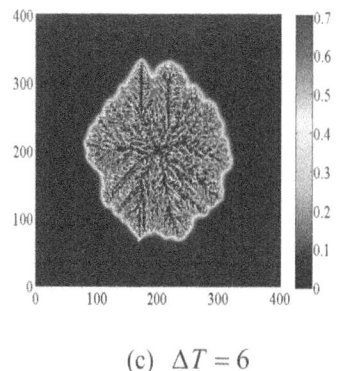

(a) $\Delta T = 2$　　(b) $\Delta T = 4$　　(c) $\Delta T = 6$

图 2　不同过冷度等轴晶生长形貌及溶质浓度分布

图 2 为模拟的同一时刻(t=0.005s)不同过冷度下等轴晶的生长形貌及溶质浓度分布。右侧衬度条表示溶质浓度值。图 2(a)、(b)、(c)过冷度分别为 2K、4K、6K。可以看到,过冷度不同,枝晶生长形貌有

差异,当过冷度较小时,择优方向的一次枝晶生长最快且易于分辨,并伴有二次枝晶的生长。随着过冷度的增大,如图2(b)所示,一次枝晶逐渐消失,二、三次枝晶迅速生长。当过冷度继续增大时,迅速生长的二、三次枝晶逐渐粗化,择优方向的一次枝晶彻底消失,整体呈近似圆形。溶质浓度方面,过冷度越大,最高溶质浓度越大,枝晶内部的偏析程度也越严重,这是由于枝晶间激烈的竞争,使得枝晶臂间更加致密,相应的溶质浓度也会升高的缘故。

3.3 不同扰动振幅下等轴晶生长模拟

图3为模拟的同一时刻不同扰动振幅下等轴晶生长形貌及溶质浓度分布。当界面无扰动时,如图3(a)所示,枝晶生长具有明确的主轴,随扰动的增大,二、三次枝晶的迅速生长,使得主轴消失,晶体的分枝越来越复杂化,且其生长速度加快,如图3(b)、(c)所示。二三次枝晶间隙形成溶质偏析,由于生长速度的加快,溶质浓度逐渐增高。

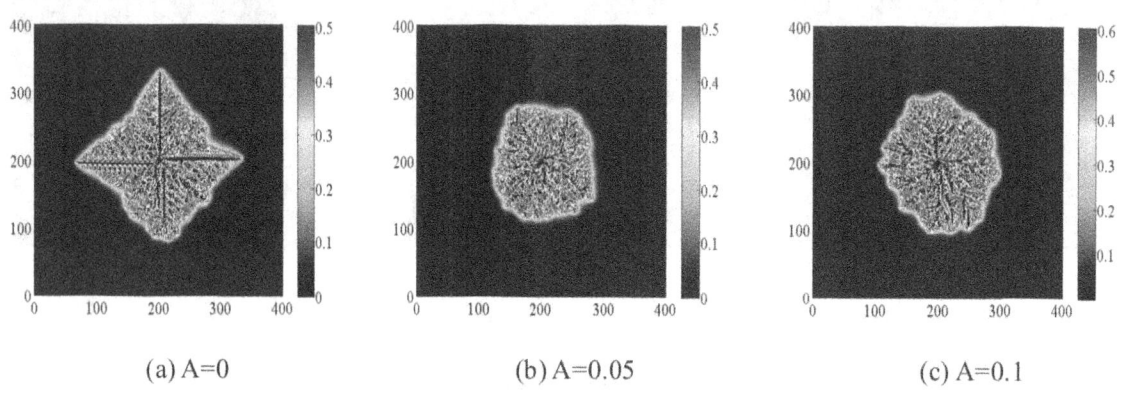

(a) A=0 　　　　(b) A=0.05 　　　　(c) A=0.1

图3　不同扰动振幅下等轴晶生长形貌及溶质浓度分布

4 结论

综合考虑了界面能各向异性、界面扰动、过冷度、溶质扩散等因素后,采用元胞自动机法对焊接熔池凝固过程中等轴晶的生长进行了模拟。模拟结果表明:

(1)建立的模型能够合理的模拟出等轴晶的生长过程。凝固过程中,枝晶沿着各自的择优方向快速生长,同时伴有二、三次枝晶的竞争生长。

(2)讨论了不同过冷度和扰动振幅对等轴晶生长的影响。过冷度不同,枝晶生长形貌有差异,过冷度越大,二、三次枝晶迅速生长并粗化;扰动振幅主要影响枝晶的分枝,扰动越大,分枝越明显,生长速度越快。

(3)分析了溶质浓度的分布规律。溶质主要富集在二、三次枝晶臂的间隙中,形成溶质偏析;随过冷度和扰动振幅的增大,富集的溶质浓度逐渐增高。

参考文献：

[1] 许庆彦,柳百成. 铸造合金凝固组织的计算机模拟与预测 [J]. 稀有金属材料与工程,2003,32(6):401–406.

[2] 朱鸣芳,于金,洪俊构. 金属凝固显微组织的计算机模拟 [J]. 中国工程科学,2004,6(5): 8–15.

[3] Mohsen A Z, Hebi Y, Sergio D F. Comparison of cellular automaton and phase field models to simulate dendrite growth in hexagonal crystals [J]. Journal of Materials Science and Technology, 2012, 28(2): 137–146.

[4] Pavlyk V, Dilthey U. Simulation of weld solidification microstructure and its coupling to the macroscopic heat and fluid flow modeling [J]. Modeling and Simulation in Materials Science and Engineering, 2004, 12(1): 33–45.

[5] Yin H, Felicelli S D, Wang L. Simulation of a dendritic microstructure with the lattice boltzmann and cellular automaton methods [J]. Acta Materialia, 2011, 59(8): 3124–3136.

[6] 黄安国,余圣甫,李志远. 焊缝金属凝固组织元胞自动机模拟 [J]. 焊接学报,2008,29(4): 45–59.

[7] 马瑞,董志波,魏艳红. 镍基合金焊缝凝固组织演变过程模拟和仿真 [J]. 焊接学报,2010,31(7): 43–49.

[8] 占小红,董志波,魏艳红,王勇. Ni–Cr 二元合金焊接熔池中柱状枝晶生长模拟 [J]. 2009,19(8): 1431–1436.

[9] 赵玉珍. 焊接熔池的流体动力学行为及凝固组织模拟 [D]. 北京: 北京工业大学,2004,50–56.

[10] 李继红，汪强，吴伟刚，张敏. Fe–C 二元合金凝固过程枝晶偏析的数值模拟研究 [J]. 材料导报, 2013,27(6): 152–155.

[11] 陈晋. 基于胞元自动机法的凝固过程微观组织数值模拟 [D]. 南京:东南大学,2005,22–23.

基于刚塑性模型的二维金属塑性成形有限元模拟

滕子浩，廖敦明，陈　涛，王新云

(华中科技大学　材料成形及模具技术国家重点实验室)

摘　要：基于刚塑性有限元法,采用罚函数法与反正切摩擦模型,分别用三角形单元和四边形单元对二维金属塑性成形过程进行了模拟与研究,开发了相应的模拟软件。针对任意复杂形状的模具,采用 Delaunay 三角形网格划分算法对网格进行重划,运用盒子树法快速查找新旧网格节点位置关系,并给出了相应的算例,将模拟得到的结果与商业软件 DEFORM 计算结果进行对比,二者基本一致,验证了计算程序的正确性。

关键词：刚塑性模型；金属塑性成形；有限元法

The Finite Element Simulation of Two Dimensional Metal Forming Based on the Rigid-Plastic Model

TENG Zi-hao, LIAO Dun-ming, CHEN Tao, WANG Xin-yun

(State Key Laboratory of Materials Processing and Die&Mould Technology University of Science and Technology)

Abstract：Based on the rigid-plastic finite element method, the penalty function method and the arctangent frictional model, the simulation of two dimensional metal forming process was done by separately using the triangular element and quadrilateral element, and the corresponding simulation software was developed. For molds with any complex shape, the Delaunay triangle meshing algorithm is utilized for the re-meshing and the box tree method is employed to quickly find the position of nodes between new and old grid, then the relevant calculation example is presented. The numerical results are compared with the commercial software DEFORM, the good agreement demonstrates the accuracy of this calculating program.

Key words: the rigid-plastic model; metal forming; the finite element method (FEM)

第一作者：滕子浩(1991—)，男，博士生。锻造有限元数值模拟。tengzihao@hust.edu.cn。

通讯作者：廖敦明(1973—)，男，教授，博士生导师。材料成形 CAD/CAE 研究。liaodunming@hust.edu.cn。

PP/DBDPE/OMMT/Sb$_2$O$_3$ 复合材料的
制备与阻燃性能的研究

卢　红[1]，**陆桂焕**[1]，**谢襄漓**[2]，**王林江**[3]

（1. 桂林理工大学 材料与科学工程学院；2.桂林理工大学 化学与生物工程学院；

3.桂林理工大学 材料与科学工程学院 有色金属及材料加工新技术教育部重点实验室）

摘　要：本文用有机蒙脱石（OMMT）部分取代十溴二苯乙烷制备聚丙烯/十溴二苯乙烷/三氧化二锑（PP/DBDPE/Sb$_2$O$_3$）、聚丙烯/十溴二苯乙烷−有机蒙脱石/三氧化二锑（PP/DBDPE/OMMT/Sb$_2$O$_3$）复合材料，采用 XRD、SEM 对材料的结构和形貌进行分析和表征；采用 TGA 和锥形量热仪对复合材料的热稳定性和阻燃性进行测试和表征。TGA 测试表明，对组成为 PP/22.5%DBDPE/7.5%Sb$_2$O$_3$ 的复合材料，当用 50%有机蒙脱石取代十溴二苯乙烷后，复合材料的初始分解温度和热分解温度分别提高了 8℃、13℃。在 600℃的残留量达到 13.71%，提高了 40.62%。锥形量热测试结果分析表明，有机蒙脱石部分取代十溴二苯乙烷后，PP/ DBDPE/OMMT/ Sb$_2$O$_3$复合材料的热释放率峰值（PHRR）最小为 374kw/m^2，与 PP/22.5%DBDPE/7.5%Sb$_2$O$_3$复合材料的 PHRR（459kw/m^2）相比，下降了 19%。说明有机蒙脱石部分取代十溴二苯乙烷对聚丙烯具有较好的协同阻燃效果。

关键词：聚丙烯；蒙脱石；十溴二苯乙烷；三氧化二锑；阻燃

Preparation and Flame Retardant Performance of PP/ DBDPE/OMMT/Sb$_2$O$_3$ Composites

LU Hong[1], LU Gui-huang[1], XIE Xiang-li[2], WANG Lin-jiang[3]

（1. College of Materials Science and Engineering, Guilin University of Technology; 2. Ministry−province Jointly−constructed Cultivation Base for State Key Laboratory of Processing for Non−ferrous Metal and Featured Materials; 3. Key Laboratory of New Processing Technology for Nonferrous Metals and Materials, Ministry of Education）

Abstract：In this paper, PP/DBDPE/Sb2O3 and PP/DBDPE/OMMT/Sb$_2$O$_3$ composites were prepared by melt blending method via using the organic montmorillonite (OMMT) partly replace the DBDPE. XRD and SEM were used to characterize the structure features and of OMMT and composites. TGA and cone calorimeter were used to characterize the thermal stability and fire retardance of the composites. The data from TGA indicated that the onset temperature of PP/ 11.25%DBDPE/ 11.25%OMMT/ 7.5%Sb$_2$O$_3$ composites 8℃ higher than PP/ 22.5%DBDPE/ 7.5%Sb$_2$O$_3$ composites and decomposition temperature is higher 13℃ than that. The amount of nonvolatile residue were

基金项目：国家自然基金（41272064）

第一作者：卢　红，女。矿物材料与绿色建材。137761187@qq.com。

通讯作者：王林江，男，教授，博士生导师。wlinjiang@163.com。

13.71% which was higher 40.62% than that while remains at 600℃. Cone calorimeter analysis indicated that the PHRR of the PP/ DBDPE/OMMT/Sb_2O_3 composites is 374 kW/m^2 minimum value after OMMT partly replacing DBDPE and deleasing 19% when compare to the PHRR of PP/ 22.5% DBDPE/ 7.5% Sb_2O_3 composites. It concludes OMMT and APP had a good synergistic effect on PP, after OMMT partly replacing DBDPE and adding APP.

Key words: polypropylene, montmorillonite; decabromodiphenyl ethane; antimonous oxide; fire retardancy

共沉淀法制备水滑石－蒙脱石复合材料

李存军 [1,2]，谢襄漓 [3]，王林江 [4]

（1.桂林理工大学 广西矿冶与环境科学实验室中心；2.桂林理工大学 材料科学与工程学院 4；

3. 桂林理工大学 化学与生物工程学院；4.有色金属及材料加工新技术教育部重点实验室）

摘　要：以内蒙古钙基蒙脱石为原料，将其钠化改型获得钠基蒙脱石；以钠基蒙脱石片层为模板，采用共沉淀法合成了水滑石，水滑石在蒙脱石片层上成核生长，制备了水滑石–蒙脱石复合材料。采用 X 射线衍射分析(XRD)、扫描电子显微镜(SEM)等技术研究了复合材料。结果表明，水滑石分布在蒙脱石片层上，两者紧密结合形成复合材料。复合材料中蒙脱石和水滑石均以晶态形式存在，但由于水滑石的生长受到蒙脱石颗粒聚集的空间限制，复合材料中的水滑石晶粒较小，结晶性较低。另外，由于部分水滑石在蒙脱石层间成核生长，破坏了蒙脱石片层的有序性，致使复合材料中蒙脱石的结晶度下降。

关键词：水滑石；蒙脱石；共沉淀

Preparation of Hydrotalcite－Montmorillonite Composite by Coprecipitation Method

LI Cun-jun[1, 2], XIE Xiang-li[3], WANG Lin-jiang[4]

(1. Guangxi Scientific Experiment Center of Mining, Metallurgy and Environment; 2.College of Materials Science and Engineering, Guilin University of Technology; 3. College of Chemistry and Bioengineering, Guilin University of Technology; 4. Key Laboratory of Nonferrous Materials and New Processing Technology, Ministry of Education)

Abstract: Sodium-montmorillonite was obtained by modification of Inner Mongolia calcium-montmorillonite. Hydrotalcite-montmorillonite composite was prepared by using coprecipitation method. The layers of montmorillonite played as templates for nucleation and growth of hydrotalcite. Hydrotalcite-montmorillonite composite was studied by X-ray diffraction (XRD), scanning electron microscopy (SEM), etc. The experimental investigation showed that hydrotalcite was composed with montmorillonite where hydrotalcite distributed on the layers of montmorillonite. Both hydrotalcite and montmorillonite were crystalloid. However, because the space for growth of hydrotalcite was restrained by aggregation of montmorillonite particles, it leaded to the poor crystallinity and small crystal grain of hydrotalcite in composite. It was also suggested that because fraction of hydrotalcite was grown between the adjacent layers of montmorillonite, the crystallinity of montmorillonite decreased.

Keywords: hydrotalcite; montmorillonite; coprecipitation

基金项目：国家自然基金(41272064)。

第一作者：李存军，男。矿物材料与绿色建材。cunjunlee@163.com。

通讯作者：王林江，男，教授，博导。wlinjiang@163.com。

层状双氢氧化物与蒙脱石在水中的剥离及组装行为研究

梁启超[1], 陈　耀[1], 谢襄漓[2], 王林江[3,4]

(1.桂林理工大学　材料科学与工程学院; 2.桂林理工大学　化学与生物工程学院;

3.广西有色金属及特色材料加工国家重点实验室培育基地; 4.有色金属清洁冶炼与综合利用广西高校重点实验室)

摘　要: 通过将阴离子粘土矿物层状双氢氧化物(LDH)与阳离子粘土矿物蒙脱石(MMT)复合,利用蒙脱石较高的比表面积、良好的热稳定性和较大的空隙度来改善LDH的结构特性,可以显著扩大其应用领域。基于水热法合成了NO_3-LDH,用离子交换法制备了醋酸根插层层状双氢氧化物(Aco-LDH)。以水为溶剂,分别将Aco-LDH和钠基蒙脱石通过机械搅拌和超声分散等技术手段进行处理,得到醋酸根型层状双氢氧化物和蒙脱石的剥离胶体液,通过混合搅拌反应制备出具有片-片组装结构的层状双氢氧化物-蒙脱石复合材料。用XRD、FTIR、TEM、AFM等手段对剥离和组装结构进行了表征。结果表明,当Aco-LDH浓度为1mg/mL时,在水中超声处理获得厚度为4.5~20nm的LDH片层;蒙脱石机械搅拌分散,5000 r/min高速离心后,层状结构特征衍射峰(001)消失,蒙脱石基本被剥离成单层状结构。组装复合物的层间距为1.51nm,接近单元LDH片层与单元MMT片层的堆叠结构厚度,表明LDH和MMT之间形成片-片组装结构。

关键词: 层状双氢氧化物; 蒙脱石; 水溶液环境; 剥离; 片-片组装结构

Study on Delaminating and Assembly Behavior of Layered Double Hydroxides and Montmorillonite in Water

LIANG Qi-chao[1], CHEN Yao[1], XIE Xiang-li[2], WANG Lin-jiang[3,4]

(1.College of Materials Science and Engineering, Guilin University of Technology; 2.College of Chemistry and Bioengineering, Guilin University of Technology; 3. Ministry-province Jointly-constructed Cultivation Base for State Key Laboratory of Processing for Non-ferrous Metal and Featured Materials; 4. Guangxi Key Laboratory of University for Clean Metallurgy and Comprehensive Utilization of Non-ferrous Metal Resources)

Abstract: The high specific surface area, good thermal stability and large hole space of montmorillonite (MMT) can improve the structure features of layered double hydroxide (LDH) which will significantly expanding its application field after preparing composite. Acetate-intercalated LDH (Aco-LDH) was prepared by ion-exchange of NO_3-LDH with sodium acetate trihydrate. The colloidal suspension of Aco-LDH and MMT was obtained by mechanical stirring and ultrasonic dispersing in water solvent, respectively. The sheet-by-sheet assembly composite was gained through mechanical stirring. The delaminating and assembly structure was characterized by X-ray diffraction (XRD),

基金项目:国家自然基金(No.41272064)。

第一作者:梁启超,男。矿物材料与绿色建材。578254994@qq.com。

通讯作者:王林江,男,教授,博导。wlinjiang@163.com。

Fourier transform infrared spectroscopy （FTIR）, Transmission electron microscope（TEM） and Atomic force microscope（AFM）. The results shown that Aco-LDH was delaminated into nanosheets with the thickness of 4.5 nm to 20 nm when the concentration is 1mg/mL; The （001） characteristic diffra cion peak of MMT disappeared after mechanical stirring and high speed centrifugation at 5000r/min, which shown that MMT was delaminated. The interlayer space of the compostie is 1.51nm, which is almost the thickness summation of one LDH sheet and one MMT sheet. The composite has sheet-by-sheet assembly structure.

Keywords: layered double hydroxides; montmorillonite; aqueous environment; delaminating; sheet-by-sheet assembly

热模拟法研究 12Cr1MoVG 再热裂纹敏感性

石云哲[1]，成　鹏[1]，朱　平[1]，王淦刚[1]，赵建仓[1]，王鹏飞[2]

(1.苏州热工研究院有限公司；2.河北沧海重工股份有限公司)

摘　要： 采用 Gleeble 热力模拟机对贝氏体组织的 12Cr1MoVG 钢进行焊接热影响区(HAZ)热模拟试验，对模拟 HAZ 组织在不同温度进行回火热处理并进行冲击试验，研究再热裂纹敏感性，验证了敏感温度下再热裂纹的实际存在。冲击试验和断口扫描结果表明，HAZ 组织主要是贝氏体，再热裂纹温度区间为 600℃~650℃。本次研究中最敏感温度点为 600℃。

关键词： 再热裂纹；热模拟

Study on Reheat Cracking Susceptibility of 12Cr1MoVG by Thermo-mechanical Simulator

SHI Yun-zhe[1], CHENG Peng[1], ZHU Ping[1], WANG Gan-gang[1], ZHAO Jian-cang[1], WANG Peng-fei[2]

(Suzhou Nuclear Power Research Institute)

Abstract: The heat affected zone (HAZ) simulation test was done to bainite 12Cr1MoVG specimen, using Gleeble Thermal-mechanical simulation machine. Then, for studying the reheat cracking susceptibility, the impact test has been done to the HAZ structures after different temperatures heat treatment. The existing of reheating cracking was proved by the presence of cracking at sensitive temperature. The result of impact test and fracture analysis showed that HAZ structure is bainite and 5%~8% ferrite, and its reheat cracking sensitive temperature range is 600~650℃ and the most sensitive point is 600℃ in our study.

Key words: reheat cracking; thermal simulation

0 引言

12Cr1MoVG 钢具有较高的热强性能、持久塑性、抗蠕变及抗破断、抗氧化能力，同时具有一定抗腐蚀的能力。工作温度可达 600℃，具有足够的可加工性、焊接性，常用于制作锅炉过热器、再热器等内部构件[1]。

12Cr1MoVG 钢焊接性能良好，但同时也具有一定的裂纹倾向。从成分上看，Cr-Mo-V 钢有一定的沉淀强化，再热裂纹敏感性倾向较大[2]。与此同时，12Cr1MoVG 钢在不同热处理制度下，其材料的金相

第一作者：石云哲(1988—)，男，工学硕士。电站金属焊接接头缺陷修复、焊接热模拟。snpisyz@163.com。

组织存在一定的差异。

　　本文主要针对两种不同组织的 12Cr1MoVG 材料，使用 Gleeble 热力模拟试验机进行焊接热模拟试验，分别进行不同温度回火后，进行冲击韧性试验，得出不同组织对再热裂纹敏感性的影响规律。

1　试验材料和试验方法

1.1　试样材料

　　试样选用的材料，一共分两组，分别取自大厚壁 12Cr1MoVG 管件的外壁和内壁。由于大厚壁管件制造热处理过程中，外壁和内壁降温速度等热处理情况不同，最终形成的组织略有不同。分别为：铁素体＋珠光体，贝氏体＋少量铁素体，如图 1 所示，两组试样编号分别命名为 A、B。两组试样的化学成分如表 1 所示。

表 1　两组试样的化学成分(Wt.%)
Tab.1　chemical composition of two group specimens

编号	C	Si	Mn	Cr	Mo	V	Ni
A	0.116	0.321	0.518	1.11	0.303	0.192	0.185
B	0.129	0.301	0.518	1.11	0.318	0.234	0.124

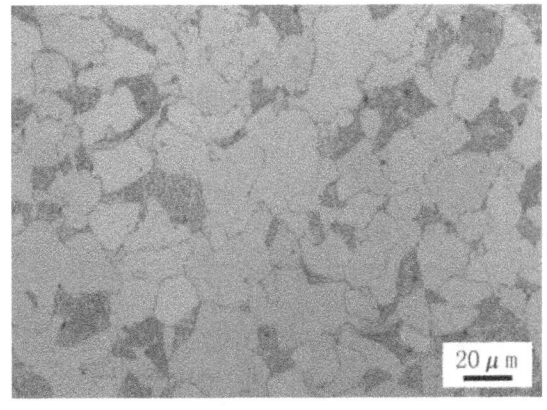 

(a)A 组试样金相组织：铁素体＋珠光体　　　　　(b)B 组试样金相组织：贝氏体＋少量铁素体

图 1　两组试样的金相组织
Fig.1　microstructure of two group specimens

1.2　试验方法

　　HAZ 热模拟试验在 Gleeble 热力模拟试验机上进行，由于实际电站现场焊接中 12Cr1MoVG 管多为大厚壁管件，所以按照 Rykalin-3D 三维传热进行计算。预热温度、峰值温度、焊接线能量均参考电站现场焊接过程中的实际情况进行选取，然后根据传热模型计算出相应的焊接热循环参数。

　　最终选用的 HAZ 热模拟试验的具体参数如表 2 所示。

表2 HAZ 热模拟试验参数

Tab.2 Parameters of HAZ thermal simulation test

参数名称	参数值	参数名称	参数值
密度(g/cm³)	7.860	线能量(KJ/cm)	25
比热容(J/g.℃⁻¹)	0.586	预热温度(℃)	200
热导率(J/cm.s⁻¹.℃⁻¹)	0.452	$t_{8/5}$(s)	14.67
最高加热温度(℃)	1300	加热速度(℃/s)	250
高温停留时间(s)	2		

HAZ 模拟过程中,通过热电偶实测的温度和程序设定曲线闭环回路,实现温度的精确控制。过程中试样的温度测量频率为 100HZ,整个过程中的热循环设定曲线和试样的实测温度曲线见图2。

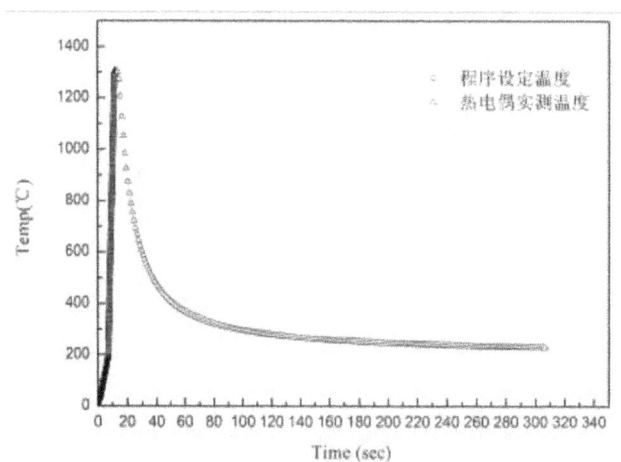

图2 热循环设定和试验实测温度曲线

Fig.2 setting curve and measured curve of Thermal cycling

两组试样经过 HAZ 模拟试验后,分别加热到 475℃、600℃、650℃、690℃、725℃、760℃温度下回火热处理,升温速度为 150℃/h,在回火温度下保温 2h,降温速度为 150℃/h,温度降至 300℃后随炉冷却。

试样经过热处理后,加工成标准 V 型缺口夏比冲击试样,进行冲击韧性试验。

2 试验结果与讨论

2.1 HAZ 模拟试验金相组织观察

两种组织的试样经过 HAZ 模拟试验后,分别对其金相组织进行观察。微观组织显示,母材为铁素体＋贝氏体的 12Cr1MoVG 试样 HAZ 组织为贝氏体＋少量铁素体,铁素体含量约为 5%~8%,如图3所示;母材为贝氏体＋少量铁素体的 12Cr1MoVG 试样 HAZ 组织为贝氏体,如图4所示。

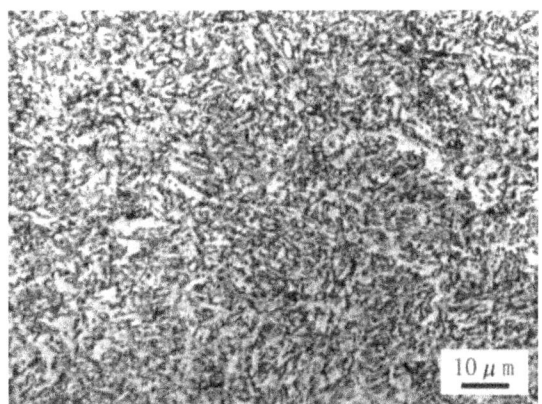

图 3　HAZ 组织：贝氏体+少量铁素体
Fig.3　HAZ structure: bainite and a little ferrite

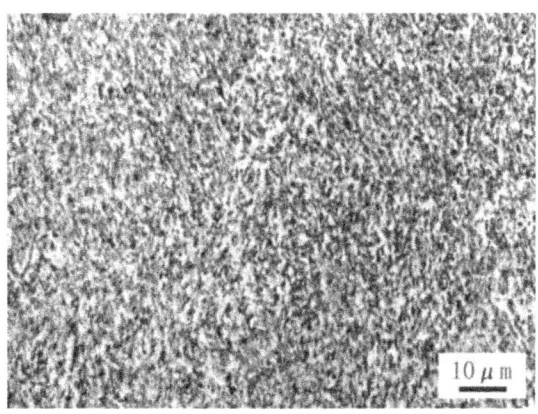

图 4　HAZ 组织：贝氏体
Fig.4　HAZ structure: bainite

2.2 冲击韧性试验结果

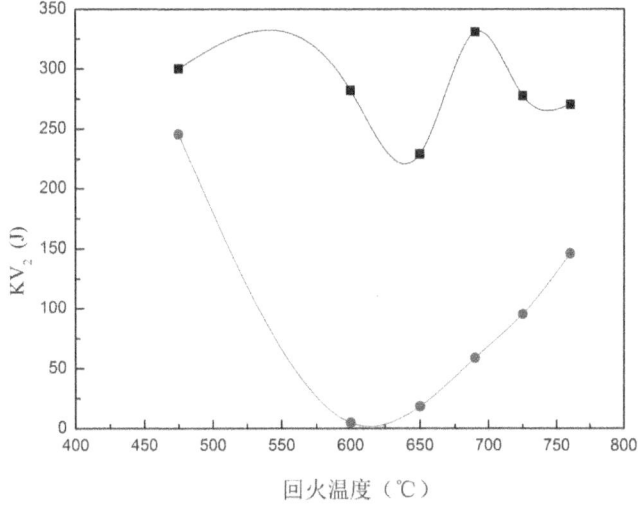

图 5　冲击韧性试验统计结果
Fig.5　statistical result of Impact toughness test

从图5可以看出,A组试样整体比B组试样在各个温度韧性相对要好，即贝氏体＋少量铁素体HAZ组织具有相对更好的防止再热裂纹的能力。

A组试样不存在明显的敏感温度区间,在475℃~760℃范围内,其冲击韧性均处于较高水平,其冲击吸收功均大于200J;B组试样的敏感温度区间为600℃~650℃,其中600℃回火状态下,HAZ粗晶区的韧性最差。回火温度升高到690℃时,HAZ冲击韧性又有所提高。

2.3 断口扫描分析

针对A、B两组试样的冲击试验结果，分别选择600℃、650℃、725℃三个温度下的冲击试样断口进行微观扫描电镜,对断裂机理进行分析研究。

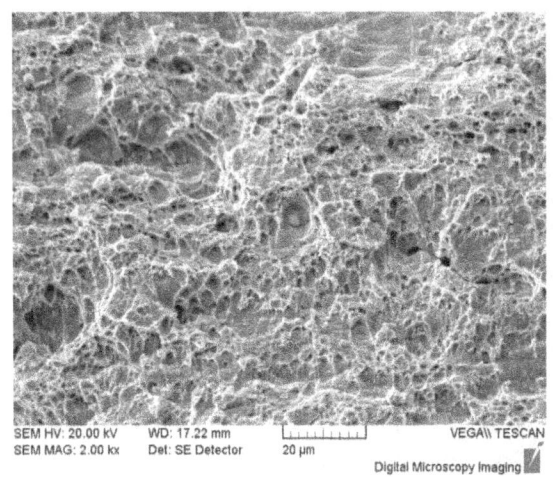

(a)A组试样断口形貌   (b)B组试样宏观形貌

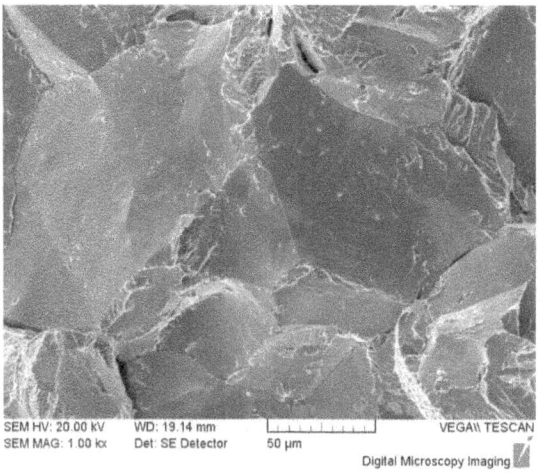

(c)B组试样断口形貌

图6 600℃试样断口扫描分析

Fig.6 SEM analysis of 600℃ specimens

A组试样属于韧性断裂,剪切唇非常明显,纤维区面积大,放射区几乎看不到。分析图6(a)扫描电镜结果可以看出,为典型的韧窝断裂[3]。

B组试样宏观断口细齐,有金属光泽,没有纤维区出现,放射区为主,几乎没有剪切唇,是典型的脆性断裂,如图 6(b)所示。对扩展区进行观察,几乎没有韧性断裂的特征,可到看到非常完整的晶粒,为典型的沿晶断裂,如图 6(c)所示。

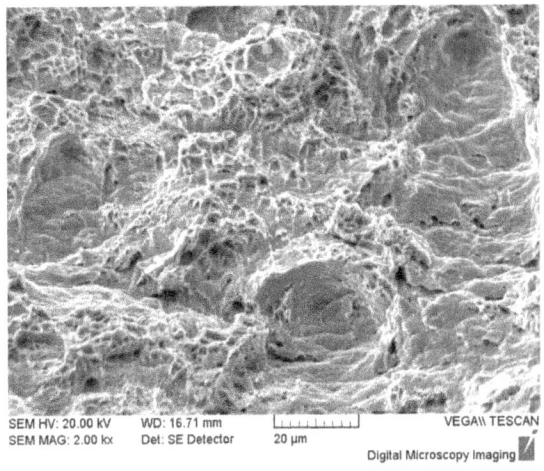 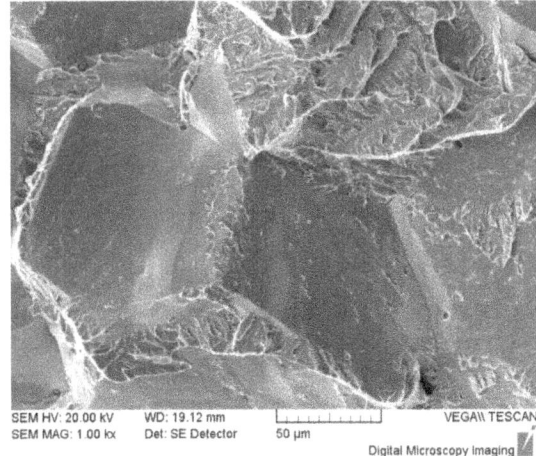

(a)A 组试样断口形貌　　　　　　　　　　　(b)B 组试样断口形貌

图7　650℃试样断口扫描分析

Fig.7　SEM analysis of 650℃ specimens

650℃ 回火后冲击试样,A 组依然是典型的韧性断裂,剪切唇非常明显,纤维区面积很大。对启裂区和扩展区观察发现,仍为韧窝形貌,但比 600℃ 回火冲击试验韧窝相对较少

因此,650℃ 回火的韧性比 600℃ 回火的韧性要低。

B 组试样断口宏观整体比较平齐,有一定金属光泽,放射区为主,几乎没有剪切唇,也是典型的脆性断裂。观察扩展区可以看到完整的晶粒,扩展区仍为典型的沿晶断裂。

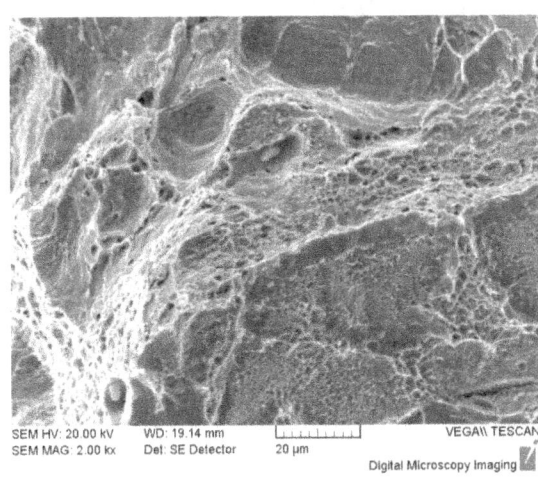 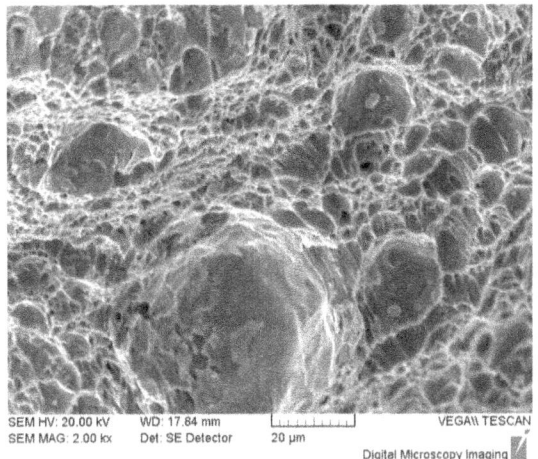

(a)A 组试样断口形貌　　　　　　　　　　　(b)B 组试样断口形貌

图8　725℃试样断口扫描分析

Fig.8　SEM analysis of 725℃ specimens

对 725℃ 回火 A 组和 B 组冲击试样进行观察。A 组依然是典型的韧性断裂。扩展区形貌为典型的韧窝,比 650℃ 时的扩展区具有更多更典型的韧窝,因此表现出更好的冲击韧性。如图 8(a)所示。

B 组试样断口宏观比较粗糙,启裂区和扩展区有沿晶断裂的倾向,在扩展区有少量的韧窝存在,如图 8(b)所示。因此 725℃回火状态下,韧性比 600℃、650℃要高。

不同回火温度冲击试样的微观观察结果,直观反映了晶界形变能力的不同,也反映了对再热裂纹的敏感程度。

焊接接头的再热裂纹与材料在一定温度下的沉淀硬化有关，也与接头金属在一定温度下蠕变塑性低、不足以适应由应力松弛而引起的附加变形有关。金属晶界区形变能力降低是产生再热裂纹的前提,材料对应于不同温度都有相应的变形能力。当外加形变量小于变形能力时,材料是安全的,但当材料变形能力小于外加变形量时,出于不能顺应变形要求就会发生破坏。

不同回火温度冲击试样的微观观察结果,直观反映了晶界形变能力的不同,也反映了对再热裂纹的敏感程度。

A 组试样 HAZ 组织为贝氏体 +5%~8%铁素体,B 组试样 HAZ 组织为贝氏体。A 组试样整体比 B 组试样在各个温度韧性相对要好，即贝氏体+ 5%~8%铁素体 HAZ 组织具有相对更好的防止再热裂纹的能力。

A 组试样均为韧性断裂,从启裂区和扩展区都能观察到典型的韧窝形貌，表明 HAZ 处的贝氏体和铁素体组织在经过不同温度回火,虽然晶粒强度升高,但晶界处强度依然很高,具有足够的形变能力,抵抗再热裂纹的能力比较高。不存在明显的敏感温度区间,在 475℃~760℃范围内,其冲击韧性均处于较高水平,其冲击吸收功均大于 200J。但在 650℃回火时的启裂区和扩展区的韧性明显相对较低,韧窝也相对较少,该温度下的冲击韧性值也相对最低,表明该温度下晶界处强度相对最弱,随着回火温度升高,强度又逐渐恢复。

B 组试样在 475℃回火下的冲击韧性值也非常理想;然而在 600℃、650℃时的断裂均为脆性断裂,宏观断口比较齐平,有金属光泽,微观观察发现均为典型的沿晶断裂,晶界处有一定的析出物。这符合再热裂纹的发生和断裂机理，也说明 HAZ 出的纯粒状贝氏体组织在 600℃、650℃时的晶界处强度非常低,无法满足晶界形变的需求,在外加变形的需求下,晶界非常容易产生滑移,从而产生开裂。在 725℃时,晶界有沿晶脆断的趋向,但有少量韧窝存在,冲击韧性值也相对较高,晶界处强度逐渐恢复。说明贝氏体组织在回火过程中,晶界处强度对回火温比较敏感,尤其在 600℃~650℃时,晶界强度非常低,非常容易产生再热裂纹缺陷,因此 600℃~650℃为 B 类试样的敏感温度。

3　结论

(1)HAZ 组织中 5%~8%的铁素体能降低再热裂纹的敏感性。贝氏体 HAZ 组织再热裂纹敏感性比较高。

(2)本试验中,贝氏体 HAZ 组织敏感温度范围为 600℃~650℃。

参考文献:

[1] 姜求志,王金瑞,等. 火力发电厂金属材料手册 [M]. 北京: 中国电力出版社,2001.

[2] 陈忠兵,赵彦芬,赵建仓,等. 厚壁 12Cr1MoVG 钢焊接接头裂纹分析及其控制 [J]. 中国电机工程学报,2012,32(35): 137-143.

[3] 钟群鹏,赵子华. 断口学 [M]. 北京: 高等教育出版社,2006.

粘土矿物在有机溶剂中的高产率剥离及蒙脱石／层状双氢氧化物复合材料制备

陈　耀[1]，梁启超[1]，谢襄漓[2]，王林江[3,4]

(1.桂林理工大学 材料科学与工程学院；2.桂林理工大学 化学与生物工程学院；
3.广西有色金属及特色材料加工国家重点实验室培育基地；4. 有色金属清洁冶炼与综合利用广西高校重点实验室)

摘　要：本文用离子交换法制备了十六烷基三甲基溴化铵插层的蒙脱石，用水热法制备了十二烷基磺酸钠插层的层状双氢氧化物。在此基础上，以氯仿为剥离介质，通过机械搅拌和超声处理，分别获得了蒙脱石和层状双氢氧化物的层板剥离溶胶，再将二者进行插层组装，制备出蒙脱石/层状双氢氧化物复合材料。用X射线衍射分析、傅里叶红外光谱、扫描电子显微镜、原子力显微镜等测试手段对样品的晶型和微观形貌进行了表征。结果表明，在粘土浓度为2mg/ml时，超声处理20min后，蒙脱石的(001)晶面以及层状双氢氧化物的(003)和(006)晶面的衍射峰基本消失，层状结构被破坏，剥离蒙脱石的片层厚度为8nm左右，层状双氢氧化物的片层厚度降低到10nm左右。组装后的蒙脱石/层状双氢氧化物复合材料特征峰的d值达到2.52nm，且兼有蒙脱石和层状双氢氧化物的特征峰，形成蒙脱石层板和层状双氢氧化物层板交替排列结构。

关键词：蒙脱石；层状双氢氧化物；层板剥离；组装

Efficient Exfoliation of Clay Minerals in Organic Solvent and Preparation of Montmorillonite/layered Double Hydroxide Composites

CHEN Yao[1], LIANG Qi-chao[1], XIE Xiang-li[2], WANG Lin-jiang[3,4]

(1. College of Materials Science and Engineering, Guilin University of Technology;　2.College of Chemistry and Bioengineering, Guilin University of Technology; 3. Ministry-province Jointly-constructed Cultivation Base for State Key Laboratory of Processing for Non-ferrous Metal and Featured Materials; 4. Guangxi Key Laboratory of University for Clean Metallurgy and Comprehensive Utilization of Non-ferrous Metal Resources,)

Abstract: In this paper, hexadecyl trimethyl ammonium bromide was intercalated into montmorillonite by ion exchange method, sodium dodecyl sulfate was intercalated into layered double hydroxides by a hydrothermal method. Montmorillonite and layered double hydroxides were respectively exfoliated through mechanical stirring and ultrasonic process on this basis, using chloroform. Then montmorillonite/layered double hydroxide composite materials was synthesized

基金项目：国家自然基金(No.41272064)。
第一作者：陈　耀,男。矿物材料与绿色建材。sdcyao168.com。
通讯作者：王林江,男,教授,博导。wlinjiang@163.com。

through inserting LDHs sheets into montmorillonite. The crystal structure and microstructure of the samples were characterized by X-ray diffraction (XRD), Fourier transform infrared(FT-IR), Scanning electron microscopy (SEM) and Atomic force microscope (AFM). Results show that, when the concentration of clay minerals in dispersion is 2 mg/ml, after sonicated 20 min, the diffraction peak of (001) plane and (003)(006) plane disappeared, indicating that the layered structure of ontmorillonite and layered double hydroxide were destroyed. And the nano-sized pieces of montmorillonite and layered double hydroxide were exfoliated with a thickness of 8 nm and 10 nm, respectively. The lattice spacing of composites increased to 2.52nm. The X-ray diffraction pattern of the precipitate shows Bragg reflections corresponding to both montmorillonite and layered double hydroxide indicating that the structure of composites was alternately stacked layer-by-layer.

Key words: montmorillonite; layered double hydroxide; platelets exfoliated; assemble

Al-5Zn-0.03In-xEr 模拟阳极合金组织与性能研究

李　航，魏　兵，许征兵*，曾建民

（广西大学 有色金属材料及其加工新技术教育部重点实验室）

摘　要：熔炼得到了不同 Er 含量的 Al-5Zn-0.03In-xEr（x=0,0.2,0.4,0.6,0.8,1,4,7）阳极合金，利用扫描电镜(SEM)、能谱(EDS)对合金的物相、组织进行分析和观察，利用极化曲线和电化学阻抗谱(EIS)研究不同成分合金在 3.5%NaCl 溶液中的电化学性能。对极化曲线拟合得到了腐蚀电流密度和腐蚀电位，对阻抗谱进行了等效电路拟合。实验结果表明：随着 Er 含量的增加，铝阳极合金晶粒得到细化，枝晶间析出相增多；Er 含量低于 1%时，合金开路电位和腐蚀电位变化不大，腐蚀电流密度较大，Er 含量超过 1%时，电位正移，腐蚀电流密度降低。阳极极化程度较小，未出现钝化现象；EIS 谱中阻抗弧半径增大，腐蚀速度减缓，阳极合金在中高频扫描区域出现容抗弧，低频出现感抗弧，Er 元素的添加只改变阻抗的大小，并不改变其类型，其电化学动力学机制不变，阳极合金的溶解由孔蚀引起。

关键词：铝阳极合金；电化学阻抗；极化曲线；拟合

Research on Microstructure and Properties of Simulative Al-5Zn-0.03In-xEr Anode Alloy

LI Hang, WEI Bing, XU Zheng-bing*, ZENG Jian-min

(Key Lab. of Nonferrous Materials and New Processing Technology, Guangxi University)

Abstract: Al-5Zn-0.03In-xEr（x=0,0.2,0.4,0.6,0.8,1,4,7)anode alloys with different Er content were melted. The microstructure and phase studies were carried out by means ofscanning electron microscope (SEM) and energy dispersive (EDS). The electrochemical property of alloys in 3.5% NaCl solution were investigated with Tafel polarization and electrochemical impedance spectroscopy (EIS). Corrosion current density and corrosion potential were obtained by fitting of Tafel polarization curve and the EIS was simulated by using the equivalent circuit model. The results show that with the increase of Er content, the grain was refined, the amount of inter-dendritic precipitates increased. When the content of Er was less than 1%, the open circuit potential and corrosion potential of the alloy didn't change a lot and corrosion current density was large . However, when the Er content exceeded 1%, the potential became positive and the corrosion current density decreased. The polarization degree of anodes were not strong, and the radius of electrochemical impedance spectroscopy enlarged and corrosion rate also decreased. The capacitive loop appeared at the range of high frequency and inductive loop at low frequency. The addition of Er only effected the size of EIS but not the electrochemical dynamics mechanism. The dissolution of alloys were caused by pitting.

Key words: aluminum anode alloy; tafel polarization curve; EIS; simulation

基金项目：国家自然科学基金资助（项目号51401057）；广西自然科学基金资助（2013GXNSFBA019252）；广西教育厅科研项目资助（No. 2013YB011）；新型铝合金材料及先进加工理论与技术系统性课题资助（桂科能 13-051-02,13-A-02-03）；"铝合金材料先进加工技术"八桂学者专项经费资助。

第一作者：李 航，男。有色金属材料加工。skylihang@163.com。
通讯作者：许征兵，男。有色金属材料加工理论与工艺，。51happiness@163.com。

超音速电弧喷涂 Ti-Al 涂层的组织与性能研究

魏 兵，李 航，许征兵*，曾建民

(广西大学 有色金属材料及其加工新技术教育部重点实验室)

摘 要：本文采用超音速电弧喷涂技术，在 45# 钢喷涂 Ti-Al 粉芯丝材，分别用不同电弧喷涂工艺参数获得相应的涂层。通过涂层的组织结构分析、硬度和结合强度等测试研究电弧喷涂工艺参数对涂层的影响，并对电弧喷涂工艺参数进行了优化。结果表明，涂层的显微组织呈典型的层状结构，层与层之间存在少量的夹杂氧化物和孔隙。确定优化后的工艺参数为喷涂电流 220A，喷涂电压 25V，涂层厚度 0.6mm。随着喷涂电流的增加，涂层硬度增大，结合强度升高；随着喷涂电压增加，涂层硬度减小，结合强度降低；并且随着涂层厚度的增加，结合强度呈下降趋势，其最大结合强度为 32.8MPa。

关键词：高速电弧喷涂；组织结构；结合强度

Study on Microstructure and Properties of High-velocity Arc Spraying Ti-Al Coating

WEI Bing, LI Hang, XU Zheng-bing, ZENG Jian-min

(Key Lab. of Nonferrous Materials and New Processing Technology, Guangxi University)

Abstract: In this study, the coating were deposited on 45 steel substrate by high-velocity arc spraying with cored wires. They were sprayed by using different arc spraying technological parameters to obtain the corresponding coating respectively. Effects of arc spraying technological parameter on the coating were studied by the microstructure analyzing, hardness and bonding strength testing and the arc spraying process was optimized by orthogonal experiment design. The results show that the microscopic structure of coatings appears typical lamellar structural feature between which a small quantity of oxides and porosities exist. The optimal parameters of the arc spraying were spraying current 220A, spraying voltage 25V, coating thickness 0.6mm. With the increasing of the spraying current, the hardness and bonding strength of the coating increase. The increasing of the spraying voltage could decrease the hardness and bonding strength of the coating. The increase of coating thickness decreases the bonding strength. The highest bonding strength of the coating is 32.8MPa.

Keywords：high-velocity arc spraying; microstructure; bonding strength

基金项目:南宁市科学研究与技术开发项目(项目号 GX20140260)；"铝合金材料先进加工技术"八桂学者专项经费资助。
第一作者:魏 兵,男。表面工程。weibing0917@163.com。
通讯作者:许征兵,男。有色金属材料加工理论与工艺。51happiness@163.com。

核电压力容器锻件热处理过程的组织转变规律

李传维，韩利战，刘庆东 ，骆晓萌，顾剑锋 *，张伟民

(上海交通大学 材料科学与工程学院材料改性与数值模拟研究所)

摘　要：对核电压力容器大型锻件,采用热模拟,结合 OM、SEM 和硬度等表征手段,研究该材料连续冷却过程中的组织转变规律。结果表明,当冷速小于 1℃/s 时,先共析铁素体中的碳扩散至未转变的奥氏体中并在低温转变为下贝氏体马氏体岛状组织;当冷速在 1℃/s~5℃/s 范围时,过冷奥氏体先后分解转变为粒状贝氏体,上贝氏体和下贝氏体;过冷奥氏体在冷速为 5℃/s~10℃/s 范围内形成贝氏体铁素体和马氏体组织。核电压力容器用钢在比较慢的冷速条件下仍有马氏体相变形成孤岛状马氏体,这主要与奥氏体向铁素体转变时碳在未转变的奥氏体中富集有关。本文还研究了不同淬火状态在650℃、5h 回火后的组织并与实际锻件取样位置进行了对比，证实采用热模拟方法研究核电压力容器锻件的的组织转变规律是可信的。

关键词：核电压力容器用钢；连续冷却；相变规律

The Microstructure Evolution Mechanism of Reactor Pressure Vessel Forging During Heat Treatment

LI Chuan–wei, HAN Li–zhan, LIU Qing–dong, LUO Xiao–meng, GU Jian–feng, ZHANG Wei–min

(Institute of Materials Modification and Modeling, School of Materials Science and Engineering, Shanghai Jiao Tong University)

Abstract: The thermal simulation combined with OM, SEM and hardness methods were used to investigated microstructure transformation of thenuclear reactor pressure vessel (RPV)heavy forgings in heat treatment,particularly, the continuous cooling transformation of quenching process. The results show that at the cooling rates of 1℃/s or below, ferrite initially formed and continuously rejecting C into the untransformed austenite, which transforms to C-rich lower bainite at lower temperature, resulting in ferrite-bainite dual-phase microstructures. At the cooling rates between 1℃/s and 5℃/s, the successive transformation products are bainite ferrite, upper bainite and lower bainite. The upper bainite and martensite dual-phase microstructures are formed at the range of 5℃/s to 10℃/s with a lower Ms. The martensite islands were found in ferrite and lower bainite at lower cooling rates in RPVsteel due to the carbon reject to the untrasformed austenite. The microstructures of different cooling rates after tempering at 650℃, 5h and compared with the sampling point of the heavy forging. It is confirmed that the thermal simulation methods is the reliable for predicted the RPV forging microstructure.

Key words: reactor pressure vessel steel, continuous cooling, phase

基金项目:国家重点基础研究发展计划资助项目(2011CB012904);中国博士后科学基金(2013M541517)。

第一作者:李传维,男。大型锻件热处理及其数值模拟。li–chuanwei@sjtu.edu.cn。

通讯作者:顾剑锋,男。材料改性及数值模拟。gujf@sjtu.edu.cn。

0 引言

核电材料通常基于安全性的考虑多采用保守的材料设计[1,2],在目前第三代商用核电反应堆压力容器设计中仍采用综合性能优异的 SA 508 Gr.3 钢,这种材料在核电压力容器上的应用已有近 40 年的历史[3]。随着核电安全的要求越来越高,核电压力容器锻件呈现出大型化、一体化的趋势,某些些锻件调质前的重量可达到进 500 吨,壁厚达到近 300mm,这给核电压力容器锻件的制造提出了新的挑战。

在过去,不同的研究者对大型锻件冷却过程的组织转变规律进行了研究,H.pous-rpmero 研究了改材料的奥氏体晶粒长大规律及其连续冷却过程中的组织转变规律[4, 5]。K. D. Haverkamp[6]研究了不同壁厚位置在不同时间的冷却速度并建立了组织和性能分布,结果表明,在不同取样位置,由于冷却速度各异,造成组织差异较大,最终的性能沿着壁厚的分布呈开口向上的抛物线分布,即越远离壁面,材料的强度越低,冲击韧性越低,韧脆转变温度越高。T. Enami[7, 8]的研究结果也得到了相似的结论。尽管这些研究已经得到定量的关系,但大多数研究都是从材料研究的角度,并没有和实际锻件的组织分布建立联系。

采用热模拟的方法来研究大型锻件的组织和性能分布是一种经济高效的方法, 它不仅可以实现对大型锻件心部组织的模拟,还能为数值模拟提供可靠性参考,为优化数值模拟方法提供基础,本文采用数值模拟的方法计算锻件在淬火过程中典型取样位置的冷却曲线,然后采用热模拟炉模拟锻件不同取样位置的温度变化历史,观察其淬火后的微观组织形貌,建立沿壁厚的方向的组织分布规律,本文还研究不同取样位置回火后的微观组织分布情况,验证了热模拟方法的准确性,为后续开展锻件的组织分布及性能分布的数值模拟提供了基础。

1 试样制备和试验方法

研究用低碳低合金 SA508 Gr.3 钢的化学成分如表 1 所示,初始组织为其正火 + 回火组织,其基本组成为铁素体和贝氏体,其中铁素体含量约 54%;平均原奥氏体晶粒度为 25μm 左右,采用膨胀实验测得的 Ac1 温度和 Ac3 温度分别为 702℃和 834℃ 。

表 1 实验钢的化学成分(质量分数,%)

Tab.1 chemical composition of the investigated steel (wt.%).

Steel	C	Mn	Ni	Mo	Cr	Si	Cu	Al	Fe
SA508 Gr.3	0.17	1.41	0.82	0.51	0.13	0.17	0.03	0.011	Bal.

本文以核电压力容器封头为研究对象,首先采用本研究所开发的 THERMAL-PROPHET 软件[9]计算了典型取样位置在淬火冷却过程中的冷却曲线,计算其在转变温度区间的平均冷却速率,根据计算得到的冷速区间,设计模拟实验选用的冷却速度,采用 Gleeble 热模拟试验机模拟其不同取样位置的冷却过程,为了缩短实验时间,其奥氏体化过程均以 60℃/min 加热到 900℃,保温 5h,然后以不同的速度冷却至室温,模拟其淬火过程。为了实际锻件调质后的组织进行对比分析,在观察其室温组织后,还在 650℃回火 5h 模拟其回火过程并观察其回火后的室温组织。微观组织观察采用金相,扫描电子显微

镜相结合的方法,样品采用机械抛光的方法制备,腐蚀剂为 4%的硝酸酒精溶液。

2 试验结果与讨论

2.1 锻件不同部位冷却速度的确定

图 1a 为选用锻件的锻造成形后的截面图,采用本研究所开发的 THERMAL-PROPHET 软件结合 MSC. Marc 计算了淬火冷却过程中的温度分布,并绘制其典型位置的冷却曲线(图 1b),从图 1b 中可以计算其在转变温度区间的平均冷却速率约为 0.1℃/s~5℃/s。为了保证实验结果在更宽的冷速范围内适用,本文设计模拟实验选用的冷却速度为 0.05℃/s~10℃/s。

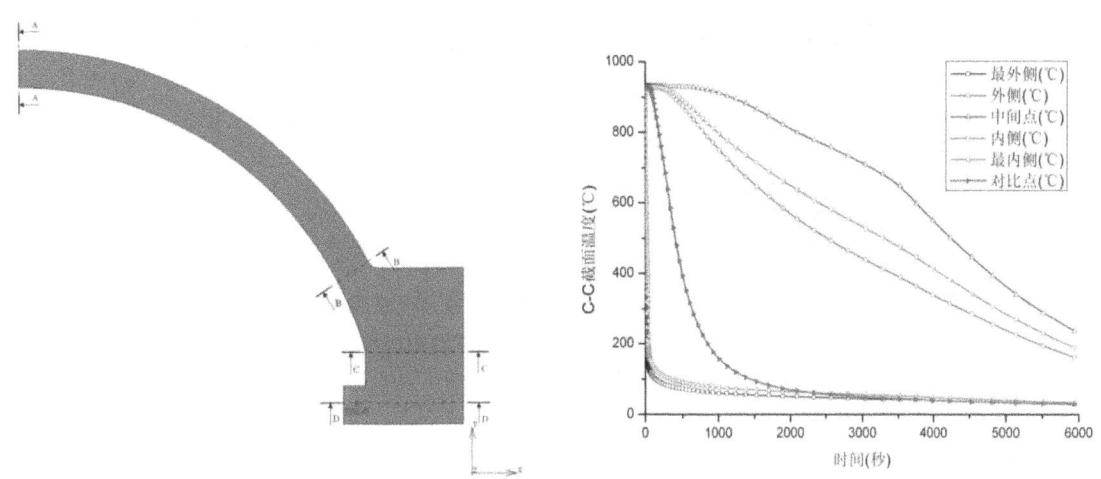

图 1 所取锻件的截面及其 C-C 面的冷却速率

Fig.1 The 1/2 cross section graphof the forging in (a) and (b)is the cooling rate distribution of C-C section.

2.2 模拟淬火后的组织

图 2、图 3 是 SA508 Gr.3 钢不同冷速的金相和扫描电镜照片。从图中可以看出在不同冷却速度下,奥氏体发生分解转变后得到的的组织差别较大,在以 0.05℃/s~0.1℃/s 冷却时,结合图 2a、b 可以发现,淬火的室温组织主要为先共析铁素体(F)和退化的珠光体组织(P)以及下贝氏体组织(BL)的混合组织,除此以外,还有少量富碳的马氏体组织存在(图 3a)。在以 0.5℃/s 冷却时,金相组织中仍然有少量的先共析铁素体(图 2c),此时先共析铁素体体积较小,主要的组织依然为铁素体(F)和贝氏体组成,并没有发现珠光体转变。当冷速增加到 1℃/s 冷却时(图 2d,图 3b),主要发生贝氏体转变,奥氏体分解后得到的室温组织为粒状贝氏体组织,同时有部分奥氏体在贝氏体转变区未来得及转变被保留到马氏体点以下,发生马氏体转变形成少量马氏体组织或 M/A 岛状组织。在冷却速度增加至 5℃/s 时(图 2e,图 3c),主要得到贝氏体组织,但此时的贝氏体组织与 1℃/s 冷却组织相比更为细小。当冷速为 10℃/s 时(图 2f,图 3d),已能观察到明显的马氏体组织,得到的室温组织为上贝氏体和马氏体的混合组织。随着冷却速度的提高(从锻件心部至表面),组织分布依次为,铁素体+退化珠光体+贝氏体→铁素体+贝氏体→贝氏体→贝氏体+马氏体,其硬度从 210.23 HV_{10} 逐渐增大 294.97 HV_{10},这与文献报道[5]中的结果一致。与这些结果稍有不同的是,本文发现在马氏体转变在所有的冷却速度下均存在,

在较慢的冷速下,马氏体往往以岛状组织出现,这些岛状组织中碳的富集,使得其转变温度低于 Ms 点,从而得到马氏体组织,由于高碳的马氏体组织硬度大,脆性高[10],这些马氏体组织对材料的性能有何种影响,还需进一步深入的研究。

2.3 回火后的组织

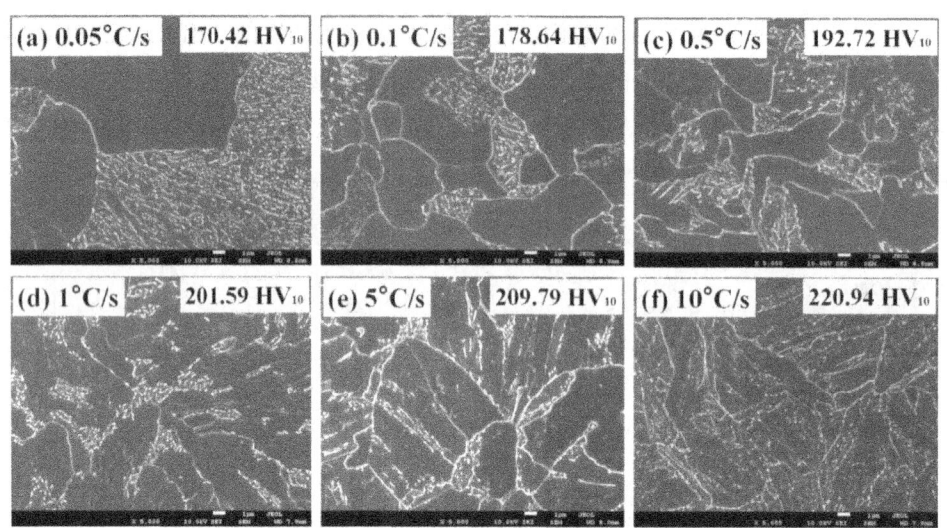

图2 不同冷速淬火后得到的室温组织金相照片

Fig.2 The OM micrographs of specimenunderdifferentcooling rate states.

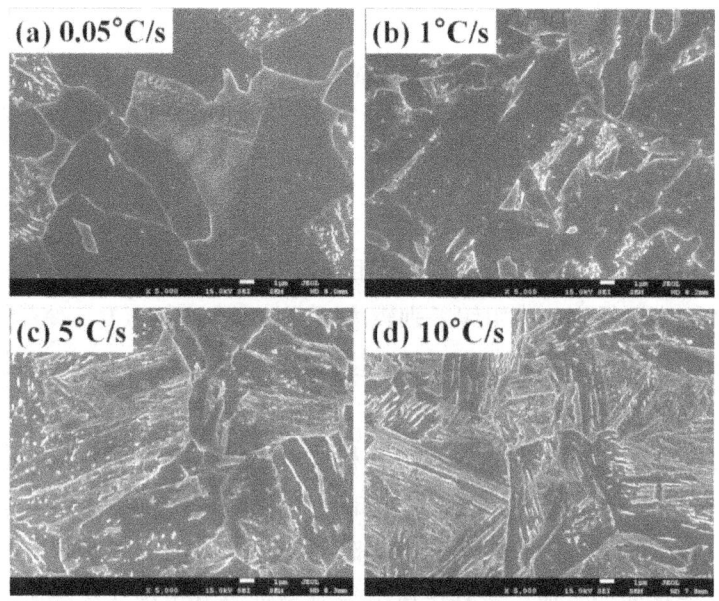

图3 不同冷速淬火后得到的室温组织 SEM 照片

Fig.3 The SEM micrographs of sampleunderdifferentcooling rates.

回火后的组织如图 4 所示,从图中可以看出,尽管经过 650℃ 5h 回火,淬火遗留下的组织特征依然明显,特别是在较慢冷速下形成的先共析铁素体,除了有少数分布均匀的析出物外,并无其他变化,原贝氏体组织经过回火后碳化物分布均匀,在 0.5℃/s 冷却后的组织中也发现了同样的情况。经过 1℃

/s 冷却得到的粒状贝氏体中的主要变化是贝氏体板条间的富碳 M/A 岛状组织的分解转变，这些碳化物堆积在原 M/A 区域，与较慢冷速下得到的先共析铁素体不同，由于贝氏体铁素体中碳的过饱和度较高[11]，在贝氏体铁素体中也能发现少量的碳化物。冷速为 5℃/s 和 10℃/s 时，组织中的碳化物分布趋于均匀，特别是 10℃/s 时，组织中的碳化物均匀细小，弥散分布在整个视场内。随着淬火冷速的提高，材料在回火后的硬度也逐渐提高，其中硬度最大值出现在 10℃/s，约为 220.94HV$_{10}$ 这比最小值大了约 50HV$_{10}$，这说明冷速对材料的性能影响较大。

2.4 模拟组织与锻件实际组织的对比

由于实际锻件质量大，浇注温度高，不可避免的会出现不同程度的偏析，为了将模拟结果与实际锻件的组织分布对比，本文分别对正、负偏析区的组织进行了硬度实验和扫描电镜分析。图 5a 为偏析的金相照片，中间用红色半透明标示的竖直方向的带状区域为正偏析区，宽度约为 700μm，两侧为负偏析区。图中蓝色线表示对应位置的硬度分布，可见正偏析区的硬度大于负偏析区，约为 240 HV$_{0.2}$，负偏析区的硬度约为 205 HV$_{0.2}$。对正、负偏析区进行大载荷的硬度分析，其硬度稍小，约为 236.37 HV$_{10}$ 及 194.64 HV$_{10}$，这与图 4c、d 模拟实验结果相近。正、负偏析区组织的扫描电镜照片如图 4b、c 所示，可见实际锻件的奥氏体晶粒和组织较为粗大，其负偏析区的组织隐约可见为贝氏体组织，与图 4c、d 中组织相近；其正偏析区的组织为下贝氏体的回火组织，其板条依然可见，其硬度较高，未能在图 4 中找到对应的组织。

实际锻件中的组织不均匀，是由于其中的合金元素偏析造成的，在文献[7,12]中采用 EPMA 的方法对正负偏析区的合金元素浓度分布进行了分析，结果表明正偏析区的 Mn、Mo、Ni 含量均偏高，使得其转变温度降低，获得的组织较为细密。由于热模拟实验没有考虑实际偏析的情况，若排除偏析的影响，可见该取样位置的实际冷速约为 0.5℃/s ~1℃/s，这佐证了数值模拟所预测的取样位置的冷却速度是可信的，也说明采用热模拟的方法对锻件的组织预测结果是符合实际情况的。推而广之，该锻件心部的冷速约为 0.1℃/s，可见其组织约为图 4a、b 所示，为块状铁素体和下贝氏体的回火组织，其硬度约为 180HV10 左右。

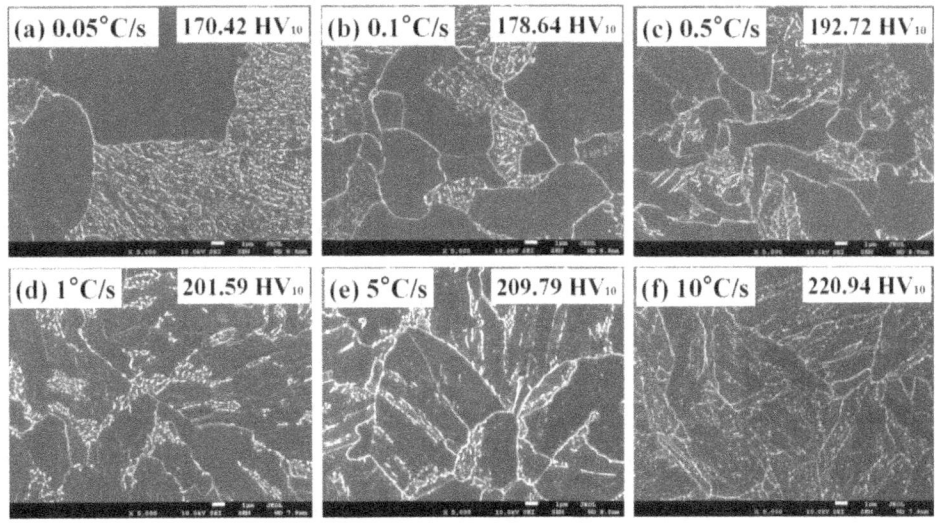

图 4　不同冷速淬火后得到组织经过模拟回火后的组织

Fig.4　The SEM micrographs of differentcooling rates sample undergoes 650℃, 5h tempering.

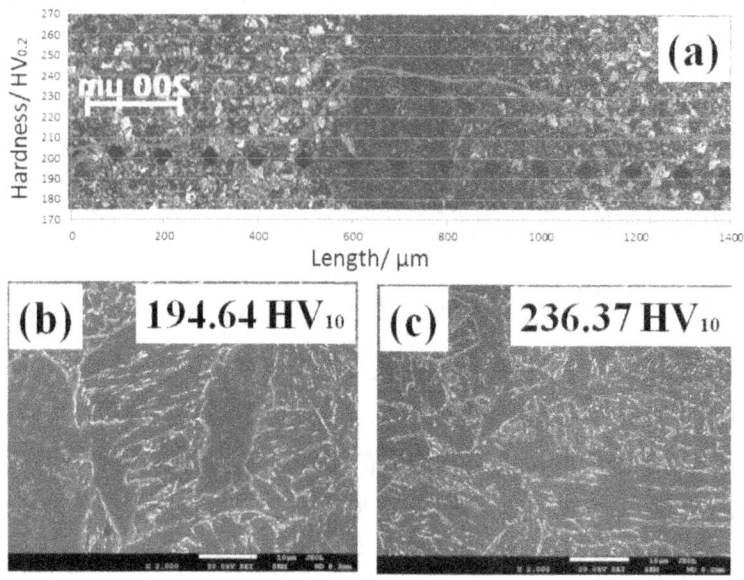

图 5　实际锻件取样位置的金相组织及硬度分布
(a)金相照片和硬度分布　(b)正偏析区的组织　(c)负偏析区的组织
Fig.5　microstructure and hardness distribution of the sampling point in forging contain some macrosegregation.

3　结论

　　通过数值模拟的方法,预测了锻件淬火冷却过程的温度分布,并提取最大截面处的冷却曲线,获得了锻件冷却过程中的冷却速率范围,采用热模拟方法获得了不同冷却速率下的室温组织,并模拟回火工艺,观察回火后的组织并与实际锻件的组织进行了对比,结果表明,锻件表层的冷却速度约为6℃/s,心部(1/2T)的的冷却速度约为0.1℃/s,取样位置(1/4T)冷却速度约为0.5~1℃/s;淬火后的组织分别为贝氏体与马氏体混合组织,粒状贝氏体组织,多边形铁素体与贝氏体的混合组织;回火后这些组织依然可以辨认, 其中富碳部分回火转变较为明显, 铁素体组织基本保持原貌;回火后的硬度分布在170.42 HV$_{10}$ 至 209.79 HV$_{10}$ 之间,这与实际锻件调质后的结果基本一致,说明采用热模拟方法能够较好的体现实际锻件的组织分布。

参考文献

[1] Zinkle, S. J. and G. S. Was.Materials challenges in nuclear energy　[J]. Acta Materialia, 2013. 61 (3): p. 735–758.

[2] Yvon, P. and F. Carré. Structural materials challenges for advanced reactor systems　[J]. Journal of Nuclear Materials, 2009. 385(2): p. 217–222.

[3] Z. Q. Sheng, H. X., F. peng. The microstructures os A508–3 steel after heat treatment　[J]. Nuclear Power Engineering, 1988. 9(1): p. 49–53.

[4] Pous–Romero, H., et al.. Austenite grain growth in a nuclear pressure vessel steel　[J]. Materials Science and Engineering: A, 2013. 567(0): p. 72–79.

[5] Pous–Romero, H. and H. K. Bhadeshia. Continuous Cooling Transformations in Nuclear Pressure

Vessel Steels [J]. Metallurgical and Materials Transactions A, 2014. 45(11): p. 4897-4906.

[6] Haverkamp, K., et al.. Effect of heat treatment and precipitation state on toughness of heavy section Mn-Mo-Ni-steel for nuclear power plants components [J]. Nuclear engineering and design, 1984. 81 (2): p. 207-217.

[7] Ohashi, N., et al., Manufacturing process and properties of nuclear RPV shell ring forged from hollow ingot [J]. Nuclear engineering and design, 1984. 81(2): p. 193-205.

[8] Enami, T., et al.. Effects of cooling rates and tempering conditions on the strength and toughness of Mn--Ni--Mo, Cr--Mo steel plates, 1974, Kawasaki Steel Corp., Chiba, Japan.

[9] Xiao-hui,, L., et al.. Development and application of heat treatment simulation software: Thermal Prophet [J]. Heat Treatment of Matels, 2012. 37(2): p. 115-118.

[10] Lan, L., et al.. Correlation of martensite - austenite constituent and cleavage crack initiation in welding heat affected zone of low carbon bainitic steel [J]. Materials Letters, 2014. 125: p. 86-88.

[11] Kim, S., et al.. Correlation of the microstructure and fracture toughness of the heat-affected zones of an SA 508 steel [J]. Metallurgical and Materials Transactions A, 2000. 31(4): p. 1107-1119.

[12] Pickering, E.J.. Macrosegregation in Steel Ingots: The Applicability of Modelling and Characterisation Techniques [J]. ISIJ International, 2013. 53(6): p. 935-949.

形貌可控纳米 ZnO 的电沉积制备

于梅花

(广西大学 材料科学与工程学院)

摘 要：以 $Zn(NO_3)_2 \cdot 6H2O$ 和 KCl 的混合溶液为电解液，采用三电极恒电位体系在 ITO 上电沉积制备了形貌可控的一维 ZnO 纳米材料。通过旋转涂布 $(CH_3COO)_2Zn \cdot 2H_2O$ 溶液，先在 ITO 表面生长了一层 ZnO 种子层。该种子层可以增加 ZnO 的密度，使 c 轴取向更加明显。通过场发射扫描电子显微镜(FESEM)，系统地分析了电沉积参数(电沉积时间、$Zn(NO_3)_2 \cdot 6H_2O$ 浓度、电沉积电压、电沉积温度、热处理温度)对电沉积制备纳米 ZnO 形貌的影响，并研究了其生长机理。

关键词：纳米 ZnO；电沉积；可控制备

Synthesis of ZnO Nanorod Arrays by Electrochemical Deposition Method

YU Mei-hua

(School of Materials Science and Engineering, Guangxi University)

Abstract: ZnO nanorod arrays were synthesized on ITO by electrochemical deposition method using the three-electrode configuration at a constant potential, which the mixture of zinc nitrate and potassium chloride was used as the liquid electrolyte. By spin coating zinc acetate solution to prepare ZnO seeds and then deposit ZnO nanorod arrays on the seeds layer, which increased the density of ZnO nanostructure and improved the preferred growth along the c axis. The effect of electrodeposition parameters (electrodeposition time, the concentration of electrolyte, electrodeposition voltage, electrodeposition temperature, heat treatment temperature) on the morphology had been investigated by field emission scanning electron microscope (FESEM) systematically and the growth mechanism had been analysized in details.

Keywords: nano ZnO; electrodeposition; controllable preparation

---

第一作者:于梅花,女。材料制备及显微分析。mhy@gxu.edu.cn,。

低合金高强高韧马氏体钢的静态软化行为

赵艳君[1]，马本莉[1]，曾建民[1]，孟庆雪[2]

(1.广西大学 材料科学与工程学院；2.邯郸钢铁集团有限责任公司)

摘　要：20SiMn3NiA 为新开发的低合金马氏体钢,具有高强高韧的特点。研究了该合金在 900℃ ~1000℃之间用双道次压缩,1.0~1000s 道次间隔时间内的静态软化行为。当道次间隔时间为 1.0s 时,变形温度为 900℃和 950℃的静态再结晶率分别为 6.48%和 7.43%，而 1000℃的静态再结晶率达到了 84.48%,静态再结晶已经接近完成。20SiMn3NiA 钢在变形量为 0.3,变形速率为 $1.0s^{-1}$ 时,静态再结晶激活能为 448 kJ/mol。

关键词：双道次压缩；静态再结晶；激活能

Static Softening of Martensite Steel of Low Alloy with High Strength and High Toughness

ZHAO Yan-jun[1], MA Ben-li[1], ZENG Jian-min[1], MENG Qing-xue[2]

(1. School of Materials Science and Engineering, Guangxi University, Nanning Guangxi 530004, China;
2. Handan Iron and Steel (Group) Co. Ltd.)

Abstract: 20SiMn3NiA is a new type of low alloy martensite steel with high strength and high toughness. The interrupt hot compression tests were performed from 900 to1000℃ using gleeble 1500.The load-free time were increased from 1.0s to 1000s and its static softening behaviors during this period were investigated. When the passes interval is 1.0s, the static recrystallization fractions at 900℃ and 950℃ deformation temperature are 6.48%, 7.43%, respectively. However the static recrystallization fractions are 84.48%, nearly full static recrystallization, when the deformation temperature is 1000℃. The steel static recrystallization active energy is 448 kJ/mol when the true stress is 0.3 and the true strain rate is $1.0s^{-1}$.

Key words: double-hit compression; static recrystallization; activiation energy

基金项目:广西科学研究与技术开发项目(桂科重 14122001-3)。

第一作者:赵艳君,女。新金属材料开发。zhaoyanjun71@gmail.com。

热挤压变形对镍基高温合金组织和热变形行为的影响

何国爱[1,2]，杨　川[1]，刘　锋[1,2]，司家勇[1]，江　亮[1,2]

(1.中南大学 粉末冶金国家重点实验室；2.中南大学 粉末冶金研究院)

摘　要：采用热模拟试验机对热等静压态和热挤压态的材料，在温度为 1000℃~1100℃的条件下，对一种新成分的镍基粉末冶金高温合金用 3000 吨的挤压机进行热压缩实验。采用光学显微镜、扫描电镜(SEM)、电子探针(EPMA)以及电子背散射衍射技术(EBSD)，分析对比了热挤压前后的组织变化。结果表明，通过热挤压变形后，材料的显微组织主要以细小而均匀的等轴晶粒，且分布也较为集中，热等静压态的原始颗粒边界基本被消除了，材料的晶粒尺寸由热等静压态的 $11.29\mu m$ 减小到热挤压态的 $8.32\mu m$。此外，分析结果表明，通过热挤压变形后，在不同变形条件下的压缩峰值应力也有大幅升高；挤压前后材料的变形激活能分别计算为 1012.9kJ/mol 和 1146.9kJ/mol。

关键字：镍基粉末高温合金；热挤压；电子探针；电子背散射衍射；热模拟；激活能

Effects of Hot Extrusion on Microstructures and Hot Deformation Behaviors of Ni Based Superalloy

HE Guo-ai[1,2], YANG Chuan[1], LIU Feng[1,2] *, SI Jia-yong[1], JIANG Liang[1,2]

(1. State Key Laboratory of Powder Metallurgy, Central South University;
2. Powder Metallurgy Research Institute, Central South University)

Abstract：A new Nickel based P/M superalloy was extruded using a 3000-ton extrusion machine and the hot compression tests were conducted under the temperature range from 1000 to 1100℃. The optical and electron microscope, EPMA and the technique of electron backscatter diffraction were used to examine the microstructural features before and after hot extrusion. The results showed that the alloy exhibited fine-uniform-equixial grain and the prior powder boundaries were eliminated after hot extrusion. The grain was refined from the as-HIPed with an average grain size of $11.29\mu m$ to the as-HEXed with $8.32\mu m$. The peak stress for as-HEXed was higher than that of as-HIPed and the deformation activity energy was determined as 1146.9 kJ/mol and 1012.9 kJ/mol, respectively.

Key words: powder metallurgy nickel based superalloy; hot extrusion; EPMA; EBSD; hot simulation; activity energy

---

基金项目：国家高技术研究发展计划资助(资助编号：2012AA03A514)；国家自然科学基金资助(项目号：51401242,61271356,51205031)；
研究生自主探索项目(项目号：2015zzts031)。

第一作者：何国爱，男，博士生。粉末冶金镍基高温合金。heguoai@csu.edu.cn。

通讯作者：刘　锋，男，讲师，博士。粉末冶金镍基高温合金。liufeng@csu.edu.cn。

正负张力对大规格棒材内部组织致密性影响规律研究

张　龙，刘青松，沈　潮，曹　燕，黄贞益

（安徽工业大学 冶金工程学院）

摘　要： 在轧制生产过程中，尽管大规格棒材进行了多道次变形，但其内部组织依然非常疏松，难以达到通过轧制改善内部组织致密性的目的。为此，本文通过引入刚塑性可压缩材料变形理论，以相对密度大小表示轧件心部疏密程度，建立起大规格棒材心部疏松缺陷三维轧制有限元模型，分析了不同张力对轧件心部密度、轧制力、宽展的影响。研究结果表明，负张力轧制能有效改善大规格棒材心部疏松缺陷，但负张力也会增大轧制力、轧制力矩以及宽展，因此采用负张力轧制改善大规格棒材内部组织致密性时需要综合轧机能力和孔型充满度等因素确定负张力大小。

关键词： 相对密度；张力；大棒材；有限元

Study on the Influence of Positive and Negative Tension on the Compactness of Big Bars' Internal Organization

ZHANG Long, LIU Qing-song, SHEN Chao, CAO Yan, HUANG Zhen-yi

(School of Metallurgical Engineering, Anhui University of Technology)

Abstract: Despite multi-stages deformation carry out on big bars, the internal organization of which is still extremely loose in the rolling process, and it is hard to attain the purpose of improving the compactness of internal organization by rolling. Thus, the present study introduced the rigid-plastic compressible material theory, assessed the compact level with relative density, established a three-dimensional finite element rolling model of big bar with loose defect in the core section, and analyzed the influence of different tension on the density of the core, the rolling force and the spread. The results showed that, the negative tension rolling process can improve the compactness of the big bars' core section. However, it also increases rolling force, rolling torque and the spread. Therefore, factors such as the mill capacity, the pass-filling degree and so on should be put into consideration before using negative tension rolling to improve the compactness of big bars' internal organization.

Keywords: relative density; tension; big bar; finite element

第一作者：张　龙，男。轧制工艺及数值模拟。laxn@foxmail.com。

薄壁大平面钛合金铸件的铸造模拟及工艺优化

张　晨，崔新鹏，周　黔，南　海

（北京百慕航材高科技股份有限公司）

摘　要：以 GE 防火墙为例，基于 ProCAST 对薄壁大平面钛合金铸件铸造过程中的温度场、流动场进行了立式静止浇注、卧式静止浇注及离心浇注等多方案的数值模拟，结果显示铸件大平面水平放置不利于金属充型和排气，而采用大平面倾斜放置的离心浇注方案能保证铸件完整无缺陷成型。通过模拟确定了最优工艺，并且实际铸造结果与模拟结果吻合得很好，验证了利用模拟进行工艺优化的可行性。

关键词：数值模拟；钛合金；薄壁大平面；工艺优化

Simulation and Optimization of the Titanium Casting with Thin Large Planes

ZHANG Chen, CUI Xin-peng, ZHOU Qian, NAN Hai

(Beijing BAIMTEC MATERIAL CO., LTD, Beijing, 100095; Beijing Institute of Aeronautical Materials,)

Abstract: The temperature field, fluid field of the GE Firewall Ti-Alloy casting with thin large planes during the pouring and solidification process were simulated based on ProCAST. The simulation plans of vertical gravity casting, horizontal gravity casting and centrifugal casting were carried out. The results showed that it's difficult for exhausting and filling to place the large plane of the casting horizontal. However, It was easy to get the integral casting with no metallurgical defects through the centrifugal casting with the large plane inclined. The actual casting results were well-fitted with the simulation results, which proved the accuracy of the simulation. Finally, the optimized method was confirmed through the simulation.

Key words: numerical simulation; Ti-alloy; thin large planes; optimization

第一作者:张　晨,男。钛合金精密铸造。zhangchen@baimtec.com。

基于铸型边界单元的铸造热变形数值模拟

陈　涛，廖敦明，庞盛永，周建新，殷亚军，滕子浩，曹　流

（1.华中科技大学 材料成形及模具技术国家重点实验室）

摘　要：铸件在成形过程的热应力变形问题一直是铸造行业中难以完全解决的问题。为了实现近净成形或者精净成形这一铸造的未来发展方向，越来越多的专家和学者采用数值模拟方法来预测铸件的热变形并改进铸造工艺。传统的铸造热应力有限元数值模拟常常需要对铸件和铸型的热应力场同时进行计算，不仅计算量大，而且复杂的铸件与铸型边界条件容易造成计算不收敛的问题。本文提出采用铸型边界单元近似处理复杂铸件与铸型边界条件的方法，可以大大减少计算量，同时保证满足实际工程应用的数值精度。对两个实际生产的铸件同时进行生产实验与数值模拟，结果表明本文提出的方法能够正确模拟出实际铸件的变形情况，为实际生产提供理论指导与工艺改进的方向。

关键词：铸造热变形模拟；有限元法；铸型边界单元

Deformation Simulation of Castings Based on Mold Surface Element Method

CHEN Tao, LIAO Dun-ming, PANG Sheng-yong, ZHOU Jian-xin,
YIN Ya-jun, TENG Zi-hao, CAO Liu

(1. State Key Laboratory of Materials Processing and Die & Mould Technology, Huazhong University of Science and Technology)

Abstract: Deformation of casing during solidification process puzzles many engineers and scientists for years. In order to attain the goal of near-net forming by casting, numerical simulation is a powerful tool. The traditional methods compute the thermal stress of both the casting and the mold. It suffers the problem of massive calculation and failure of convergence. This paper proposed an improved Mold Surface Element Method, which is able to simulate the complicated interaction between casting and mold in efficiency and provide satisfied precision for engineering. Two practical casting products are used to verify the proposed method. The simulation results agree well with the results observed in practical products. The proposed method is believed to benefit the production practice and provide theoretical guidance.

Key words: mold surface element method; thermal stress simulation of casting; casting deformation

基金项目：教育部新世纪优秀人才支持计划(NCET-09-0396)；国家数控重大专项(2012ZX04010-031,2011ZX04014-052,2012ZX04012-011)；湖北省自然科学基金(2011CDB279)；湖北自然科学基金创新群体项目(2010CDA067)。

第一作者：陈　涛，男。铸造热应力数值模拟。cht1399@foxmail.com。

通讯作者：廖敦明，男。铸造热应力数值模拟 & 铸造工艺 CAD。liaodunming@hust.edu.cn。

X65 管线用钢焊接热模拟研究

刘文艳，徐进桥，袁桂莲，黄治军，郑江鹏

(武钢研究院)

摘　要：对 X65 管线钢焊接热影响区粗晶区进行了不同热输入模拟焊接热循环试验。结果表明：t8/5=5s 时，热影响区粗晶区组织主要为板条贝氏体，维氏硬度仅为 224HV，说明试验管线钢淬硬性低，在不预热和后热的条件下出现冷裂纹的可能性小。热影响区粗晶区没有发生明显的硬化和软化现象。t8/5 在 5~100s 范围内，X65 钢-20℃粗晶区的冲击韧性平均值可达到 185J 以上，相当于 28.6mm 试板，初始温度 20℃，线能量范围为 14~79kJ/cm。因此 X65 钢对焊接热输入具有较好的适应性，适合焊接的线能量范围较宽。

关键词：X65；粗晶区；热输入；线能量

Studies in Simulated Heat-affected Zones of X65

LIU Wen-yan, XU Jin-qiao, YUAN Gui-lian, Huang Zhi-jun, ZHENG Jiang-peng

(Rearch and Development Center, Wuhan Iron & Steel(Group) Corp)

Abstract: Experiments were carried out upon the influence of the heat-input on microstructures and properties in simulated coarse grain heat-affected zone of X65. The results showed that when the maximum cooling rate reaches 60℃/s (namely t8/5=5s) the microstructures are lath-bainite and Vickers hardness is 224 HV, which displays that the test steel has a small hardening quenching tendency and a good cold cracking resistance ability. The results of the impact toughness and Vickers hardness in simulated CGHAZ showed that hardening and softening doesn't exist and the impact toughness at -20℃ reaches over 185 J when t8/5 is chosen from 5 to 100 seconds. So X65 has a broad range of heat-input, that is to say, the line energy can be selected between 14 and 79kJ/cm for a 28.6 mm thick plate when its initial temperature is 20℃.

Key words: X65; coarse grain heat-affected zone; heat-input; line energy

第一作者：刘文艳，女。钢种的焊接性及热模拟。liuwenyan2006@aliyun.com。

PLD 过程中 InAl 靶材温度场分布及其相结构稳定性

覃嘉媛[1], 黄 瀛[1], 阮孟财[2], 韦小凤[1], 符跃春[1], 何 欢[1], 陆书龙[2], 边历峰[2], 沈晓明[1]

(1.广西大学 材料科学与工程学院,有色金属及材料加工新技术教育部重点实验室;
2.中国科学院 苏州纳米技术与纳米仿生研究所,纳米器件及应用重点实验室)

摘 要：采用数值模拟的方法,分析了脉冲激光沉积(PLD)制备 InAlN 薄膜过程中激光烧蚀 InAl 靶材表面温度场的分布, 计算了激光功率密度和重复频率对温度场的影响以及在一定条件下靶材表面温度场的分布和烧蚀深度,并通过对比分析激光烧蚀前后靶材的 XRD 测试结果研究其相结构的稳定性,分析结果可为利用 PLD 制备高 In 组分 InAlN 薄膜提供指导。

关键词：PLD; 激光烧蚀; 温度场分布; 相稳定性

Temperature Field Distribution of InAl Target and its Phase Stability during PLD Process

QIN Jia-yuan[1], HUANG Ying[1], RUAN Meng-cai[2], WEI Xiao-feng[1], FU Yue-chun[1], HE Huan[1], LU Shu-long[2], BIAN Li-feng[2], SHEN Xiao-ming[1]

(1. Key Laboratory of New Processing Technology for Non-ferrous Metals and Materials, Ministry of Education, School of Materials Science and Engineering, Guangxi University; 2. Key Laboratory of Nanodevices and Applications, Suzhou Institute of Nano-Tech and Nano-Bionics, Chinese Academy of Sciences)

Abstract：In this work, numerical simulation for the temperature field distribution of laser-ablated InAl target during the pulsed laser deposition （PLD） process of InAlN films is carried out. Based on the dependence of temperature field distribution on laser power density and repetition frequency, temperauture field distribution and ablated depth under some conditions are analyzed. The phase stability of InAl target is also investigated by comparing the XRD measurements before and after the laser-ablation. The results would be useful for the PLD preparation of InAlN thin films with high indium concentration.

Key words：PLD; laser ablation; temperature field distribution; phase stability

基金项目：国家自然科学基金项目(61474030);中国科学院重点实验室开放基金项目(15ZS06);有色金属及材料加工新技术教育部重点实验室开放基金项目(GXKFZ-04, GXKFJ09-25, GXKFJ11-09);广西理工科学实验中心重点项目(LGZX201105)。
第一作者：覃嘉媛,女。宽带隙半导体材料的制备与表征。1549496234@qq.com。
通讯作者：沈晓明,男。半导体光电功能材料与器件。docsjh@gxu.edu.cn。

低碳微合金钢的变形行为及流变应力模型

赵宝纯，李桂艳，刘凤莲，谢广群

(鞍钢股份有限公司技术中心)

摘　要：利用 Gleeble-3800 热力模拟试验机对低碳微合金钢进行了高温单道次压缩实验，获得了实验钢在不同变形温度和变形速率下的流变应力曲线。分析了变形参数对实验钢变形行为和变形抗力的影响。结果表明：实验钢在较高温度或较低应变速率条件下，容易发生再结晶。在其他条件一定时，试样的热变形抗力随温度的升高而降低，随着应变速率的增大而增大。基于流变应力曲线，获得了峰值应变和发生动态再结晶的临界应变等特征值，并通过回归分析，得出实验钢的热加工方程。基于确定出的特征值，结合实验钢变形过程中经历的不同阶段，建立了实验钢高温变形抗力的分段函数数学模型，该模型计算结果与实验值进行了比较。

关键词：低碳微合金钢；动态再结晶；变形抗力；数学模型

Hot Compression Deformation Behavior and Flow Stress Modeling of a Low Carbon Microalloyed Steel

ZHAO Bao-chun, LI Gui-yan, LIU Feng-lian, XIE Guang-qun

(Iron and steel research institute of AnGang Group, Anshan)

Abstract：Single pass hot compression test was conducted on a low carbon microalloyed steel by using a Gleeble-3800 thermo-mechanical simulator. And the flow stress curves of the tested steel deformed at various temperatures and strain rates were obtained, which permitted the analysis of deformation behavior and deformation resistance. The results show that dynamic recrystallization occurs at low strain rate or high temperature. With other processing parameters unchanged, the deformation resistance increases with increasing strain rate while decreases with increasing temperature. Based on the flow stress curves, characteristic values for the steel, such as peak strain and critical strain for the onset of dynamic recrystallization, can be obtained. By regression method, the hot working equation can be formulated. The flow stress curves were segmented according to the characteristic point, which facilitates the modeling for the high temperature deformation resistance of the tested steel. And the results from the mathematical model and the experimental ones were compared.

Key words: low carbon microalloyed steel; dynamic recrystallization; flow stress; mathematic modeling

---

第一作者：赵宝纯，男。材料加工过程模拟与设计。baochunz@163.com。

原位水解制氢多功能燃料电池包的研发

唐　美[1]，詹海鸿[2]，许征兵[1]，曾建民[1]，何国强[1*]

（1.广西大学 材料科学与工程学院；2.广西有色金属集团有限公司广西冶金研究院）

摘　要：随着传统化石资源的枯竭、环境的日益恶化，世界各国纷纷加大力度开发和利用新能源，如太阳能、氢能、风能、地热能等。氢能具有来源广泛、清洁环保、可储存和可再生等优点，被视为21世纪最具发展潜力的清洁能源。质子交换膜燃料电池（PEMFC）是氢能利用的绝佳方式，它是一种将燃料的化学能通过电化学反应直接转化为电能的发电装置，具有发电效率高、低碳环保和可再生等优点。我们基于传统石化重整制备氢气不易存储与运输、加氢站设施少、杂质气体难以去除等问题，结合铝水解制氢原理和广西铝资源优势，提出铝/铝合金水解原位制氢的设想。通过调节制备参数和优化装置设计，研制氢气现制现用的多功能燃料电池一体化装置，并提出废铝制氢及反应产物循环利用的方案。该燃料电池一体装置的具有很强的的环境适应性，在0℃至100℃温度区间均可正常工作，污水、冰雪、尿液、血液、酒等液体均能作为制氢原料；能长时间持续发电而无需充电；同时它具有静音、便携、可移动、易维护、成本低、绿色环保等优点。该燃料电池包可用于手机、平板电脑、笔记本电脑、玩具等电子产品充电，还可广泛应用于偏远或特殊地区作业、野外旅游、应急供电、军事等领域。有望成为中国首个商品化的可用于笔记本电脑充电、急救供电、野外旅游等领域的便携式多功能燃料电池包，该燃料电池包的研制对便携式燃料电池的商品化和产业化具有重要意义。

关键词：原位水解制氢；燃料电池；商品化

In-situ Water Splitting Hydrogen Production for Multifunctional Fuel Cells Pack

TANG Mei[1], ZHAN Hai-hong[2], XU Zheng-bin[1], ZENG Jian-min[1], HE Guo-qiang[1*]

（School of Materials Science and Engineering, Guangxi University; Guangxi Non-ferrous Metals Group Co., Ltd, Guangxi Research Institute of Metallurgy）

Abstract: With the depletion of traditional fossil resources, environment worsening, many countries'government have increased efforts to develop and utilization of new energy sources, such as solar, hydrogen, wind, geothermal energy and others. Hydrogen energy as a extensive sources, clean and environmental friendly, storable and renewable and other advantages, is regarded as the most potential for development of clean energy in the 21st century. Proton exchange membrane fuel cell (PEMFC) is a excellence way to the use of hydrogen energy, it is a energy generating device which makes the

基金项目：广西科学研究与技术开发计划课题（桂科攻 14122007-31）及 ZD2014007、XGZ130955 等项目。

第一作者：唐　美，女。第一性原理计算。　hytangtangtang@126.com。

通讯作者：何国强，男。新能源材料。　heguoq@163.com。

chemical energy through electrochemical reaction to direct conversion into electrical energy, with the advantages of higher power generation efficiency, low carbon environmental protection and renewable. Base on the disadvantages of traditional petrochemical reforming hydrogen production, such as difficult to storage and transport, less hydrogenation station, difficult to remove the impurity gas, combined with the mechanism of aluminum hydrolysis hydrogen production [1-3] and aluminum resources advantage of Guangxi province, the idea of aluminum / aluminum alloys in-situ hydrolysis hydrogen production is proposed. By adjusting the preparation parameters and optimize the device design, we develop synchronization of hydrogen Production and application for multifunctional fuel cells device, and puts forward a program to using of waste aluminum for hydrogen production and reaction products renewable. The fuel cell device has strong adaptability to the environment, can normally work from 0 oC to 100 oC range, and the sewage, ice and snow, urine, blood, wine and other liquid can be a raw materials for hydrogen production. It can keep working long time continuous to provide power but without charge process. Meanwhile, it is no noise, portable, mobile, easy maintenance, low cost, green environmental protection. The fuel cells device can be used to charge mobile phones, panel PCs, notebook PCs, toys and other electronic products, but also widely be used in remote or special area and region, outdoor travel, emergency power supply, military and other fields. It is expected to become the first commercialization multifunctional portable fuel cells pack in China, which is used for notebook PCs charging, emergency power supply, outdoor travel and other fields. the development of the fuel battery pack for portable fuel cell commercialization and industrialization has important significance. It is significant to develop the portable fuel cells commercialization and industrialization.

Key words: in-situ water splitting hydrogen production; fuel cells; commercialization

铝 / 钢 CMT 熔钎焊接头强度的数值模拟研究

付　参，林　健，雷永平，符寒光，吴中伟

（北京工业大学 材料科学与工程学院）

摘　要：为实现汽车轻量化，采用铝钢一体化车身框架是一条有效途径，提高铝钢连接接头强度是改善一体化车身安全性的关键。目前研究表明铝钢 CMT（冷金属过渡）熔钎焊接头有较好的剪切强度，但其剥离强度仅为剪切强度的 1/50。本研究针对低碳镀锌钢与 AA6061 铝合金的 CMT 熔钎焊接头的剥离过程建立有限元模型，对剥离时板材受力情况及界面层破坏模式进行数值模拟，探讨剥离强度低于剪切强度的原因。

关键词：冷金属过渡焊；剥离强度；数值模拟；铝钢异种材料

Investigation on Numerical Simulation of Strength of Al/Steel CMT－brazed Joint

FU Can, LIN Jian, LEI Yong–ping, FU Han–guang, WU Zhong–wei

（College of Materials Science and Engineering, Beijing University of Technology）

Abstract: Aluminum-steel hybrid structure used in vehicle body is a promising method for automotive lightweight. The key to improve the safety of hybrid structure is increasing the strength of aluminum-steel joint. The reported studies indicate that aluminum-steel CMT melt-brazed joint has good shear strength, while peel strength is very low which is only fifty percent of shear strength.

The finite element model of peel process of galvanized low carbon steel and AA6061 aluminum dissimilar materials CMT melt-brazed welding joint is established in this study. The peel strength and failure mode in peel process are simulated. Considering this, the reasons why peel strength is lower than shear strength are discussed.

Key words: CMT; peel strength; numerical simulation; aluminum-steel dissimilar materials

基金项目：国家自然科学基金资助项目（51005004，51275006）；北京市自然科学基金资助项目（3132006）；
　　　　　北京市教委项目科研项目（面上）（KM2012100050010）。
第一作者：付　参，男。铝钢异种金属焊接。S201309036@emails.bjut.edu.cn。
通讯作者：林　健，男。铝钢异种金属焊接。linjian@.bjut.edu.cn。

热喷涂扁平粒子的形成及其物理模拟

李 辉,周 正,贺定勇,雷永平

(北京工业大学 材料科学与工程学院)

摘 要：热喷涂涂层是由大量加热至熔化或半熔化粉末颗粒撞击到基材后经过变形扁平铺展后堆积形成的,单个颗粒的形成具有时间和空间的独立性,因此,单个颗粒的形成过程是认识热喷涂涂层形成机制及其性能的基础。近几十年来,国内外大量热喷涂领域的研究集中于热喷涂粒子的形成机制及其特点,但由于单个粒子形成时间在数个微秒内,尺度在几十至几百个微米,很难进行直接观察,一些研究者基于相似性原则尝试进行毫米级扁平颗粒的物理模拟,得出了一些有价值的发现。本文主要综述了近二十年来在本领域的研究进展,并对热喷涂领域的物理模拟工作进行了展望。

关键词：热喷涂；扁平粒子；物理模拟

Formation of Thermal Spray Splat and Its Simulation

LI Hui, ZHOU Zheng, HE Ding-yong, LEI Yong-ping

(College of Materials Science and Engineering, Beijing University of Technology)

Abstract: Thermal spray coating is formed by the accumulation of large quantity of molten or semi-molten powder particles impinging onto the substrate that are quickly deformed and flattened. The single splat is separately distributed in time and space and therefore the formation of one splat is known to be the basic of the understanding of formation mechanism of thermal sprayed coating. In recent years, a large number of work focuses on the formation and the behavior of thermally sprayed particles, which is however difficult to be characterized, owing to the fact that an individual particle is flattened in a few microseconds scaling in dozens micrometers. Some researches attempted the physical simulation on the splat by using millimeter-size particle, and some interesting findings were reported. This paper summarizes the current progress on this issue and the development trend is also proposed.

Key words: thermal spray; splat; simulation

基金项目：国家自然科学基金资助项目(50805002,5121007)。

第一作者：李 辉,男。热喷涂及材料连接。hui.li@bjut.edu.cn。

毫米级羟基磷灰石粒子的微观组织及其形成

李　辉，陈　伟，谷佳宾，贺定勇，雷永平

（北京工业大学 材料科学与工程学院）

摘　要：等离子喷涂羟基磷灰石涂层是应用于人工植入体的一种生物活性涂层，涂层中包含非晶态组织及分解组织。为了充分揭示羟基磷灰石涂层组织转变特点，实现对涂层组织及性能的准确控制，本研究采用悬熔的方式设计了毫米级熔化颗粒的撞击实验，物理模拟撞击形成的毫米级扁平粒子，对不同熔化状态下的扁平粒子进行了微观组织分析，探讨了羟基磷灰石扁平粒子的形成机制。

关键词：热喷涂；羟基磷灰石；扁平粒子

Microstructure of Millimeter-size Hydroxyapatite Splat and Its Formation Mechanism

LI Hui, CHEN Wei, GU Jia-bin, HE Ding-yong, LEI Yong-ping

（College of Materials Science and Engineering, Beijing University of Technology）

Abstract: Plasma sprayed hydroxyapatite coating has been successfully applied to artificial implant as a bioactive coating, which contains large amount of amorphous calcium phosphate and other decomposed phases. In order to reveal the transformation mechanism of hydroxyapatite and to better manipulate the microstructure of the coating as well as its properties. This study designed an impact of millimeter-size molten particle to substrate using levitation melting, and the flattened splat was obtained by physical simulation. The effect of the melting state of particle was considered and the microstructure was characterized. The formation mechanism of flattened hydroxyapatite particle was discussed.

Key words: thermal spray; hydroxyapatite; simulation

基金项目：国家自然科学基金资助项目（50805002，5121007）。

第一作者：李　辉，男。热喷涂及材料连接。hui.li@bjut.edu.cn。

毫米级镍扁平粒子的微观组织特征及其传热过程

李 辉，王本鹏，栗卓新，贺定勇，雷永平

(北京工业大学 材料科学与工程学院)

摘 要：扁平粒子是热喷涂涂层的基本形成单元，也是近几十年来关注的热点。为了更好地理解扁平粒子的形成，本论文设计并制备了毫米级镍熔滴坠落撞击实验，对形成的镍扁平颗粒的微观组织和传热凝固过程进行了数值仿真。实验研究了颗粒温度、撞击速度及基体温度对扁平颗粒的影响。研究发现，过热的镍扁平颗粒撞击至基材后有可能发生基材熔化，随着基体温度的提升，这一现象更为明显，而增大撞击速度有可能导致颗粒发生飞溅，降低颗粒与基体的界面温度，这对于认识喷涂颗粒与基体的结合有指导意义。

关键词：热喷涂；扁平颗粒；传热

Microstructural Characteristics and Heat Transfer of Flattened Millimeter−size Nickel Droplet

LI Hui, WANG Ben−peng, LI Zhuo−xin, HE Ding−yong , LEI Yong−ping

(College of Materials Science and Engineering, Beijing University of Technology)

Abstract: The splat is the fundamental unit of thermally sprayed coatings, which has been one of the hottest issues for decades. In order to study the splat formation, an experiment was designed and conducted, in which a millimeter-size nickel metal droplet fell freely and impacted on aluminum and stainless steel substrate. The microstructural characteristics of the splat and the heat conduction and solidification processes during the flattening process have been studied numerically and experimentally. The effect of the droplet temperature, impact velocity as well as the substrate temperature was investigated. The phenomenon of substrate melting was observed after the spreading of nickel droplet, which became more pronounced when the initial substrate temperature increased. Increasing the impact velocity of droplet resulted in a decrease in the interfacial temperature between droplet and substrate.

Key words: thermal spray; flattened behavior; heat transfer

基金项目：国家自然科学基金资助项目(50805002,5121007)。

第一作者：李 辉，男。热喷涂及材料连接。hui.li@bjut.edu.cn。

钢／铝异种材料界面结合的热模拟研究

余本红，林 健，赵海峰，雷永平，吴中伟

（北京工业大学 材料科学与工程学院）

摘 要： 使用热模拟试验方法研究热镀锌低碳钢和铝合金在热力共同载荷作用下的界面结合过程。研究了压强、温度、时间关键参数对接头力学性能的影响，并且对不同参数下获得的焊接接头的界面层进行了显微分析。研究结果表明，当压强为 263.2Mpa、保持时间为 60s 时，随着温度的升高，接头的拉剪力先减小后增大。当温度为 390℃时，界面层主要由 Al-Zn 共晶物组成，并且剪切力最大；当温度为 430℃时，界面层中 Al-Zn 共晶物消失，Fe-Al 金属间化合物开始生成，并且剪切力发生下降；当温度为 490℃时，界面层全部由 Fe-Al 金属间化合物组成，接头拉剪力开始增大。390℃下，较大的压强不利于形成界面结合，得到接头的剪切力较低。延长保持时间使界面反应更充分，有利于提高接头力学性能。

关键词： 热模拟；钢/铝接头；金属间化合物；剪切力

Research on Interfacial Bonding of Dissimilar Materials of Steel and Aluminum by Thermal Simulation

YU Ben-hong, LIN Jian, ZHAO Hai-feng, LEI Yong-ping, WU Zhong-wei

(College of Materials Science and Engineering, Beijing University of Technology)

Abstract: The welding of hot dip galvanized low-carbon steel and aluminum alloy are manufactured by thermal simulation testing machine. The experiment can get many welding joints with different mechanical properties by controlling the parameters such as temperature, pressure and time. The layer of welded joints obtained from different process parameters are researched. Research shows that with the temperature increasing, the shear strength of the joint is first decreased and then increased. When the pressure is 263.2MPa and the holding time is 60s, at the temperature of 390℃,the interface layer consists of eutectic Al-Zn and with the maximum tensile-shear force. At the temperature of 430℃, the eutectic Al-Zn is missing and the Fe-Al intermetallic compound is appearing. At the temperature of 490℃, the interface layer consists of Fe-Al intermetallic compound. Increasing pressure will have great influence on Al-Zn eutectic reaction. The influence is that the mechanical properties will decrease. Extending the holding time will make the interfacial reactionmore fully and increase the mechanical properties of joint.

Key words: thermal simulation; steel/aluminums joint; intermetallic compound;tensile-shear force

基金项目： 国家自然科学基金资助项目（51005004，51275006）；北京市自然科学基金资助项目（3132006）；
北京市教委项目科研项目（面上）（KM2012100050010）。
第一作者： 余本红（1993— ），男，硕士。钢铝异种材料焊接。:yubh@emails.bjut.edu.cn。
通讯作者： 林 健（1979— ），男，副教授，硕导。汽车车身的轻量化研究。 linjian@bjut.edu.cn。

热处理对 TC18 钛合金电子束焊接接头组织性能的影响

张思聪[1,2]，雷永平[1]，刘 昕[2]

(1.北京工业大学 材料科学与工程学院；2.北京航空制造工程研究所,高能束流加工技术重点试验室)

摘 要：采用普通退火、去应力退火两种焊后热处理工艺对 TC18 钛合金电子束焊接接头进行焊后热处理,分别测量其接头室温拉伸和冲击性能,并对其显微组织和拉伸、冲击断口进行光学和扫描电镜观察。实验表明,随着退火温度的升高,普通退火的组织发生了再结晶和晶粒长大的过程,相比去应力退火, 相含量减少, 相含量增多,体现为去应力退火后的接头强度优于普通退火,接头塑性低于普通退火。

关键词：TC18 钛合金；电子束焊接；焊后热处理；力学性能

A Study on the Influences of Heat Treatment on the Structure and Performance of TC18 Titanium Alloy Electron Beam Welded Joint

ZHANG Si-cong[1,2], LEI Yong-ping[1], LIU Xin[2]

(1.College of Materials Science and Engineering, Beijing University of Technology; 2.Science and Technology on Power Beam Processes Laboratory, Beijing Aeronautical Manufacturing Technology Research Institute)

Abstract: The tensile and impact properties of TC18 titanium alloy electron beam welded joint at room temperature were measured separately by two different post weld heat treatment, including normal annealing and stress relief annealing. The microstructure, tensile and impact fracture morphology were studied under optical and scanning electron microscope. It is shown in the experiment that the structure of joint using normal annealing went through the process of recrystallization and grain growth with improve of temperature. Compared with the one treated by stress relief annealing, the phase of the former joint decreases significantly while its phase increased dramatically. The result is that the strength of the joint adopting normal annealing was better than that of stress relief annealing, while its plasticity was the opposite.

Key words: TC18 titanium alloy; electron beam welding; post weld heat treatment; mechanical property

---

第一作者：张思聪,男。高强钛合金电子束焊接。191885976@qq.com。

钛合金激光焊接接头的显微组织研究

罗　霞[1]，徐　周[1]，王晓南[1*]，张　敏[2]，陈长军[2]

（1.苏州大学 沙钢钢铁学院；2.苏州大学 机电工程学院激光加工中心）

摘　要：研究了 2mm 厚钛合金激光焊接接头的显微组织和疲劳性能。研究结果表明：激光焊缝区显微组织为粗大的柱状 β 晶粒和针状马氏体；热影响区中的显微组织为细小的 α′ 针状马氏体，原始 α 相和 β 相，且热影响区组织明显比焊缝区组织细小；母材的显微组织为典型的 α+β 双态组织，β 相分布于 α 相等轴晶的晶界上。当应力比 R=0.1、循环基数为 10^7 时，母材的条件疲劳强度明显高于焊缝。

关键词：TC4 钛合金；激光焊接；显微组织；针状马氏体；疲劳性能

Study on Microstructure of Titanium Alloy Welded Joints Using Fiber Laser

LUO Xia[1], XU Zhou[1], WANG Xiao-nan[1*], ZHANG Min[2], CHEN Chang-jun[2]

（1.Shagang School of Iron and Steel, Soochow University; 2.Laser Processing Research Center, School of Mechanical and Electrical Engineering, Soochow University）

Abstract: Microstructure and fatigue properties of titanium alloy welded joints were studied. The result showed that the microstructure of fusion zone (FZ) was coarse columnar-shaped β phase and needle-like matensite. The microstructure of heat-affected zone (HAZ) was a mixture of fine needle-like matensite, primary α phase and β phase. The microstructure of base material (BM) consists of intergranular β phase and equaixed α phase. When the stress ratio was R=0.1 and circulation base was 10^7, the fatigue strength of the BM was significantly higher than that of the FZ.

Keywords: titanium alloys; laser welding; microstructure; needle-like matensite; fatigue behavior

基金项目：国家自然科学基金资助项目（No.51305285）、江苏省基础研究计划（自然科学基金）（No.BK20130315）；江苏省光子制造科学与技术重点实验室开放基金资助项目（No.GZ201304）。

第一作者：罗　霞（1985— ），女，博士，讲师。表面处理及激光焊接。xluo@suda.edu.cn。

通讯作者：王晓南（1984— ），男，博士，讲师。高性能钢铁材料及激光焊接。wxn@suda.edu.cn。

The Thermal-Mechanical Coupled FEM Analysis on Continual Tube Rolling MPM Deformation Process

LIU Bing, LI Peng-xue

(School of Mechanical Engineering, University of South China)

Abstract: Based on the characteristics of MPM deformation process, a thermal-mechanical coupled model of this process was established by the three-dimensional elastic-plastic finite element method. Some important parameters such as the rolled product size in each stand, the temperature variation curves of the key nodes on the tube, the rolling force and torque of each roller, and the equivalent stress and strain of the tube were analyzed using this model. The analysis results are as follows: The external diameter and wall thickness are alternate reduced between the pass bottom position and the roll gap position and tend gradually to finished size; The temperature of nodes, contacting with mandrel or rolls, decrease for heat transfer between tube and mandrel, tube and rolls, rube and surroundings, while the temperature of internal nodes increase for plastic work and decrease following for conduction within tube; The flow law of metal is that the metal in the bottom of pass is radial compressed, and flow towards the roll gap position or along the axial direction. Comparison between simulation solutions and experiment results shows a good agreement, which means that this model is capable of simulation MPM deformation process as well as forecasting product quality.

Key words: MPM; finite element model; continual tube rolling; simulation

1 Introduction

Steel tube especially of seamless steel pipe plays a major role at the national economy development. Compared with other forms of tube rolling, continuous tube rolling train gets development for its high production and good quality. Therein, MPM(Multi-Stand Pipe Mill), which use 2-adjustable-roll and a retained mandrel, get widely attention for such advantages: high stability of mandrel, little lateral metal flow, wide production dimension, uniform deformation, as well as low rolling energy consumption[1].

During the MPM deformation process, the tube gets a three-dimensional inhomogeneous deformation, by the action of the pass which is composed of two rolls and one mandrel. The deformation is very complicated because of the characteristics of pass rolling, such as nonlinearity of geometric and material, non-simultaneity of contact, and the velocity difference of each point in the pass. In addition, the process should satisfy the stable condition of multi-stand rolling. Therefore, the thermal-mechanical coupled finite element model possesses guiding significance for actual production of seamless steel

第一作者:刘 兵,男。机械工程材料。13789377318@139.com。

pipe to understand and grasp the rolling state, to analysis the force rule during rolling process[2].

2　FEM Model Foundation

2.1 Basic Equation of Thermal–Mechanical Coupled FEM

Energy conservation equation for the continuous medium, whose volume is V, density is ρ, boundary is S, is written as[3]:

$$\int_V \left\{ \rho \left[\overline{Q} - \frac{dU}{dt} \right] + \sigma_{ij} \frac{\partial v_i}{\partial x_j} \right\} dV = \int_S H dS \quad (1)$$

Where σ is stress, v is velocity field, U is given internal energy, $\overline{Q}$ is volume heat flux, H is heat flow rate per unit area on the boundary S, x_j is coordinate.

2.2 Friction Model Determination

There are three types of friction models, including sliding coulomb friction model, shear friction model, and stick-slip friction model, are often adopted in the contact problems. The modified sliding coulomb friction model was select in this finite element model[3].

$$\sigma_{fr} \leq -\mu \sigma_n \frac{2}{\pi} \arctan \left(\frac{v_r}{v_{vcnst}} \right) B \quad (2)$$

Where, σ_{fr} is shearing stress, σ_n is equivalent shearing stress, μ is coulomb friction factor, B is unit vector of the relative sliding velocity, $B = \frac{v_r}{|v_r|}$, v_r is relative sliding velocity, v_{vcnst} is the critical value of the relative sliding velocity.

2.3 Thermal Boundary Condition Determination

During the MPM process, three boundary conditions all exist on tube surface including heat conduction, thermal convection and thermal radiation. Considering that the heat loss of convection occupies the little proportion during the process, the thermal convection and radiation can be merged into one condition, which is written as[4]:

$$q = h(T - T_\infty) \quad (3)$$

Where, q is heat flux, T is surface temperature of the tube, T_∞ is ambient temperature, h is equivalent heat transfer coefficient, which include convective heat transfer coefficient h_1 and radiate heat transfer coefficient h_2

$$h_2 = \varepsilon \sigma (T + T_\infty)(T^2 + T_\infty^2) \quad (4)$$

Where ε is blackness coefficient, σ is Boltzmann constant.

Considered comprehensive factors, h was set to 0.17kW/(m2·℃).

The contact heat transfer between rolls, mandrel and tube can be described as follow:

$$q_r = h_r(T - T_d) \quad (5)$$

Where q_r is heat flux density, h_r is the coefficient of contact heat transfer, which was set to 20 kW/(m²·℃), T_d is the contact temperature between the roll, mandrel and tube.

2.4 FEM Model Foundation

A FEM model of MPM was established based on the practical rolling process of a certain factory. The model was composed of workpiece, six rolls (including one idle roll), one mandrel and one pusher, as shown in fig.1. Main parameters of rolls were given in table 1.

Tab. 1 The main parameters of rolls

pass	1	2	3	4
ECC(Eccentricity/mm)	5.6	0	0	0
R1(the radius of pass/mm)	191.55	184.6	182.3	182
S(roll gap value/mm)	40	40	40	40
Φ(the diameter of rolls/mm)	850	850	850	750
Velocity(rotational velocity of roller/rpm)	6.0	8.8	10.6	15.9

The updated Lagrange algorithm was adopted in this model, combined with the flow rule of Prandtl-Reuss and Von Mises yield criterion. The workpiece was assumed to be elastic-plastic deformable body, rolls and mandrel were assumed to be rigid body. The friction coefficient between workpiece and rolls was set to 0.2, and 0.07[5] between workpiece and mandrel for the lubricated mandrel. The initial temperature of the workpiece was set to 1050℃. The material of the workpiece was 20#, and the capillary tube size was Φ410mm× 23mm. Considering calculation time, the distance between rolling stands was assumed to be 500mm and 1600mm for the workpiece in length. In addition, the workpiece was evenly divided into 100 layers in length, 2 layers in radius, and 72 shares in circumference. The workpeice contained totally 14400 elements and 21816 nodes. A pusher [6] was established in order to be bitten smoothly for the tube.

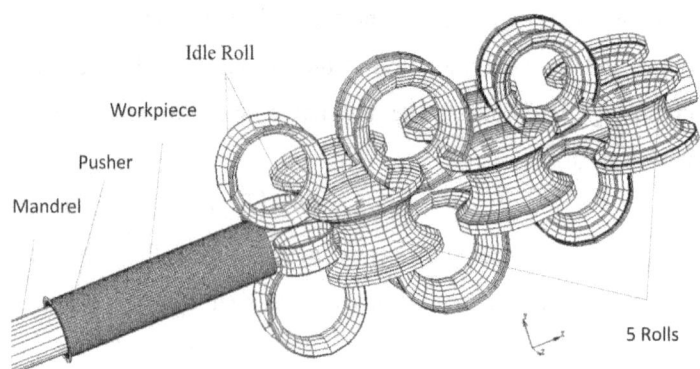

Fig. 1 the FEM model

3 Result Analysis and Modeling Verification

3.1 The Tube Shape Analysis

The circumferential variation rule of external diameter and the wall thickness in each stand outlet were illustrated in Fig. 2. The external diameter is alternate reduced between the pass bottom position and the roll gap position. After all, all circumferential points tend to finished size. The average radius value was 183.04 mm, and the error was 0.65%. On account of the geometric characteristics of the pass, the tube in roll gap is not in contact with the rollers. Therefore, the overfill phenomenon is liable to occur, and the tube will thicken here, which is called lateral spread. The simulation result showed that in third stand outlet, the thickness of tube in roll gap position was larger than that in second stand outlet. This proved that there is lateral spread phenomenon in stands. The finished thickness of steel tube was 10.80 mm, the error was 3.67%. The tube shapes in each stand outlet were shown as Fig. 4.

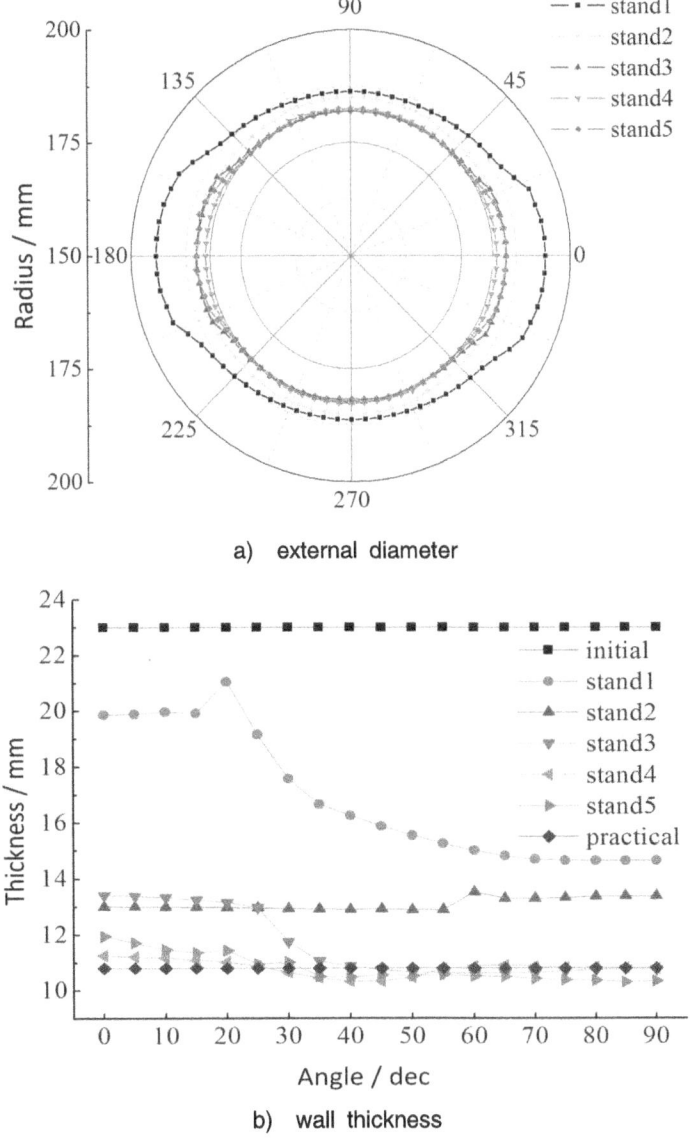

a) external diameter

b) wall thickness

Fig.2 The circumferential variation rule of external diameter and the wall thickness in each stand outlet

3.2 Temperature Analysis

Fig. 3 showed the tube temperature variation curves. The characteristic nodes were shown in Fig. 3a). Node 3 contacted with pass bottom on odd stands, while node 9 contacted with pass bottom on even stands. The temperature of node 1, node 4 and node 7 decreased for contacting with the mandrel. The temperature of node 3, node 6 and node 9 decreased for two reasons. One is contacting with the rollers, the other is convection and radiation existed with surrounding environment. The temperature of node 2, node 5 and node 8 increased for plastic work and decreased following for conduction within tube.

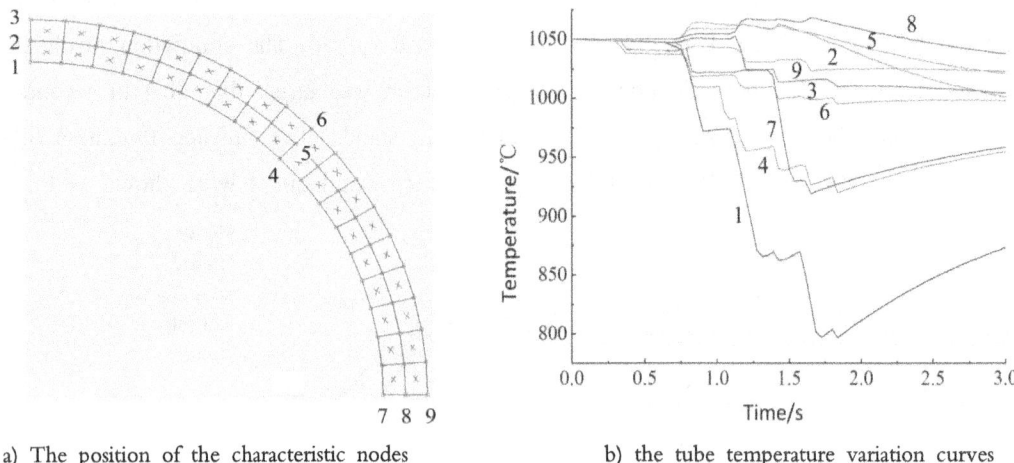

a) The position of the characteristic nodes   b) the tube temperature variation curves

Fig.3 The tube temperature variation curves

3.3 Stress and Strain Analysis

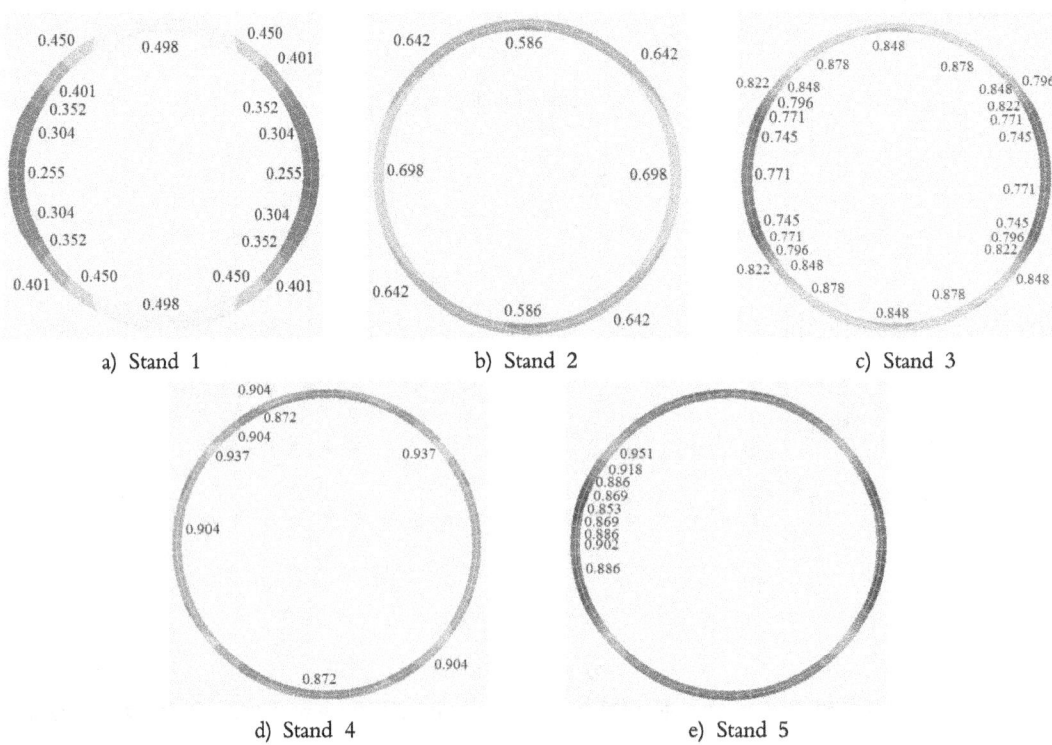

a) Stand 1   b) Stand 2   c) Stand 3

d) Stand 4   e) Stand 5

Fig.4 The equivalent strain distribution at the outlet of each stand

Fig. 4 showed the equivalent strain distribution at the outlet of each stand. The maximum value is located in pass bottom position, while minimal value is situated in roll gap position. The reason is that both external diameter and thickness, especially first three stands, reduce in pass bottom. But the equivalent strain will tends to be uniform for the staggered arrangement of pass bottom and roll gap. The equivalent strain of latter two stand outlet increased rarely , because that the last pass is mainly applied to form the same curvature at the circumference direction, the passes is same, and the reduction amount is small.

To analyze the metal flow law in the deformed area, the nephograms of each stress distribution above the first stand rolling area was shown in Fig.5, and the cylindrical coordinates was adopted here. The metal in the bottom of pass is at three-direction compressive stress state (σ_r , σ_θ, σ_l are negative), while the metal in roll gap is at the three-direction tensile stress state (σ_r , σ_θ, σ_l are positive). Therefore, the flow law of metal can be described that the metal in the bottom of pass is radial compressed, and flow towards the roll gap position or along the axial direction.

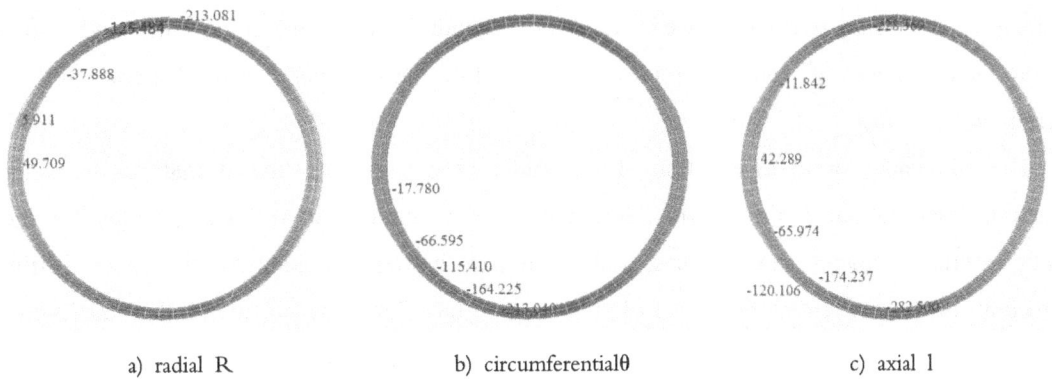

a) radial R   b) circumferentialθ   c) axial l

Fig.5   The nephograms of each stress distribution above the first stand rolling area

3.4 Rolling Force and Torque Analysis

The rolling force and torque curves were shown in fig.6. With the reduction amounts gradually deceased along the rolling direction, the value of force and torque are also decreased. In the figures, curve 1 to curve 5 was corresponding to stand 1 to stand 5. Fig.6c) showed the comparison diagram between simulation and measurement, and the results showed a good agreement, which meant that the thermal-mechanical coupled model is capable of simulation MPM deformation process.

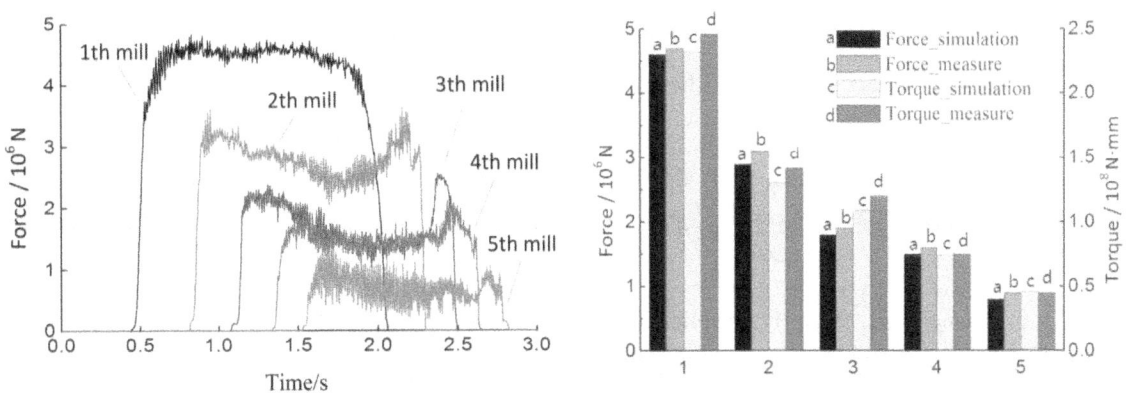

a) The rolling force curves   b) the comparison diagram between simulation and measurement

Fig.6   The rolling force and torque curves

4   Conclusions

A thermal-mechanical coupled model has been established by the nonlinear finite element method and the simulation according to 5-stand-MPM rolling process of a certain factory has been completed. The analysis results are as follows:

1. The tube shape analysis showed that, on account of the geometric characteristics of the pass, the external diameter and wall thickness are alternate reduced between the pass bottom position and the roll gap position and tend gradually to finished size.

2. The temperature analysis showed that the temperature of nodes, contacting with mandrel or rolls, decrease for heat transfer between tube and mandrel, tube and rolls, rube and surroundings, while the temperature of internal nodes increase for plastic work and decrease following for conduction within tube.

3. The equivalent strain distribution at the outlet of each stand showed that the maximum value is located in pass bottom position, while minimal value is situated in roll gap position for the reason that both external diameter and thickness reduce in pass bottom. In addition, the stress distribution in the direction of radius, circumference and axis reveal the flow law of metal that the metal in the bottom of pass is radial compressed, and flow towards the roll gap position or along the axial direction.

4. Comparison between simulation solutions and experiment results shows a good agreement. Therefore, this foundation is valuable to optimize the process parameters and analysis the product quality.

References:

[1]   Wang X J, Xu S C. Continuous Rolling Theory to Steel Tube. China: Metallurgical Pr., 2005

[2]   Miguel A C, Marcela B G, Eduardo N D. Finite element analysis of steel rolling processes. Computers & Structures, 2001. 79(22–25): 2075–2089

[3]   MSC.MARC2013. User Manual. MSC Software Corporation. 2013

[4]   Wang F X, Du F S, Yu H. the thermal–coupled FEM analysis on 3–roll continual tube rolling PQF deformation process. Advanced Materials Research, 2011,189–193:1670–1674

[5]   Hong H P, Kang Y L, Feng C T. Three–dimensional thermo–mechanical coupled FEM simulation for hot continuous rolling of large–diameter mandrel bar. J Mater Sci Technol. 2003, 19(suppl): 228–230

[6]   Liu J S, Zhang S H, Xiao H. Application of MSC.Marc in material processing engineering. China: Water Conservancy and Hydropower Pr., 2010

金属粉末热等静压成形过程数值模拟模拟与验证

汪　敏，周　顺，周建新，殷亚军，史玉升

(华中科技大学 材料成形与模具技术国家重点实验室)

摘　要：本文以 Shima 屈服准则描述粉末本构特性，进行热等静压高温高压条件下的粉末变形模拟。模拟结果显示：外包套变形较大，驱动内部钛合金致密化，而内部粉末变形较小。工件变形趋势以及工件变形量和模拟结果基本相符，误差控制在 4%以内。工件各部位最终相对密度和模拟结果基本一致。引起误差的主要原因是模型基于连续介质而忽略了颗粒重排的微观过程。讨论了各个阶段致密化机理。实验结果表明，数值模拟可以为金属粉末 HIP 工艺以及模具设计提供参考。

关键词：热等静压；有限元；数值模拟；Shima 模型

Numerical Simulation and Verification of Metal Powder Under Hot Isostatic Pressing

WANG Min, ZHOU Shun, ZHOU Jian-xin, YIN Ya-jun, SHI Yu-sheng

(State Key Laboratory of Material Processing and Die & Mould Technology, Huazhong University of Science and Technology)

Abstract: This work describes powder constitutive properties using Shima yielding criterion, simulates powder deformation in the process of hot isostatic pressing. Simulation results show that the thin can has large deformation and drives powder densification. The internal core has almost no deformation. Experimental results show that the trend and size simulation error is less than 4%. There is not much different in the value of densification between experiment and simulation results but the simulation value in low density areas is smaller than actual values, this is because the simulation process ignores some microscopic behaviors such as movement and rearrangement of powder particles. Mechanism of densification at different stages was discussed. The simulation model can give a useful method to analysis the stainless steel powder HIP process.

Key words: HIP; finite element method; numerical simulation; shima model

0 引言

粉末热等静压技术起源于美国巴特尔(Battelle)研究所。是一种在高温、高压同时作用下使粉末得

基金项目：材料成形与模具技术国家重点实验室自主研究项目(2014-02)；国家自然科学基金项目(No.51475181)。

第一作者：汪　敏，男。材料成形过程计算机数值仿真技术。wangmin.luck@foxmail.com。

通讯作者：周建新，男。材料成形过程模拟仿真及数字化技术。zhoujianxin@hust.edu.cn。

以压制烧结的技术[1,2],使传统的粉末成形工艺与烧结工艺同时进行,可消除粉末内部的孔隙,该技术综合了粉末冶金和冷等静压的优点,其制件强度拥有优良的力学性能和热工艺性能,且制件各向晶粒组织均匀,能成形具有优异性能的复杂零部件。由于其制件的这种特性,热等静压可满足航天航空以及核工业等重要领域的需求[3,4]。在热等静压成形过程中,粉末高温高压致密化成形的变形量较大。热等静压成型过程是材料,接触,热传导等多方面的高度非线性问题,因此很难把握粉末致密化过程的力学变化[5,6]。传统试错法制造产品时,通过试验结果修正设计方案,需反复尝试,直到制造满意的产品。该方法一般依赖设计者的经验,难以保证产品的质量,制造成本高,尤其是对于成形过程中的复杂现象、变形规律等科学问题难以得到可视化的认识。因此,难以满足现代制造业的飞速发展的要求。计算机数值模拟技术为设计与制造提供预测与方案优化,已经在现代材料加工工程中起着越来越重要的角色。为了节约成本、提高生产效率,有必要通过数值模拟的方式研究粉末体成形的力学特征。

由于粉末体的可压缩性,使得粉末成形过程的模拟与不可压缩材料的成形过程模拟有显著区别。早期学者主要结合多孔材料屈服理论,结合刚塑性有限元法模拟粉末压制成型过程中的形状、相对密度、应力应变等的变化情况。其中代表性的学者主要有:Mori[7]、Olevsky[8]、Shima 和 Oyane[9]等人。上世纪八十年代之后,随着数值模拟技术的不断发展,对粉末热等静压过程的数值模拟逐步成为研究热点[10]。在这一期间内,粉末热等静压数值模拟技术也得到飞速发展。在粉末体宏观变形方面,以 Sanchez 为代表的学者考虑到包套刚性作用的影响,修正了粘塑性模型中认为粉末材料内部各处压力均等于外部压力的假设,结合数值模拟结果与实验测量数据迭代修正屈服方程以及粉末材料变形过程中的流动应力模型,更精确地表征粉末体热等静压过程中的相关参数的变化[11]。在微观方面,以 Kim 为代表的学者将颗粒边界扩散模型、晶界扩散模型、晶格扩散模型等与蠕变模型结合,研究热等静压下粉末致密化行为[12]。在国内,粉末热等静压数值模拟技术起步较晚,只有少数单位进行了初步研究,其中华中科技大学李少波根据 HIP 的致密化数学模型建立 HIP 图,实现了 TZP 陶瓷材料 HIP 致密化过程的微观模拟[13],北京科技大学贺俊基于 Shima 模型对陶瓷粉末冷等静压过程进行了有限元的数值模拟。金属粉末热等静压计算经过国内外学者数十年的研究,一些经典模型如 Shima 模型得到修正,数值模拟结果更接近实际情况。本工作基于刚塑性有限元理论,系统介绍了金属粉末热等静压过程中的数值建模过程,并结合商业软件 MSC.MARC 与实验结果讨论数值模拟的准确性。

1 有限元模型的建立

金属粉末热等静压过程涉及到外部致密金属包套的刚塑性变形和内部非致密金属粉末的致密化过程。对于致密的金属包套,通常采用刚塑性有限元的相关理论分析其变形行为,而对于非致密的金属粉末在研究其变形时,通常将其视为可压缩的连续介质。在整个成型过程中存在着几何非线性、物理非线性、接触非线性及材料非线性等特征,整个成型过程十分复杂,为了方便在数学上进行处理,在数值模拟过程中,我们定义了如下的假设:

(1) 由于粉末热等静压过程中变形较大,不计材料的弹性变形;

(2) 材料的变形流动服从 Levy-Mises 流动法则;

(3) 材料各处成分组织均匀是匀质各向同性;

(4) 致密的金属包套在变形过程中材料满足体积不可压缩条件;

(5) 由于表面变形力较大,不计体积力和惯性力;

1.1 刚塑性有限元的基础公式

致密的金属包套材料在发生塑性变形时,满足如下的基本方程

(1) 速度 - 应变速率关系 $\dot{\varepsilon}_{ij} = \dfrac{1}{2}\left(u_{i,j} + u_{j,i}\right)$;

(2) 体积不可压缩条件 $\dot{\varepsilon}_V = \dot{\varepsilon}_{kk} = 0$;

(3) Levy-Mises 应力应变速度本构方程 $\dot{\varepsilon}_{ij} = \dot{\lambda}\sigma'_{ij}$;

(4) 速度边界条件及外力边界条件;

根据第一变分原理,在满足上述所有条件的一切可动容速度场中,真实速度场使能量泛函取最小值,因此可以得到满足刚塑性问题的能力泛函表达式如下:

$$\pi = \int_V \overline{\sigma}\dot{\overline{\varepsilon}}\, dV - \int_{S_F} F_i u_i\, dS$$

仅仅利用第一变分原理进行有限元模拟求解, 获得既满足边界条件又满足体积不可压缩条件的动可容速度场是很困难的,在实际的求解过程中,基于体积不可压缩条件,通常的做法是对上公式进行改造,采用 Lagrange 乘子法或罚函数法将体积不可压缩条件引入到泛函中,得到满足致密金属包套变形过程中体积不可压缩条件的新泛函。这里我们介绍采用罚函数法构造的新泛函:

$$\pi = \int_V \overline{\sigma}\dot{\overline{\varepsilon}}\, dV - \int_{S_F} F_i u_i\, dS + \frac{K}{2}\int_V \left(\dot{\varepsilon}_v\right)^2 dV$$

其变分形式为:

$$\delta\pi = \int_V \overline{\sigma}\delta\dot{\overline{\varepsilon}}\, dV - \int_{S_F} F_i \delta u_i\, dS + \frac{K}{2}\int_V \dot{\varepsilon}_v \delta\dot{\varepsilon}_v\, dV$$

对泛函求极值,并由于 δv_i 的任意性,则:

$$\sum_{i=1}^M \frac{\partial \pi^m}{\partial v_i} = \frac{\partial \pi_D^m}{\partial v_I} + \frac{\partial \pi_P^m}{\partial v_I} + \frac{\partial \pi_{S_F}^m}{\partial v_I} = 0$$

$$\frac{\partial \pi_D^m}{\partial v_I} = \int_V \frac{\overline{\sigma}}{\dot{\overline{\varepsilon}}} P_{IJ} v_J\, dV$$

其中: $\dfrac{\partial \pi_P^m}{\partial v_I} = \displaystyle\int_V K C_J v_J C_I\, dV$

$$\frac{\partial \pi_{S_F}^m}{\partial v_I} = -\int_{S_F} F_J N_{JI}\, dS$$

矩阵 P 、C 、的详细表达式参考文献[13,14]。

与致密金属包套不同,金属粉末在热等静压的变形过程中具有可压缩性,不仅会产生塑性变形,同时也伴随有体积收缩。因此,剪切应力和静水压力都会影响金属粉末的屈服行为,同时由于金属粉末致密化过程中粉末相对密度也会不断发生变化,粉末变形过程中的屈服准则与密度相关,许多研究者是通过对粉末烧结的方式获取多孔材料,以研究多孔介质的屈服特性。与致密材料的 Mises 屈服准则相比,粉末材料的屈服准则表达式中包含了应力张量第一不变量和相对密度的影响.目前,各学者提出的屈服准则一般形式可以写成如下形式,

$$f\left(J_1, J_2', \rho\right) = AJ_2 + BI_1^2 = \delta Y_0^2 = Y_R^{\,2}$$

式中 A, B, δ 为相对密度的函数, Y_R, Y_0 分别表示粉末烧结材料和相同成分完全致密材料的屈服强度。

对应于这种屈服函数,得到流动法则如下: $\dot{\varepsilon}_{ij} = \dfrac{\partial f}{\partial \sigma_{ij}} \dot{\lambda}$

为了保证有限元模拟过程的进行,上式中关于相对密度函数的系数 通常都要通过实验进行获取,目前各国学者所提出的适用于粉末材料变形的屈服准则的主要差异也来至于此。文献[10]对相关模型进行了总结。本工作中主要讨论 Shima 和 Oyane 的早期研究模型,即满足:

$$A = 2 + \rho^2, \delta(\rho) = 2\rho^2 - 1, B = 1 - \dfrac{A}{3}$$

至此,我们得到了适用于粉末等静压变形的屈服准则和本构方程,进而推出适用于粉末等静压变形的能量泛函表达式:

$$\int_V Y_R \delta\dot{\bar{\varepsilon}}_R dV - \int_{S_F} F^T \delta u dS = 0$$

1.2 摩擦条件的处理

在粉末热等静压过程中,金属粉末与包套相互接触的区域存在着剧烈的摩擦,正是这种摩擦力的作用使粉末材料发生变形,摩擦力的大小和方向与许多因素相关,如何准确的考虑摩擦力将直接影响模拟结果的准确性。目前在塑性加工模拟中通常都是采用反正切摩擦模型。反正切模型是由 C.C.Chen 和 S.Kobayashi 提出的。该模型不仅可以较为理想的反映出摩擦力的情况。同时该模型可以避免成型过程中速度中性点处摩擦力突然更换方向的问题。该模型的表达式如下:

$$f_s = -mk \dfrac{u_s}{|u_s|} = -mk\left(\dfrac{2}{\pi} \tan^{-1} \dfrac{u_s}{u_0}\right)$$

从上公式我们可以发现为了确保数值模拟工作的正常进行,公式中引入 u_0。显然 u_0 的引入将会对数值结果的可靠性产生重要影响。SHIRO[15]在他的专著中指出:为了获得较为真实的摩擦效果, u_s/u_0 的比值应当大于 10,但该比值过大时,中性点附近摩擦值的突变将会给数值模拟带来困难。文章中指出当 u_s 的数量级为 0.1 时, u_0 的建议值取 $10^{-3} \sim 10^{-4}$。

从上文的介绍可知,求解刚塑性有限元问题是以变分原理为基础的,变分原理以能量积分的形式将塑性偏微分方程组的求解问题变成了求能量泛函的驻值问题。基于罚函数法的刚塑性不完全广义变分原理建立起来的能量泛函中包含应变能率、体积变化惩罚和外力功率对能量泛函的贡献。那么当求解的问题中摩擦不可避免的需要考虑的时候,摩擦功率对能量泛函的贡献就必须纳入能量方程式中。即:

$$\pi_{S_C} = \int_{S_C}\left[\int_0^{u_s} \dfrac{2}{\pi} mk \tan^{-1}\left(\dfrac{u_s}{u_0}\right) du_s\right] dS$$

2 热等静压模拟与实验

本次试验设计热等静压包套为正方体包套边长为 46mm,包套壁厚为 3mm。包套材料为 304 不锈钢。由于正方体包套对称性取 1/4 模型进行模拟分析。粉末网格划分 465 个网格,包套网格划分为 367

个网格,网格类型为四面体网格。HIP 过程加载温度为 930℃,加载压力为 120MPa,整个过程工艺采取传统同时升温升压方法。在 2h 内温度压力加到最大然后保温保压 5h 之后炉冷。HIP 制件取回后用排水法测各部分相对密度值,与模拟结果对比。试样用线切割制成小样后,用 800 到 2000 目砂纸打磨并用 0.1mm 抛光膏抛光后用丙酮洗净然后用 Kroll 腐蚀液腐蚀 10S 之后再次用丙酮浸泡并超声波震动清洗 20 分钟。最后用 Quanta 200 环境扫描电子显微镜观察 TC4 粉末热等静压后尖角处和内部显微组织。

2.1 模具变形流动分析

压坯在 HIP 过程中的变形和流动过程可通过不同时间步的位移及速度矢量来衡量。各时间步压坯位移云图如图 3.1.1:

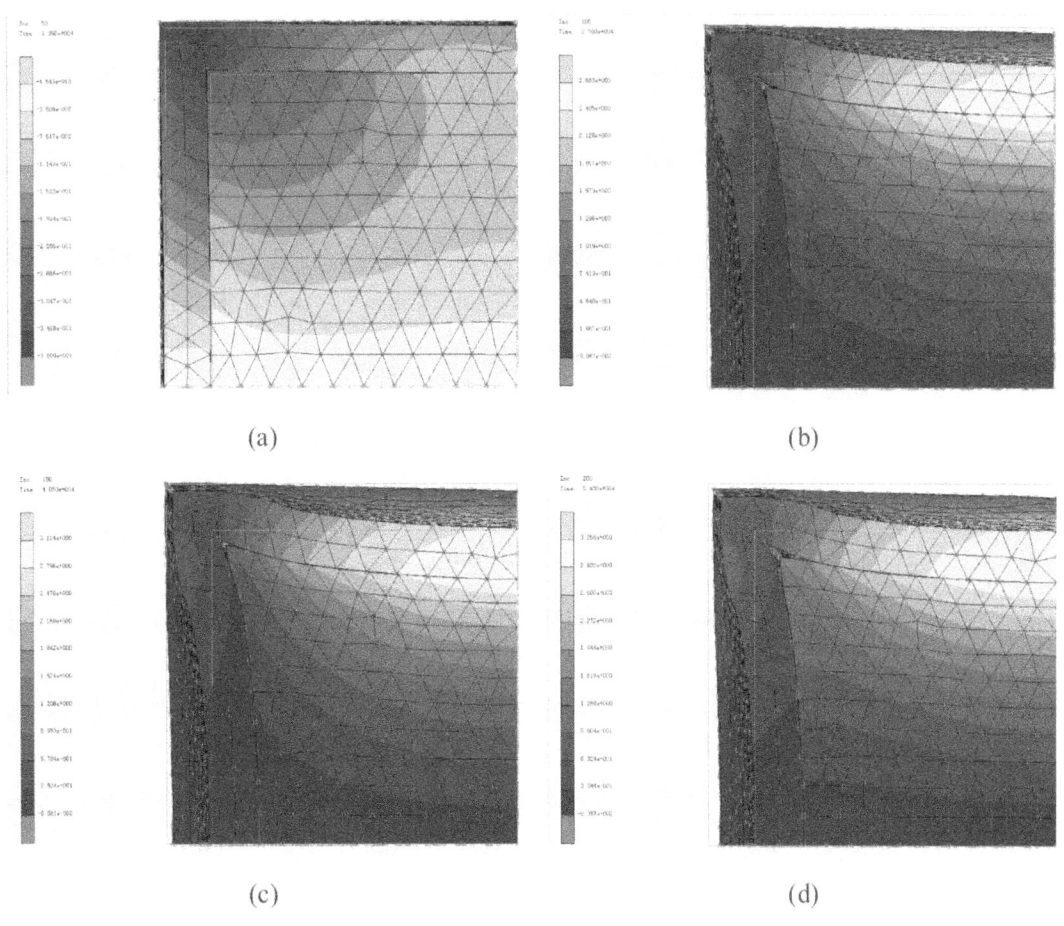

(a)   (b)

(c)   (d)

图 1　各时间步压坯位移云图

其中 a 图为 13500 秒时压坯的 X 方向位移云图,图中蓝色到黄色位移量逐渐变大。由图 a 可看出此时位移量很小且主要发生在粉末区域。位移为负值代表此时为粉末受热膨胀阶段,在这个阶段由于包套阻隔作用,再加上外加压力很小,粉末体主要表现为径向膨胀,并驱动包套的径向位移。图 b 为升温升压末期保温保压初始时期,大概 27000 秒时的位移云图,此时可看到包套已经有初步变形,粉末受压体积收缩从这个时期开始粉末的相对密度会逐渐提高。这主要是由于外加压力超过包套屈服应力,驱动粉末收缩变形,粉末颗粒重排并发生大塑性变形。图 c 为保温保压末期,降温降压开始阶段。这个时候粉末致密度已经有很大的提升,包套变形剧烈,粉末体积收缩已经到末期,此时粉末主要发生蠕变及,颗粒相互挤压的微观形变,粉末受四周压力沿切向方向往中心以及对角方向流动。颗粒间元素扩散以及颗粒挤压重结晶可有效提高抗拉强度。图 d 为降温降压阶段。这个时候粉末以及包套的变形已经到末期。随着压力,温度下降主要表现为粉末体冷却收缩变形。这个阶段粉末的相对密度会有细微提升而整个压坯的变形量不会有很大变动。从整个过程云图分析可知 HIP 过程,包套变形最为剧烈,粉末变形由外向内成减小趋势。

2.2 粉末相对密度分析

在整个粉末体中选取 2 个特征点分别位于正方体包套尖角处的粉末和粉末体内部的粉末。提取特征点整个 HIP 过程的相对密度变化数据并制成曲线图 2:

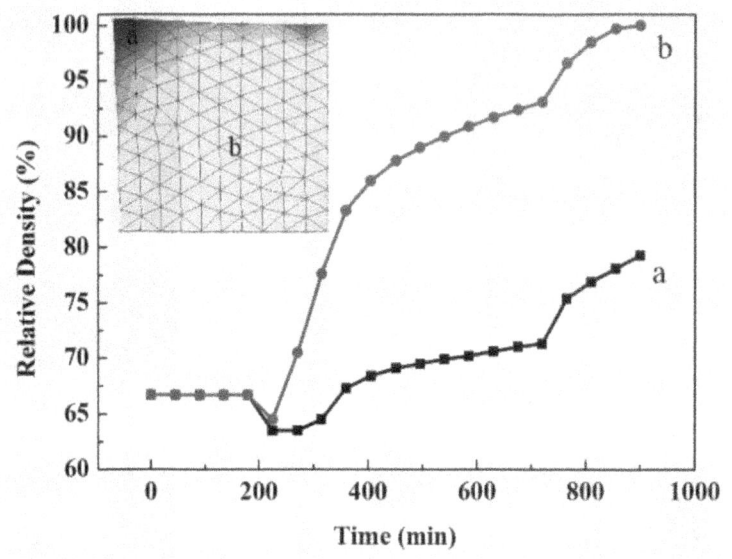

图 2　粉末相对密度随时间变化过程

在 200 分钟以前粉末相对密度变化很小，这是由于粉末虽然此时受热膨胀但是包套阻碍了粉末的膨胀整体变形小,而 200 分钟后包套由于温度压力到达临界值,整体膨胀,粉末相对密度降低。之后保温保压阶段粉末体相对密度剧烈变化有很大提高这是由于在高温高压下粉末塑性变形，颗粒蠕变以及粉末晶体的扩散共同作用,而降温降压阶段粉末体积收缩也会使得粉末相对密度有些微提升。这与上述粉末位移云图分析相互印证。HIP 后可见特征点 a 即粉末尖角处致密度仅有 0.8 左右,相对密度不高。这主要是由于尖角处粉末流动差且加压过程中压力主要作用在垂直于正方体四边方向。驱动粉末主要沿对角线向心部流动,而尖角处粉末仅受切向应力分量作用,压力小且并未受三向压力作用故而最终致密化程度差。中心部分粉末 b 受三向压力作用,粉末体积变化大,粉末流动性好,最终致密度为 0.996 几乎完全致密。

2.3 实验验证

我们对模拟件进行了实际的 HIP 制样,图 3 为热等静压前后的实物图。将 HIP 后的试样切块后用三维扫描提取实体件截面轮廓,然后从模拟结果中提取截面轮廓坐标并在 CAD 中取点描图。然后对比三维扫描和模拟的轮廓误差如图 4。

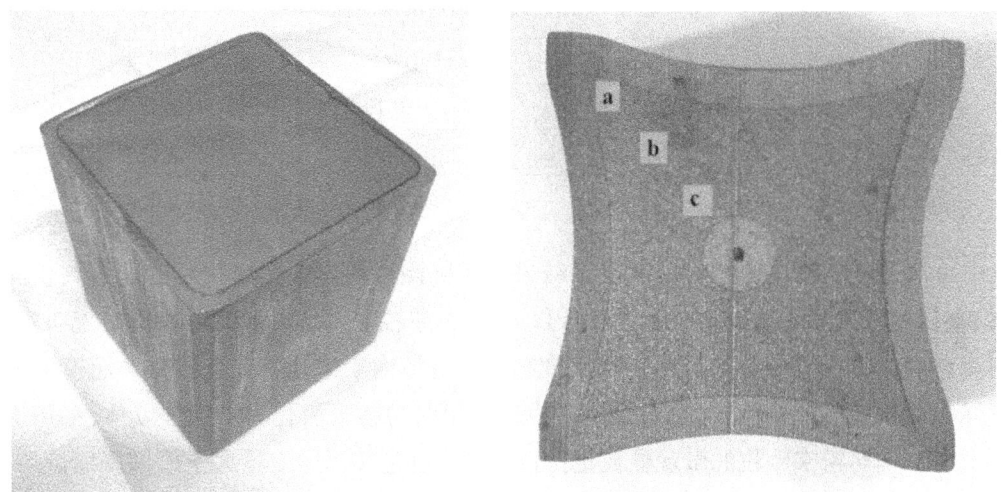

图 3　粉末热等静压前后实物图

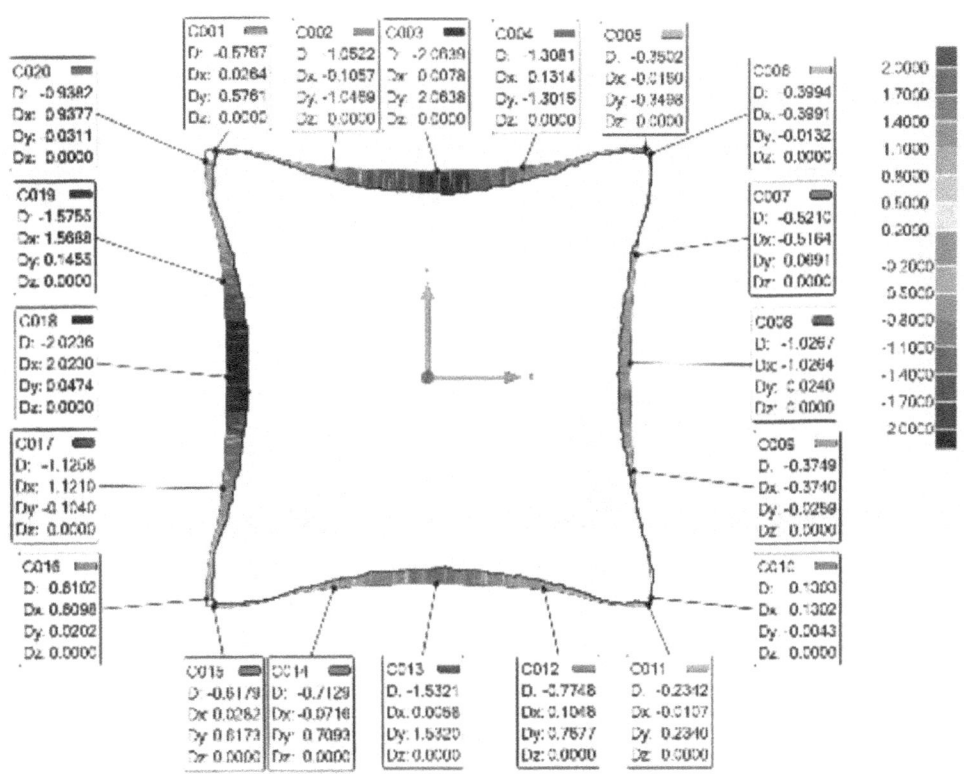

图 4　三维扫描实体轮廓与模拟轮廓误差云图

从图中可清楚看到模拟轮廓和实验得到真实轮廓的误差很小径向最大误差在 2mm 以内在边长中点处,此处的误差为 4%。而越往边角处误差越小。在尖角处误差几乎可以忽略不计。这些误差的主要来源是包套的加工误差以及模拟简化了上下盖的抽气口封焊等因素。

从 a,b,c 三个区域切取样品并用阿基米德排水法测得三个区域的相对密度与模拟结果的三个区域相对密度对比如表 2 所示(D1 为实际相对密度,D2 为模拟相对密度,模拟密度误差 =(—模拟相对密度 - 试验相对密度—/ 实验相对密度)*100%)。

表1　粉末各区域相对密度对比

分区	D1 (%)	D2(%)	相对密度误差(%)
a	0.83	0.79	4.8
b	0.98	0.95	3.1
c	0.99	0.97	2.0

从表中可清楚看到粉末体相对密度模拟结果和实际测得粉末各区域相对密度大致相同，相对密度误差在5%以内。其中a区的相对密度较低仅有0.83左右主要是由于尖角处粉末流动差，压力驱动粉末径向流动因此越靠近中心部位粉末致密度越高c区域内粉末几乎完全致密。而模拟的相对密度相对于实际相对密度偏低一点,这主要是由于模拟建模时将粉末简化为宏观连续体,采用宏观有限元建模忽略了HIP初期粉末颗粒的微观颗粒重排过程。这也是实际压缩量大于模拟压缩量的主要原因。

2.4 金相分析实验

选取粉末边角和粉末体内部两个区域的试样,制样后观察两个区域不同的组织结构。金相图如图5。

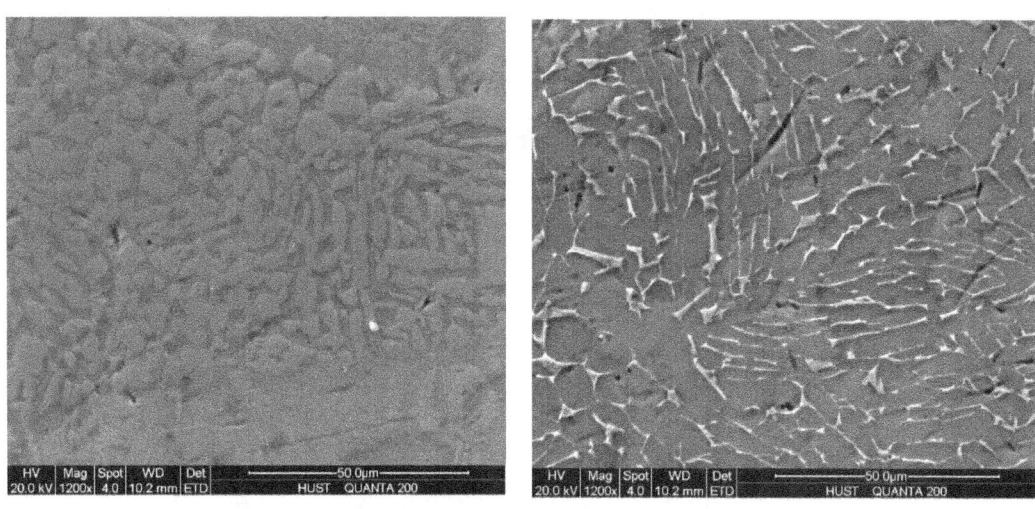

图5　尖角位置粉末金相图(左),粉末体内部金相图(右)

从图中可以看出尖角部分粉末体晶粒尺寸均匀,为细小等轴晶。这主要是由于尖角部位粉末变形量大,在HIP高温高压过程中粉末颗粒压碎并发生重结晶形成。而粉末内部晶粒主要为典型的钛合金α相(灰色)和β相(白色)。而且从尖角处粉末体金相图上可以明显看出晶间的孔隙。这也能说明尖角处粉末致密度差,未能在热等静压后达到完全致密。

3 结论

(1)对金属粉末热等静压过程设计了简单正方体包套方案,建立了粉末HIP过程有限元宏观模型并进行了粉末HIP过程的粉末流动,致密化过程分析。模拟结果表面:粉末致密化阶段主要在升温升压和保温保压阶段,此时在外加压力驱动下粉末发生塑性变形和蠕变并沿对角切线方向向内流动。整个过程粉末外部变形量大,内部变形量小。

(2)对 HIP 后试样进行三维扫描对比模拟取点描图结果表明:模拟变形尺寸结果和实验尺寸结果基本一致误差在 4%以内。对零件分区制样并用排水法测相对密度可知:模拟相对密度和实际相对密度误差在 5%以内。实际变形量略大于模拟变形量且实际相对密度也略大于模拟相对密度,这主要是由于模拟忽略了粉末微观颗粒重排过程。

(3)对粉末尖角部分和粉末体内部分别制样并观察电镜可知:粉末体尖角部分晶粒细小均匀为等轴晶组织,这是由于粉末尖角部分变形剧烈产生,颗粒发生大的变形并再结晶。粉末体内部颗粒变形较小,其组织为典型的 α+β 相组成。尖角处金相图可看到明显孔隙也能说明尖角处未达到完全致密。

参考文献:

[1] 马福康. 等静压技术 [M]. 北京: 冶金工业出版社,1992.

[2] Atkinson HV, Rickinson BA. Hot Isostatic Processing [M]. Netherlands: Kluwer Academic Publishers Group, 1991.

[3] 张义文,上官永恒. 粉末高温合金的研究和发展 [J]. 粉末冶金工业,2004,14(6):30–43.

[4] Bocanegra–Bernal M H. Hot Isostatic Pressing Technology and Its Applications to Metals and Ceramics [J]. Journal of Materials Science, 32004, 39(21): 6399–6420.

[5] Kim H S. Densification Mechanisms during Hot Isostatic Pressing of Stainless Steel Powder Compacts [J]. Journal of Materials Processing Technology, 2002, 123(2): 319–322.

[6] 韩凤麟. 热等静压(HIP)工艺模型化进展 [J]. 粉末冶金工业,2005,15(1): 12–25.

[7] Mori K, Shima S, Osakada K. Finite element method for the analysis of plastic deformation of porous metals [J]. Bulletin of the Japan Society of Mechanical Engineers, 1980, 23(178): 516–522

[8] Olevsky EA. Theory of sintering: from discrete to continuum [J]. Materials Science & Engineering R–Reports, 1998, 23(2): 41–100

[9] Shima S, Oyane M. Plasticity theory for porous metals [J]. International Journal of Mechanical Sciences, 1976, 18(6): 285–291

[10] 刘国承. 金属粉末热等静压致密化数值模拟与试验研究 [D]. 华中科技大学,2011.

[11] Sanchez L, Ouedraogo E, Dellis C, et al. Influence of container on numerical simulation of hot isostatic pressing: final shape profile comparison [J]. Powder Metallurgy, 2004, 47(3): 253–260

[12] Kim HS. Densification mechanisms during hot isostatic pressing of stainless steel powder compacts [J]. Journal of Materials Processing Technology, 2002, 123(2): 319–322

[13] 王忠雷. 三维金属体积成形过程有限元模拟若干关键技术研究与系统开发 [D].山东大学,2011.

[14] 陈文. 增量体积成形数值模拟技术及其在多道次拔长工艺设计中的应用 [D].上海交通大学,2011.

[15] KOBAYASHI, SHIRO, OH, S O O–I K, ALTAN, TAYLAN. Metal forming and the finite–element method ((Book)). 1989.

铸造过程氧化夹杂数值模拟

殷亚军，汪　敏，周建新，陈　香

(华中科技大学　材料成形与模具技术国家重点实验室)

摘　要：铸件的"洁净"程度对铸件质量的影响非常重大，铸件的力学性能与金属液中氧化夹杂物密切相关，因此对铸造过程中夹杂物数量分布、形态、碰撞和聚集长大的建模及模拟极具意义。本文在基于连续介质模型描述铸件充型凝固过程的基础上，运用夹渣物扩散模型模拟预测铸造过程中二次夹渣物的数量分布，同时对铸件不同位置二次夹杂物的粒径分布也进行了预测，了解在铸造过程中的新渣运用弹道模型研究以及在铸造充型过程中的运动情况。

关键词：氧化夹杂；连续介质模型；铸造；数值模拟

Numerical Predication for the Distribution of Oxide Inclusions during Casting Process

YIN Ya-jun, WANG Min, ZHOU Jian-xin, CHEN Xiang

(State Key Laboratory of Material Processing and Die & Mould Technology,
Huazhong University of Science and Technology)

Abstract: The importance of casting cleanliness is increasing with the demand for high quality, so various methods have been developed to analyze the number, morphology, collisions and coalescence of inclusions in casting. Based on the continuum model of classical mixture theory, a three dimensional coupled mathematical model is established for describing the thermo-solute transfer behavior of multicomponent alloy molding filling and solidification process, and a inclusion mass conservation model has been used to study the distribution of the oxide inclusions in casting, also distribution of the inclusions' diameter and number have been predicated by the inclusion population balance model with considering the effects of the Stokes collisions. And a trajectory model has been used to trace the "young" oxide films.

Key words: oxide inclusion；continuum model；casting；numerical simulation

基金项目：国家数控重大专项(2012ZX04012-011)，材料成形与模具技术国家重点实验室自主研究项目(2014-02)，
自然科学基金(No.51305149)，

第一作者：殷亚军，男。材料成形过程计算机数值仿真技术。yinyajun436@gmail.com。

通讯作者：周建新，男。材料成形过程模拟仿真及数字化技术。zhoujianxin@hust.edu.cn。

铸钢件凝固过程与热处理过程组织的快速预测技术研究

张东桥，周建新*，殷亚军，孙 飞

(华中科技大学 材料成形与模具技术国家重点实验室)

摘 要： 本文基于已有的凝固组织快速预测系统，获得铸钢件凝固过程及最终的组织分布。将此组织作为热处理的初始组织，根据 TTA 曲线与 TTT/CCT 曲线，针对某热处理工艺方案进行组织模拟。分析不同的初始组织分布与不同热处理工艺对奥氏体化过程及降温冷却过程的组织分布的影响，可以得出如下结论：初始组织对奥氏体化时间有很大影响，升温速率的快慢决定了奥氏体化时间的长短；控制冷却速度可以很好地控制铸件各个部位的组织分布。

关键词： 铸钢件；凝固过程；热处理过程；组织；预测

The Quick Prediction Technology Research of the structure for the Steel Casting Solidification Process and Heat Treatment Process

ZHANG Dong-qiao, ZHOU Jian-xin, YIN Ya-jun, SUN Fei

(Huazhong University of Science and Technology, State Key Laboratory of Materials Processing and Die & Mould Technology)

Abstract: The article can get the structure distribution of the casting solidification process based on the existing fast prediction system. According to TTA curves and TTT/CCT curves, and the casting structure as the initial structure of the heat treatment, we can simulate the structure distribution on a specific heat treatment process scheme. Analyzing the influence of the various initial structure distributions and different heat treatment process to the structure distribution of austenitization process and quenching process. We can get following conclusions. The initial structure has a great influence on the austenitizing time, heating rate speed determines the length of austenitizing time; Controlling cooling rate can easily get the desired structure distribution of the casting parts.

Key words: steel casting; solidification; heat treatment; structure; prediction

基金项目： 国家数控重大专项(2012ZX04012-011)，材料成形与模具技术国家重点实验室 2015 年重点自主研究项目(数字化智能化铸造技术及应用)

第一作者： 张东桥，男。热处理过程数值模拟。zhangdongqiao117@gmail.com。

通讯作者： 周建新，男。材料成形过程模拟仿真及数字化技术。zhoujianxin@hust.edu.cn。

基于三维离散传播辐射模型的高温合金定向凝固数值模拟

郭 钊，周建新，殷亚军，吴 凯

(1.华中科技大学 材料成形与模具技术国家重点实验室)

摘 要：在实际定向凝固炉膛内,辐射传热是其主要换热方式。为了更准确地获得高温合金定向凝固过程温度分布及其随时间的变化规律, 本文基于三维离散传播辐射模型编制相应的数值模拟程序,并运用该程序对高温合金定向凝固过程温度场进行数值模拟,重点考察了高温合金温度场分布及不同位置凝固过程的冷却曲线。结果表明:随着抽拉过程的进行,高温合金内部等温面呈不同程度的倾斜,在炉膛加热区,高温合金等温面呈现向上凸;在冷却区,高温合金等温面呈现向下凸;而在两区之间,等温线接近水平,温度分布最均匀。随着区域位置的不同,其冷却曲线变化程度也呈现不同规律,所得模拟结果与已有的研究结果吻合较好。

关键词：辐射模型；高温合金；定向凝固；数值模拟

Numerical Simulation of Superalloy's Directional Solidification Based on Three-dimensional Discrete Transfer Radiation Model

GUO Zhao, ZHOU Jian-xin, YIN Ya-jun, WU Kai

(1. State Key Laboratory of Materials Processing and Die & Mould Technology, Huazhong University of Science and Technology)

Abstract: Radiation heat transfer is the main way of heat transfer in the furnace of actual directional solidification. In order to more accurately obtain the temperature distribution and its variation law with time in directional solidification process of superalloy, the numerical software was programmed based on three-dimensional discrete transfer radiation model and then adopted to simulate the temperature field in directional solidification of superalloy. The temperature field distribution and cooling curve of different position in solidification process of superalloy were emphatically analyzed. The results show that: in the pulling process, the internal isothermal surfaces of superalloy appear different degrees of tilt. In the heating zone of furnace, the isothermal surface is convex upward; in the cooling zone, the isothermal surface is convex downward, and between the two zones, the isothermal surface is close to level. As such, in different zones, the cooling curves also present different rules. And the simulation results are in good agreement with the existing research results.

Key words: radiation model; superalloy; directional solidification; numerical simulation

基金项目：材料成形与模具技术国家重点实验室 2015 年重点自主研究项目(数字化智能化铸造技术及应用),
国家数控重大专项(2012ZX04012-011)。

第一作者：郭 钊,男。材料成形过程模拟仿真及数字化技术。zg2014@hust.edu.cn。

通讯作者：周建新,男。材料成形过程模拟仿真及数字化技术。zhoujianxin@hust.edu.cn。

拜尔法赤泥地聚物制备及抗压强度特征

王　斌[1]，朱文凤[1]，王林江[2]

（1.桂林理工大学 材料科学与工程学院；2.广西有色金属及特色材料加工国家重点实验室培育基地）

摘　要：以广西拜耳法赤泥为主要原料，添加偏高岭土、氢氧化钠、水玻璃等工艺制备赤泥地聚物。在赤泥含量50%时（赤泥和偏高岭土各占一半），水玻璃模数为1.5，养护时间24h，养护温度为70℃，得到赤泥地聚物的抗压强度为67.4MPa。通过X射线衍射分析（XRD）可看出基体均包括弥漫的衍射峰和尖锐的衍射峰，为赤泥和偏高岭土叠加的特征。红外光谱分析表明地聚合物在合成过程中AlO$_4$四面体键接在SiO$_4$四面体上，共同构成了地聚合物的三维网络状结构。场发射扫描电子显微镜分析可以看出地聚物中赤泥和偏高岭土主要是由片状结构组成，材料的基体之间结构较均匀且较致密。

关键词：赤泥；偏高岭土；地聚物；抗压强度

The Preparation and Compression Strength Analysisof Bayer Red Mud Geopolymer

WANG Bin[1], ZHU Wen-feng[1], WANG Lin-jiang[2]

（1.School of Materials Science and Engineering, Guilin University of Technology,;2. Ministry-province Jointly-constructed Cultivation Base for State Key Laboratory of Processing for Non-ferrous Metal and Featured Materials）

Abstract: In this paper, red mud geopolymer was prepared via using Guangxi Bayer process red mud as principal raw material with metakaolin, sodium hydroxide, water glass. As the sample contained 50% of red mud (red mud and metakaolin were fifty-fifty split),with modulus of water glass being 1.5, Curing time being 24h, curing temperature being 70℃ ,the compression strength of red mud geopolymer is 67.4Mpa. It can be from XRD seen that sample had diffuse diffraction peaks and sharp diffraction peaks, which is the feature of superposition of red mud and metakaolin Infrared spectroscopy showed that AlO$_4$ tetrahedron were attached to SiO$_4$ tetrahedron forming a three-dimensional network structure of geopolymer in the process of synthesis. It can be seen from SEM that red mud and metakaolin of geopolymer were made up of laminated structure, and the laminated structure is uniform and compact.

Key words: red mud; metakaolin; geopolymer; compressive strength

基金项目：国家"973"前期研究专项（2012CB722804）和国家自然科学基金（41272064）。

第一作者：王　斌，男。研究方向矿物材料与绿色建材。wangbin900418@163.com，。

通讯作者：王林江，男，教授，博导。wlinjiang@163.com，。

Con-aid 稳定剂固化土电镜扫描分析

黄纪蓉[1]，张信贵[2,3]，易念平[2,3]

(1.广西大学 材料科学与工程学院；2.广西大学 土木建筑工程学院；3.广西防灾减灾与工程安全重点实验室)

摘 要：稳定土是土木与水利工程上应用的加固材料,因其应用广泛而被大量采用。目前稳定土的方法有物理的、化学的、物理与化学两者并用等方法,物理方法往往依靠一定的机械手段通过增强密实度而获得土的强度,而化学方法则是依靠改变土粒内部的连接让土的强度增加,当物理与化学两者并用时将取得更好加固效果。Sulphonated Petroleum Products 类液体稳定剂是稳定土的化学产品。因为稳定剂-粘土-水系统产生的物理化学作用是一个十分复杂的过程,多种因素的共同作用,影响了稳定剂对土的加固效果,Sulphonated Petroleum Products 类液体稳定剂加固土质的机理依然不甚明确。通过引入加拿大 Con-Aid 液态稳定剂对南宁公路工程高液限土进行处置,从土的细观结构的新视角对其固化机理进行重新认识,为此设计了扫描电镜(SEM)试验。试验与分析说明液体稳定剂的加固效果被细观结构所控制,稳定土细观结构是土强度来源的的主要场所。

关键词：扫描电镜；稳定土；液体稳定剂；细观结构；强度

SEM Analyses for Con-Aid Liquid Stablizer Solidification Soil

HUANG Ji-rong[1], ZHANG Xin-gui[2,3], YI Nian-ping[2,3]

(1.School of Materials Science and Engineering, Guangxi University 2.School of Civil Engineering and Architecture, Guangxi University 3.Guangxi Key Laboratory of Disaster Prevention and Structural Safety)

Abstract: The stability soil is a reinforcement materials being applied extensively to civil engineering and hydraulic engineering. At present, methods of reinforcement soil are of three kinds, physicals, chemicals, physicals and chemicals methods. Physicals methods is a kind of method that strength of soil is from increasing of its density by packing particles closer together rely on mechanical tool, for example, by means of rollers, vibrators or rammers. Chemicals methods is a kind of method that strength of soil is from increasing of state of interparticles contacts and forces among particles or particle groups by chemicals reaction. Generally, density effect of soil is more highlight by combined physicals and chemicals means. Liquid stabilizer-Sulphonated Petroleum Products is the chemical reinforcing agent of stability soil. Because of reinforcing agent - clay - water system produces is a very complicated process, the physical and chemical interaction of many kinds of factors affect the reinforcement effect of soil. The mechanism of strengthening soil using the Sulphonated Petroleum Products is still made very clear. Canadian Con-Aid liquid stablizer is a

基金项目：广西重点实验室开放项目防灾与结构安全工程(2012ZDK08,2013ZDX11)；国家自然科学基金(No.51268003)。
第一作者：黄纪蓉,女。材料电镜显微分析。Hjr0592@sina.com。
通讯作者：张信贵,男。边坡工程、桩基工程、地基处理；环境岩土工程、岩土工程测试技术及土的基本性质研究。xgzhangchn@foxmail.com。

Sulphonated Petroleum Products too. In order to research Canadian Con-Aid liquid stablizer strengthened the mechanism of soil, designed many samples of different mixture proportions in laboratory, tested theirs compressive strength, workability and durability, completed experiments of Cation exchange capacity, SEM, X-ray diffraction and so on. And Canadian Con-Aid liquid stablizer is applied to the high liquid limit soil in Nanning highway projects. Form the result of the test and analysis of its characteristics, View solidification mechanism of stability soil from the structure of the soil's standpoint, we recognized that the reinforcement effect of the liquid stablizer is controlled by the ministructure of the soil, strength of the stability soil mainly stems from the ministructure of the stability soil.

Keywords: SEM; stability soil; liquid stabilizer; reinforcing agent; ministructure of the soil; strength

1 试验目的

分析稳定剂作用的机理,对未加稳定剂土和加稳定剂处理后土进行扫描电镜分析(SEM),探求稳定土在细(微)观结构层面上的变化。

2 试验步骤

(1)试验土样同以上两个试验,将取回的扰动土样风干,按最佳含水量配水,并以四种稳定剂掺入比(0、20∶100,000、50∶100,000 和 100∶100,000)将稳定剂加入水中一并配制土样,将土样养护一段时间,让其充分浸润;

(2)以最大干密度为标准,用击实法制成直径为 3.91cm,高为 8cm 的重塑土样,并养护一段时间;

(3)取出重塑样,小心将其掰成细小的试样、编号,并置于室温下让其自然阴干,再抽真空喷金做扫描电镜试验;

(4)为较好地反映出稳定剂作用后土结构变化情况,所有样品扫描区域都选在表面及表面下 5mm 厚度内。

3 试验结果分析

扫描电镜能直接揭示细观结构中的颗粒与颗粒之间关系。选取合适的扫描区域的放大倍数非常关键,扫描区域太小(放大倍数大)很难进行横向比较,扫描区域太大(放大倍数小)将看不到细观结构 - 颗粒与颗粒之间的情况,在合适的扫描区域放大倍数才能有效地反映出稳定剂作用后土细观结构变异的效果。

为较全面地考察加入稳定剂后对试样在细观结构上的改造效果,将未经稳定剂处理(稳定剂掺入量为 0)及加入不同稳定剂掺入比(20∶100,000、50∶100,000 和 100∶100,000)处理的重塑土试样分别做了四组,每组 5 个倍数(放大倍数分别为 300、500、1000、2000、5000 倍)的电镜扫描测试。SEM 照片见图 1—图 4。

对不同倍数的照片分析发现:在放大倍数为 300 和 500 时,可观察到四组试样均显示为较为密集粘土基质体结构,其上还看到附有较大的团块。由于放大倍数较低,不能获取更细节的信息,看不出其接触形式和其中的孔隙,在放大倍数为 1000 时,已能较清楚地看到扁平的团状体以面 - 面的方式叠聚而成,其中还有一些较小的孔隙,在放大倍数为 2000 和 5000 时,就能清楚地观察到团状物及其间的连接形式和孔隙的形态。团状物是由大部分扁平的片状体及少量的曲片叠聚而成,叠聚体自相紧密集聚形成一定范围的粘土基质结构,其间仅有极少的孔隙。根据前人的研究,成扁平的片状叠聚体主要是高岭石型矿物,曲片状叠聚体为伊利石—蒙脱石型矿物。因此,为了尽可能多地获取细(微)观结构信息的角度出发,反映稳定剂作用的效果,选择 2000 与 5000 的放大倍数进行比较分析。图 1 至图 4 分别为稳定剂掺入比为 0、20:100,000、50:100,000 和 100:100,000 处理后重塑土样的 300、500、1000、2000 和 5000 的 SEM 照片。

稳定剂掺入比为 0 的重塑土样 SEM 照片 d)、e)可观察到,主要由片状体以面—面接触形成,有少量以边—面形式接触,基本单元体为片状叠聚体,单元体间孔隙较多。

4 结论

由稳定剂掺入比为 20:100,000、50:100,000 和 100:100,000 的重塑土样的 SEM 照片可看出,与素土相同,也主要由片状体以面—面接触形成,基本单元体为片状叠聚体,但作为整体的粘土基质类结构上则更为致密,单元体密集排列,单元间几乎未见孔隙。可见经稳定剂处理后,土样结构发生了变异。

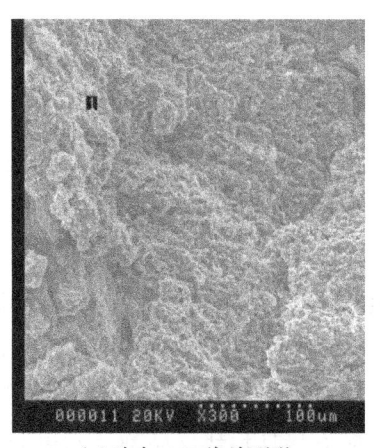

(a)放大 300 倍的形貌
(稳定剂掺入比=0)

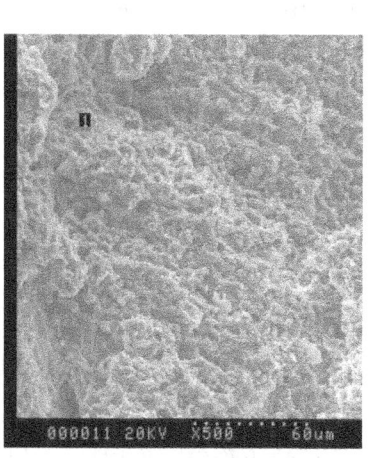

(b)放大 500 倍的形貌
(稳定剂掺入比=0)

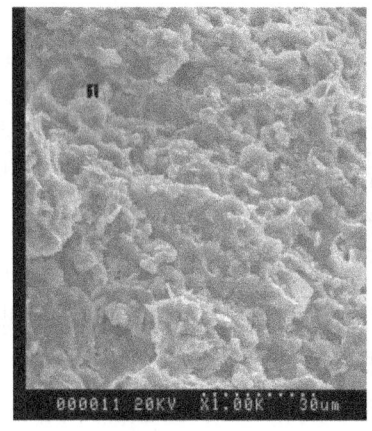

(c)放大 1000 倍的形貌
(稳定剂掺入比=0)

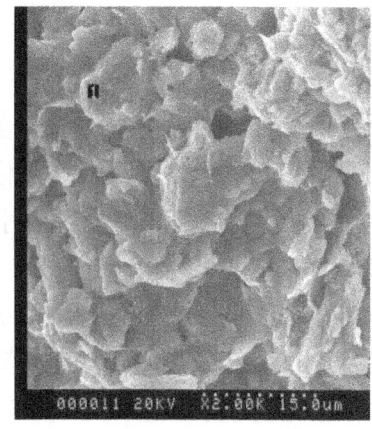

(d)放大 2000 倍的形貌
(稳定剂掺入比=0)

(e)放大 5000 倍的形貌
(稳定剂掺入比=0)

图 1 未经稳定剂处理(掺入比=0)的重塑土样(素土)的 SEM 照片

Fig.1 SEM images of Remodeling of soil samples (plain soil) without stabilizer treatment (incorporation ratio=0)

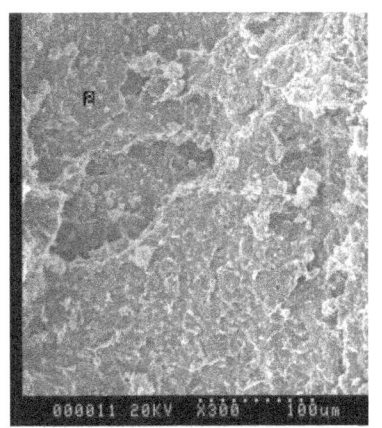

(f)放大 300 倍的形貌
(掺入比为 20:100,000)

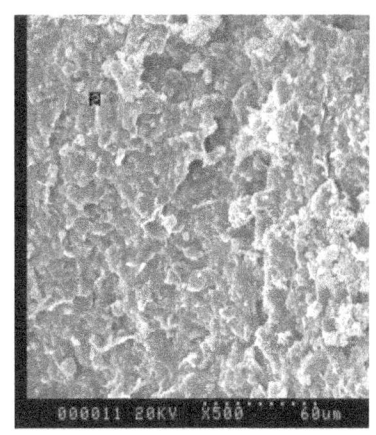

(g)放大 500 倍的形貌
(掺入比为 20:100,000)

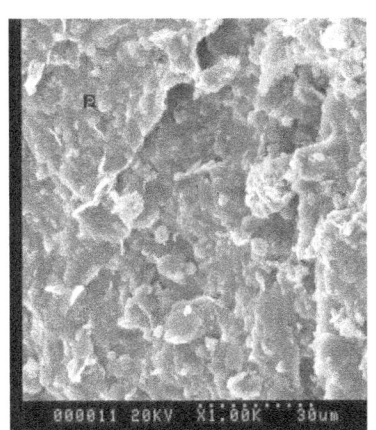

(h)放大 1000 倍的形貌
(掺入比为 20:100,000)

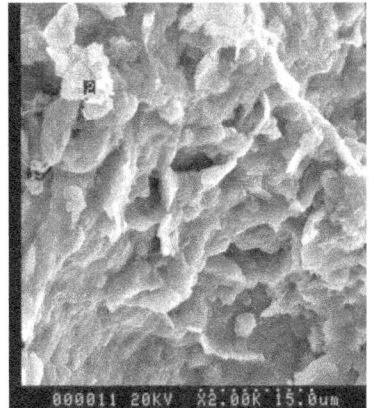

(i)放大 2000 倍的形貌
(掺入比为 20:100,000)

(j)放大 5000 倍的形貌
(掺入比为 20:100,000)

图 2 经掺入比为 20:100,000 稳定剂处理的重塑土样的 SEM 照片
Fig.2 SEM images of The reconstruction soil samples treated by 20:100000 stabilizer

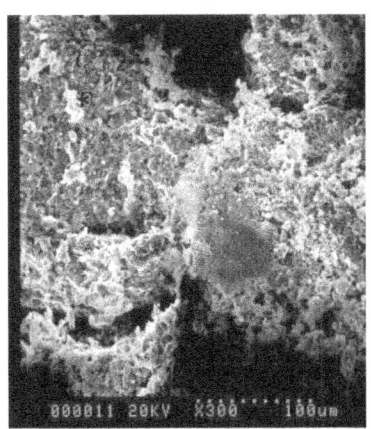

(k)放大 300 倍的形貌
(掺入比为 50:100,000)

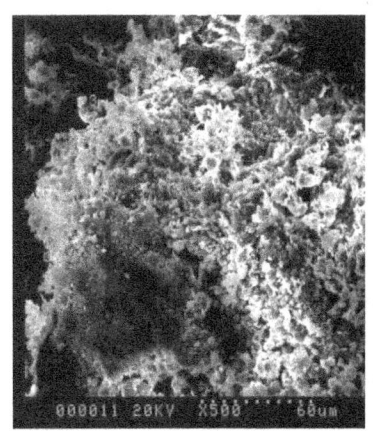

(l)放大 500 倍的形貌
(掺入比为 50:100,000)

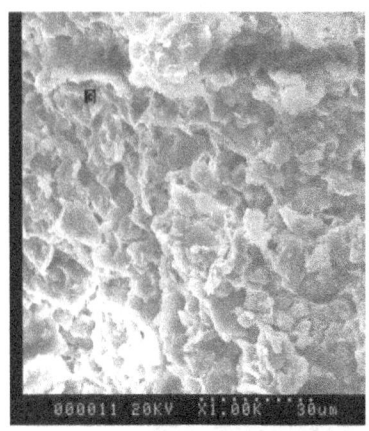 

(m)放大 1000 倍的形貌　　　　(n)放大 2000 倍的形貌　　　　(o)放大 5000 倍的形貌
(掺入比为 50:100,000)　　　　(掺入比为 50:100,000)　　　　(掺入比为 50:100,000)

图3　经掺入比为 50:100,000 稳定剂处理的重塑土样的 SEM 照片
Fig.3　SEM images of The reconstruction soil samples treated by 50:100000 stabilizer

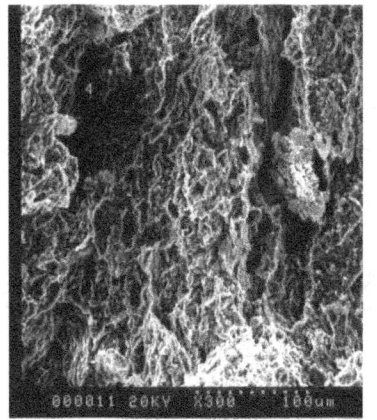 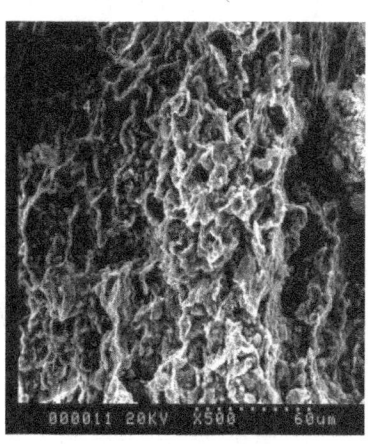

(p)放大 300 倍的形貌　　　　　　　(q)放大 500 倍的形貌
(掺入比为 100:100,000)　　　　　　(掺入比为 100:100,000)

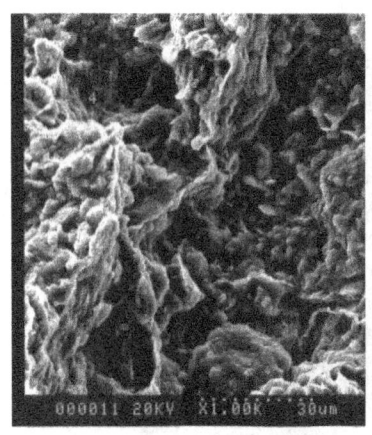  

(r)放大 1000 倍的形貌　　　　(s)放大 2000 倍的形貌　　　　(t)放大 5000 倍的形貌
(掺入比为 100:100,000)　　　　(掺入比为 100:100,000)　　　　(掺入比为 100:100,000)

图4　经掺入比为 100:100,000 稳定剂处理的重塑土样的 SEM 照片
Fig.4　SEM images of The reconstruction soil samples treated by 100:100000 stabilizer

参考文献:

[1] 谢定义,齐吉琳. 土结构性及其定量化参数研究的新途径 [J]. 岩土工程学报,1996,11(6): 651-656.

[2] 施斌. 粘性土微观结构研究回顾与展望 [J]. 工程地质学报,1996,3(3),39-44.

[3] 蒋明镜. 结构性粘土研究综述 [J]. 水利水电科技进展,1999,2(1): 26-33.

[4] 张信贵. 城市区域水土作用分析与土的结构强度研究 [D]. 南宁: 广西大学土木建筑工程学院,2002.

[5] 沈珠江. 结构性粘土试样人工制备方法研究 [J]. 水利学报,1997(1),56-72.

基于摩擦修正的含钪 Al-Zn-Mg-Zr 合金热压缩变形行为研究

李 波[1], 吴海华[1], 陈从平[1], 潘清林[2], 尹志民[2]

(1.三峡大学 机械与动力学院；2.中南大学 材料科学与工程学院)

摘 要：采用 Gleeble-1500 热模拟试验机对含钪 Al-Zn-Mg-Zr 合金进行热压缩试验，以试验所得数据(变形温度 340~500℃,应变速率 0.001~10s^{-1})为基础,基于摩擦修正后的流变应力曲线采用双曲正弦形式的修正 Arrhenius 关系对含钪 Al-Zn-Mg-Zr 合金的本构模型进行回归。结果表明：合金的流变应力随应变速率的增大而增大,随变形温度的升高而减小,其热变形机制主要以动态回复和动态再结晶为主。同时,热压缩变形过程中摩擦对于合金流变应力的影响十分显著,采取的修正方法大大降低了实验中摩擦引起的误差,比较回归模型计算的应力值与实测值其平均相对误差仅为 1.98%,因此采用 Z 参数的双曲正弦函数形式能够较为精确地预测含钪 Al-Zn-Mg-Zr 合金高温变形时的流变应力。

关键词：含钪 Al-Zn-Mg-Zr 合金；热压缩变形；流变应力；摩擦修正；本构方程

Hot Compression Deformation Behavior of Al-Zn-Mg-Zr Alloy Containing Sc Based on Friction Correction

LI Bo[1], WU Hai-hua[1], CHEN Cong-ping[1], PAN Qing-lin[2], YIN Zhi-min[2]

(1.College of Mechanical and Power Engineering, China Three Gorges University；
2.School of Materials Science and Engineering, Central South University)

Abstract: The hot compression test of Al-Zn-Mg-Zr alloy containing Sc was performed on Gleeble-1500 thermal simulation machine. According to the obtained experimental data (deformation temperatures of 340-500℃ and strain rates of $0.001 \sim 10s^{-1}$), the flow stress behavior were analyzed based on the true stress-strain corrected for friction; The constitutive model of Al-Zn-Mg-Zr alloy containing Sc was established. The result shows the flow stress of the alloy increased with increasing strain rate, and decreased with increasing deforming temperature, the main soften mechanisms were dynamic recovery and dynamic recrystallization. Meanwhile, the effect of friction on flow stress during compression process was notable and the adopted correction method could reduce the error. Comparing with experimental and predicted stress, the average relative error was 1.98%. Thus, the Arrhenius-type hyperbolic sine equation with Z parameter could accurately predict the flow stress of Al-Zn-Mg-Zr alloy containing Sc during hot deformation.

Key Words：Al-Zn-Mg-Sc-Zr alloy；hot deformation；flow stress；corrected for friction；constitutive model

---

基金项目：国家重点基础研究计划("973"计划)项目(2012CB619503)。

第一作者：李 波,男。金属塑性成形与数值模拟。liboiec@163.com。

半固态 ZCuSn10 铜合金坯料单向压缩变形本构模型研究

肖　寒[1]，陈泽邦[1]，李　勇[1]，周荣锋[1,2]，卢德宏[1]，周　荣[1]

(1. 昆明理工大学 材料科学与工程学院；2. 昆明理工大学 分析测试研究中心)

摘　要：利用 Gleeble-1500D 热/力模拟实验机对半固态 ZCuSn10 铜合金坯料进行了单向压缩实验，研究了不同应变量、变形温度、应变速率下半固态 ZCuSn10 铜合金真应力-应变曲线的规律，通过对真应力-应变曲线的回归分析得出了半固态 ZCuSn10 铜合金的本构关系模型。结果表明：在其他变形条件相同的情况下，随着应变速率的增加，变形应力增加；随着变形温度的升高，变形应力降低；随着应变量的增加，变形峰值应力无明显变化。在其他变形条件相同的情况下，在真应力-应变曲线达到峰值应力前，不同应变量的真应力-应变曲线几乎重合在一起。

关键词：铜合金；半固态；单向压缩；本构模型

Constitutive models of semi-solid ZCuSn10 copper alloy billet during uniaxial compression

XIAO Han[1], CHEN Ze-bang[1], LI Yong[1], ZHOU Rong-feng[1,2], LU De-hong[1], ZHOU Rong[1]

(1. Faculty of Materials Science and Engineering, Kunming University of Science and Technology;
2. Research Center for Analysis and Measurement, Kunming University of Science and Technology)

Abstract: The compression test of semi-solid ZCuSn10 copper alloy was conducted at Gleeble-1500 thermal simulation testing machine, and the thixo-deformation true stress-strain curves were obtained. By uniaxial compression test with different temperature, strain rate and strain, it proves that stress of semi-solid ZCuSn10 copper alloy has a close relationship with temperature, strain rate and strain. A viscoplastic constitutive model during thixo-deformation was established through regression analysis. The experimental results show that, with the deformation strain rate increasing, deformation stress increases, with the deformation temperature increasing, deformation stress induces, with the deformation strain increasing, no significant change in peak stress. Before the true stress - strain curve achieve the peak stress, true stress - strain curves of different strain almost coincide together.

Key words: copper alloy; semi-solid; uniaxial compression; constitutive equations

基金项目：云南省应用基础研究重点项目(2011FA007)；高等学校博士学科点专项科研基金资助项目(20125314120013)；云南省应用基础研究面上项目(2014FB131)；国家级大学生创新创业训练计划项目(201410674001)。

第一作者：肖　寒，男。半固态成形技术。zztixh@163.com。

0 引言

金属半固态成形技术(Semi-solid Metal Forming Processes, SSM 或 Semi Solid Processing, SSP)是通过对处于由固态向液态转变或者液态向固态转变过程中的金属采取搅拌或控制凝固过程等方法，得到一种微观组织是由液相包裹并具有一定圆整度近球形固相颗粒组成的浆料的成形加工技术[1-5]。由于半固态金属材料同时具有流动特性和触变特性，为了弄清楚半固态金属在变形过程中的流动特性与触变特性，优化变形过程和加工工艺参数，对半固态金属建立能准确反映其流变特性的本构方程是十分有意义的。目前,描述半固态金属变形的模型都是以两个理论为基础的,一个是固体骨架的变形行为用连续多孔材料的本构方程描述,半固态合金中液相的流动用达西定律描述。另外一种是半固态金属材料在变形过程中,宏观变形与固相晶粒的滑移转动变形相互协调,同时考虑固相颗粒间及液相对固相颗粒的润滑作用两方面的相互作用,并将二者进行耦合[6,7],在半固态技术本构方程的研究方面,Flemings 教授最早提出了半固态金属的流动应力与应变速率之间的关系[8]。有关半固态金属变形行为本构方程的研究目前处于探索阶段,主要是对半固态金属的压缩力学行为进行研究,通过压缩实验测得不同应变量、温度、应变速率下的材料应力 - 应变关系,采用回归分析的方法获得了半固态金属的刚粘塑性本构方程。Kang 通过引入分离系数提出了流变应力与应变速率之间的关系[9]。考虑到固相率、液体的粘度和固相颗粒间液体通道的厚度等对流变应力的影响,Daniel[10]等建立了高固相率下的 Al-4.5Cu 合金拉伸变形的本构关系模型。

本课题以采用 SIMA 法制备的半固态锡青铜坯料为研究对象,通过在半固态温度区间进行单向自由压缩试验来研究半固态铜合金在不同的变形速率下的变形行为和不同变形温度下的组织演变,以期能找出半固态铜合金的变形规律和组织演变的规律,为半固态铜合金成形工艺的选择提供参考,推进铜合金半固态加工技术在实际工业中的应用。

1 试样制备和试验方法

1.1 试样制备

试验材料为 ZCuSn10 铜合金,其化学成分如表 1 所示。采用耐驰 STA449F3 同步热分析仪进行差热分析(DSC: Differential Scanning Calorimetry)确定该合金的固相线温度为 830℃,液相线温度为 1020℃。

表 1　ZCuSn10 铜合金化学成分 Wt.%
Table.1　Chemical composition of ZCuSn10 copper alloy Wt.%

Element	Cu	Sn	Else.
Contents	88.25	10.48	1.27

SIMA 法制备 ZCuSn10 铜合金半固态坯料的过程为:第一步,铸锭的制备与轧制预变形。ZCuSn10 铜合金在1180℃浇注至金属模成型,凝固后开模取出空冷至室温,获得 ZCuSn10 铜合金铸锭。在铸锭上截取试样,机加工为 25mm×25mm×150mm 的矩形棒料作为轧制试样。第二步,半固态温度区间等温

处理。等温过程在 910℃ 下进行,保温时间为 25min。保温处理后立即水淬,得到半固态 ZCuSn10 铜合金试样。

1.2 试验方法

半固态单向自由压缩试验采用 Φ10mm×15mm 的圆柱形试样, 在 Gleeble-1500 材料热 / 力学模拟试验机上进行,为了减少压缩过程中摩擦阻力的影响,压缩试样按照 GB/T73142005《金属材料室温压缩试验方法》中有关压缩试样制备要求进行加工,压缩过程中试样端部涂有一定量的石墨润滑[11,12]。压缩实验过程中, 加热速度为 10℃ /s, 但为了避免加热系统的惯性使试样的实际温度超出预定变形温度,在加热到距预定变形温度 50℃ 时,加热速度降为 2℃ /s,加热到预定变形温度之后保温 10 s。具体的热压缩变形参数为: 实验热压缩应变量为 0.05,0.1,0.2,0.4,0.6,0.8。变形温度为 900℃ ,910℃ ,920℃ ,930℃ 。应变速率为 0.5s⁻¹,1s⁻¹,1s⁻¹,5s⁻¹,10s⁻¹。试样在半固态温度区间压缩变形后立即水冷,以保留其高温原始组织。试样沿轴线从中心剖开,经粗磨、细磨、抛光后腐蚀并采用 LEICA DMI 5000M 金相显微镜观察显微组织。

2 试验结果与讨论

2.1 单向自由压缩试验

2.1.1 温度对压缩真应力–应变曲线的影响

不同温度下的压缩真应力 - 应变曲线如图 1 所示。图 1(a)为半固态 ZCuSn10 铜合金在应变速率(s⁻¹)为 0.5,应变量 0.4,不同温度下压缩真应力 - 应变曲线,图 1(b)为半固态 ZCuSn10 铜合金在应变速率(s⁻¹)为 0.5,应变量 0.6,不同温度下压缩真应力 - 应变曲线,图 1(c)为半固态 ZCuSn10 铜合金在应变速率(s⁻¹)为 5,应变量为 0.8,不同温度下压缩真应力 - 应变曲线。可以看出,无论是应变速率(s⁻¹)为 0.5 还是 10,应变量为 0.4 、0.6 还是 0.8 的 ZCuSn10 铜合金压缩试样,压缩变形的温度都是一个重要的工艺参数,它对变形过程的特别明显。真应力 - 应变曲线整体呈现的走势都是一致的,压缩一开始时,变形应力迅速上升到最高点,随着应变的增大,变形应力又逐渐降低。由于压缩过程中外在因素的影响,真应力 - 应变曲线中间有较小的波动,但是最后都缓慢降低到一个较低的稳定值。这是由于在压缩变形刚开始时,随着压缩应变的增加,压缩应力要克服半固态 ZCuSn10 铜合金压缩试样内液相间的流动应力,液固相之间的摩擦阻力,所以在压缩一开始应力快速上升,直至达到半固态 ZCuSn10 铜合金的屈服极限,即压缩应力的最高值,这时半固态 ZCuSn10 铜合金压缩试样的变形进行到固相晶粒间的塑性变形,应力达到最高值时,说明半固态 ZCuSn10 铜合金压缩试样的固相颗粒间发生了永久塑性变形。随后, 由于压缩应力对试样的剪切作用,半固态 ZCuSn10 铜合金试样中初生(α-Cu)颗粒被液相包裹,液相的润滑作用使得滑移转动更加容易进行,降低了(α-Cu)固相颗粒间的摩擦力,引起压缩应力降低。

随着压缩的进行, 半固态 ZCuSn10 铜合金试样中的液相在外力的作用下向试样边缘的自由变形区流动,在压缩变形后期,试样心部的液相越来越少。在应力降低过程中的小波动,可能与变形应力数值的误差有关,干扰了应力数据的变化规律。

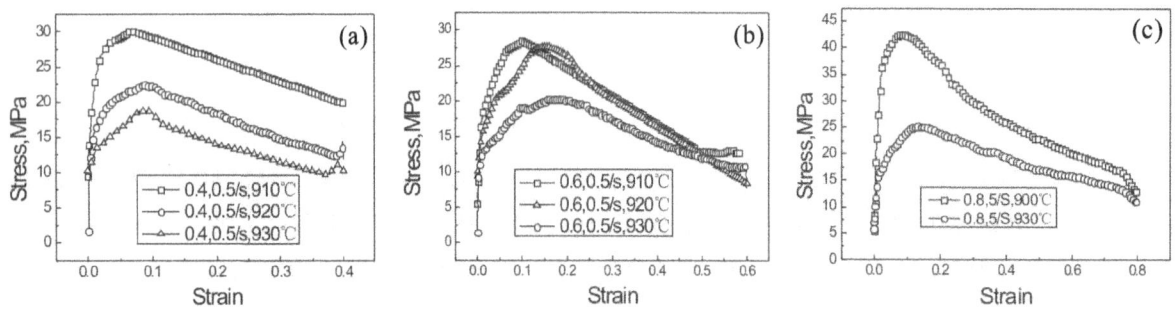

图1 ZCuSn10铜合金半固态试样在不同温度下进行单向压缩变形后真应力-真应变曲线

Fig.1 True stress-true strain curves of semi-solid ZCuSn10 copper alloy with different temperatures

由图1还可以看出,在相同应变速率和应变量下,变形温度越低,压缩时的变形应力越高,相反,变形温度越高,压缩变形应力越低。这是由于随着温度的升高,半固态ZCuSn10铜合金的液相增加,压缩变形时液相对固相骨架的变形起到一定的润滑作用,且液相增多能更好地包裹固相晶粒。压缩变形时,固相晶粒间的液相薄膜越厚,固相间几乎没有结合力,这样便使得变形越容易进行,压缩变形应力就越低。

2.1.2 应变速率对压缩真应力-应变曲线的影响

图2(a)是半固态ZCuSn10铜合金在应变量0.6和变形温度910℃,不同变形速率下的压缩真应力-应变曲线,图2(b)是半固态ZCuSn10铜合金在应变量0.6,变形温度920℃下的真应力-应变曲线。由图2(a)、(b)可以看出,按照ZCuSn10铜合金半固态压缩变形时真应力-应变曲线的变化趋势,可以将曲线分为三个阶段,第一阶段,即压缩开始应变很小时,流变应力随着应变的增加瞬间增大,并很快达到峰值应力,是瞬态激增阶段。ZCuSn10铜合金半固态组织可以看成是近球形固相颗粒和包裹在周围的液相组成的多孔材料,在压缩变形开始时,由于受到压力的作用,液相流动性较好先行流动,大量的液相流动具有一定的冲击力,这样便带动近球形的固相颗粒一起流动,流变应力要克服液相间的流动阻力和固相与液相之间的摩擦阻力,则会出现瞬态激增的现象。第二阶段,是流变应力缓慢下降阶段,随着压缩变形的进行,应力达到峰值应力之后,流变应力出现了缓慢下降阶段,这是因为在应力达到峰值应力时固相晶粒之间出现了永久性塑性变形,已经达到屈服极限,因此之后的压缩变形应力缓慢下降。第三阶段,流变应力达到稳定状态。

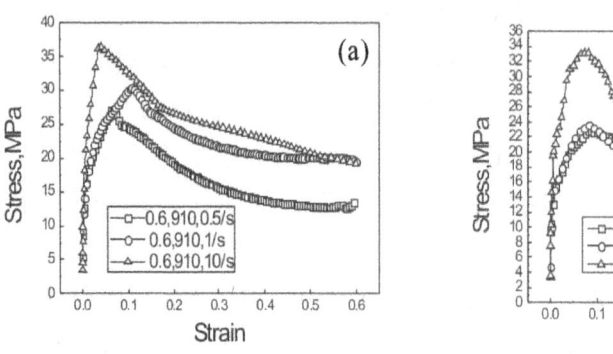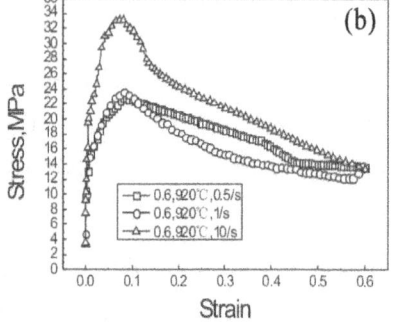

图2 ZCuSn10铜合金半固态试样在不同应变速率下进行单向压缩变形后真应力-真应变曲线

Fig.2 True stress-true strain curves of semi-solid ZCuSn10 copper alloy with different strain rate

在半固态 ZCuSn10 铜合金试样压缩变形时,应变速率是另一个重要的工艺参数,它对变形过程的影响也很明显。由图 2(a)可以看出,在应变量和变形温度相同时,应力岁应变速率的增加而增加,且都在应变很小时便达到了峰值应力。图 2(b)与图 2(a)的变化趋势是一致的。这是由于压缩变形开始时,液相最先流动,然后带动近球形的固相颗粒一起流动,这时的变形主要是液相流动和固液混合流动,这一变形阶段出现了固液协同流动的现象。但是应变速率越大,变形速度就越快,在很短的时间内,压缩变形就由液相流动和固液间的混合流动这一协同阶段过渡到固相颗粒间相互接触挤压的塑性变形阶段,固相颗粒间的塑性变形程度随着变形速率的增加而加强,从而导致流变应力随应变的增加而急剧增加,显然流变应力对应变速率比较敏感。表明半固态 ZCuSn10 铜合金是一种应变速率敏感材料。

2.1.3 应变量对压缩应力的影响

图 3 是 ZCuSn10 铜合金半固态试样在应变速率 1s^{-1},变形温度 920℃ 时不同应变量的的单向压缩真应力 - 应变曲线。图 3 可以看出,ZCuSn10 铜合金半固态压缩试样在不同应变量下的单向压缩真应力 - 应变曲线,当应变量大于 0.1 时,即应变量,0.2,0.4,0.6 时,真应力 - 应变曲线可以分为三个阶段,即压缩变形初始阶段的瞬态激增阶段,第二阶段为缓慢下降阶段,第三阶段为稳态阶段。由图 3 还可以看出,峰值应力都出现在应变为 0.1 的附近,达到峰值应力之前,不同应变量的压缩真应力 - 应变曲线几乎是重合的,这是由于压缩变形的试样状态是相同的,变形条件也是相同的。峰值应力过后的缓慢下降阶段,较小应变量如应变量为 0.1 时,压缩试样的真应力 - 应变曲线几乎没有出现应力缓慢下降阶段和之后的稳态应力阶段。这是由于应变量为 0.1 时,压缩试样的变形程度较小,还没有进入下一个变形阶段压缩变形就已经结束,所以不会再出现变形的后续两个阶段,而应变量为 0.1 时出现了一小段应力下降的曲线,这可能是压缩变形完成时,实验设备在卸载的过程中出现的误差。

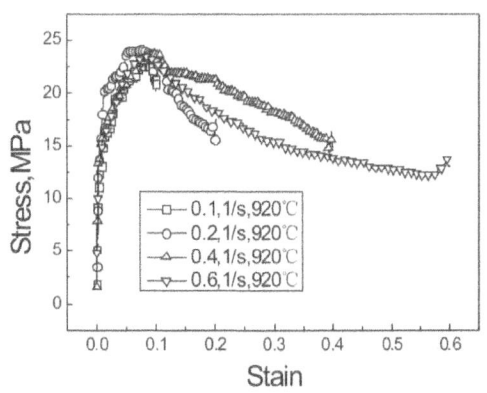

图 3 ZCuSn10 铜合金半固态试样在不同应变量下进行单向压缩变形后真应力–真应变曲线
Fig.3 True stress–true strain curves of semi–solid ZCuSn10 copper alloy with different strain

2.2 半固态 ZCuSn10 铜合金本构模型的建立

由前文所述综合分析可知,半固态 ZCuSn10 铜合金在半固态温度区间变形时,其流变应力是应变速率、应变量、变形温度和液相率的函数。对高固相分数的半固态金属本构关系模型的研究表明[54],流变应力、与应变速率、应变量呈幂函数关系,而与温度 T、液相率呈指数函数关系。同时由图 3 半固态 ZCuSn10 铜合金在不同应变量下的压缩真应力 - 真应变曲线可知:当应变量的值时,流变应力随着应变增大而迅速增大并达到峰值;当应变量的值时,流变应力缓慢减小并趋于稳定。因此,本文采用分阶段的方法建立半固态 ZCuSn10 铜合金的本构关系模型。由于当应变量的值时,变形主要以液相流动和固液混合流动为主,这一阶段流变应力受液相的影响较为明显,因此建立半固态 ZCuSn10 铜合金本构关系模型时考虑液相率的影响;而当应变量的值,即峰值应力过后的变形过程中流变应力受固相间滑移转动以及固相晶粒间塑性变形的影响,因此建立半固态 ZCuSn10 铜合金本构关系模型时不考虑液相率的影响。

2.2.1 应变量 $\varepsilon < 0.1$ 的半固态 ZCuSn10 铜合金本构关系模型

当应变量 $\varepsilon < 0.1$ 时,对流变应力 σ 与应变速率 $\dot{\varepsilon}$、轴向应变 ε、温度 T 和液相率 f_L 之间的关系假设如下:

$$\sigma = a_0 \exp(a_1 / T)\dot{\varepsilon}^{a_2}\varepsilon^{a_3}[1 - \beta f_L]^{a_4} \quad (\text{其中} f_L = (\frac{t_M - t_L}{t_M - t})^{\frac{1}{1-K}}), \quad \varepsilon < 0.1 \quad (1)$$

式中, σ 为流变应力; ε 为轴向应变; $\dot{\varepsilon}$ 为轴向应变速率;T 为变形温度; f_L 为液相率; β 为几何参数($\beta = 1.5$); t_M 为纯金属溶剂的熔点; a_0、a_1、a_2、a_3、a_4 均为常数; t_L 为合金的液相线温度;K 为平衡分配比值。

对式(1)两边取对数,把非线性回归转化成线性回归,可得

$$\ln \sigma = \ln a_0 + a_1 / \theta + a_2 \ln \dot{\varepsilon} + a_3 \ln \varepsilon + a_4 \ln(1 - \beta f_L), \quad \varepsilon < 0.1 \quad (2)$$

式(2)可以用线性方程表示:

$$y = A_0 + A_1 X_1 + A_2 X_2 + A_3 X_3 + A_4 X_4, \quad \varepsilon < 0.1 \quad (3)$$

令 $\quad y = \ln \sigma \quad X_1 = 1/T \quad X_2 = \ln \dot{\varepsilon} \quad X_3 = \ln \varepsilon \quad X_4 = \ln[1 - \beta f_L]$

$\quad A_0 = \ln a_0 \quad A_1 = a_1 \quad A_2 = a_2 \quad A_3 = a_3 \quad A_4 = a_4$

利用 SPSS 数理统计软件进行回归系数与统计检验指标的计算,回归分析的计算过程中,分别取应变为 0.01、0.05、0.1 对应的真应力,如表 2 所示。求出的回归系数与各项统计指标的值,见表 3。

表 2 $\varepsilon < 0.1$ 时,流变应力 σ 与应变 ε、应变速率 $\dot{\varepsilon}$、温度 T 的原始数据
Tab.2 Original data of stress σ, strain ε, strain rate $\dot{\varepsilon}$, and temperature T, $\varepsilon < 0.1$

$\dot{\varepsilon}$ (s-1)	$\varepsilon < 0.1$	σ, MPa		
		910℃	920℃	930℃
0.5	0.01	16.43	15.13	12.06
	0.05	25.17	28.86	13.65
	0.1	25.24	24.94	21.02
1	0.01	16.93	15.64	12.24
	0.05	25.63	21.67	13.72
	0.1	26.97	24.97	21.35
5	0.01	27.02	16.59	12.43
	0.05	29.45	23.45	14.57
	0.1	29.93	27.03	21.55
10	0.01	28.87	20.09	13.93
	0.05	29.66	27.59	17.02
	0.1	30.84	27.98	26.62

表 3 非线性回归系数

Tab.3 Nonlinear regression coefficients

标号	非标准系数	标准差	Sig. P	R
A_0	55.08	46.533	0.246	
A_1	−55294.367	51893.862	0.295	
A_2	0.079	0.016	0	0.932
A_3	0.18	0.02	0	
A_4	13.085	7.697	0.99	

通过式(2)、(3)的换算,将表 3 中的相关参数代入式(1)可得:

$$\sigma = \exp(55.08-55294.367/T)\dot{\varepsilon}^{0.079}\varepsilon^{0.18}[1-\beta f_L]^{13.085}$$

$$(其中) \quad f_L = (\frac{t_M - t_L}{t_M - t})^{\frac{1}{1-K}} \quad , \quad \varepsilon < 0.1 \tag{4}$$

2.2.2 应变量 $\varepsilon > 0.1$ 的半固态 ZCuSn10 铜合金本构关系模型

当应变量 $\varepsilon > 0.1$ 时,对流变应力 σ 与应变速率 $\dot{\varepsilon}$、轴向应变 ε、温度 T 和液相率 f_L 之间的关系假设如下:

$$\sigma = b_0 \exp(b_1/T)\dot{\varepsilon}^{b_2}\varepsilon^{b_3} \quad , \quad \varepsilon > 0.1 \tag{5}$$

式中 σ 为轴向应力;ε 为轴向应变;$\dot{\varepsilon}$ 为轴向应变速率;T 为变形温度;b_1、b_2、b_3 均为常数。

表 4 $\varepsilon > 0.1$ 时,流变应力 σ 与应变 ε、应变速率 $\dot{\varepsilon}$、温度 T 的原始数据

Tab.4 Original data of stress σ, strain ε, strain rate $\dot{\varepsilon}$, and temperature T, $\varepsilon > 0.1$

$\dot{\varepsilon}$ (s−1)	$\varepsilon > 0.1$	σ,MPa		
		910℃	920℃	930℃
	0.1	25.24	24.94	21.02
0.5	0.2	24.04	23.32	20.11
	0.4	13.78	13.58	13.15
	0.6	13.66	9.4	7.68
	0.1	26.97	24.97	21.35
1	0.2	24.45	23.15	20.16
	0.4	20.45	13.7	13.42
	0.6	19.55	13.96	13.36
	0.1	27.93	27.03	21.55
5	0.2	26.64	25.18	23.12
	0.4	24.75	22.23	21.06
	0.6	23.06	21.54	16.25
	0.1	28.43	27.98	26.62
10	0.2	27.25	26.98	25.45
	0.4	24.83	23.74	21.89
	0.6	24.65	22.18	19.76

对式(5)两边取对数,把非线性回归转化成线性回归,得

$$\ln \sigma = \ln b_0 + b_1/T + b_2 \ln \dot{\varepsilon} + b_3 \ln \varepsilon \quad , \quad \varepsilon > 0.1 \qquad (6)$$

式(6)可以用线性方程表示为:

$$y = B_0 + B_1 X_1 + B_2 X_2 + B_3 X_3 \quad , \quad \varepsilon > 0.1 \qquad (7)$$

令　　$y = \ln \sigma \quad X_1 = 1/T \quad X_2 = \ln \dot{\varepsilon} \quad X_3 = \ln \varepsilon$

$\quad\quad\quad B_0 = \ln b_0 \quad B_1 = b_1 \quad B_2 = b_2 \quad B_3 = b_3$

利用 SPSS 数理统计软件进行回归系数与统计检验指标的计算,回归分析的计算过程中,分别取应变为 0.1、0.2、0.4、0.6 对应的真应力,如表 4 所示。求出的回归系数与各项统计指标的值,见表 5。

通过式(6)、(7)的换算,将表 6 中的相关参数代入式(5)可得:

$$\sigma = \exp(-10.878 + 16036.273/T)\dot{\varepsilon}^{0.136}\varepsilon^{-0.262} \quad , \quad \varepsilon > 0.1 \qquad (8)$$

表 5　非线性回归系数

Tab.5　Nonlinear regression coefficients

标号	非标准系数	标准差	Sig. P	R
A_0	−10.878	3.171	0.001	0.874
A_1	16036.273	3782.043	0	
A_2	0.136	0.018	0	
A_3	−0.262	0.032	0	

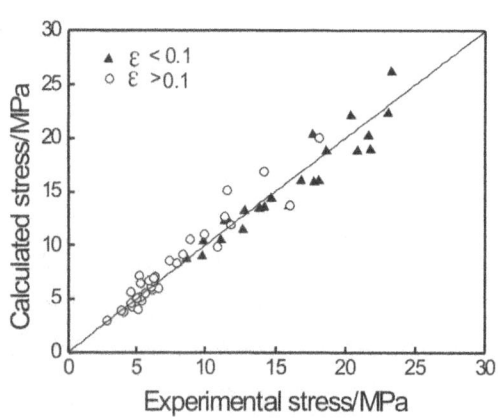

图 4　流变应力计算值与实验值的关系

Fig.4　Relation between calculated and experimental stress values

式(4)、(8)是半固态 ZCuSn10 铜合金的压缩真应力与应变速率、压缩真应变、温度以及液相分数之间的粘塑性本构方程,从表 3、表 5 可知,两个阶段的相关系数分别为 0.932、0.874。如图 4 所示,流变应力的计算值与实验值差别很小,符合较好。

3　结论

(1) 半固态 ZCuSn10 铜合金在半固态温度区间单向压缩真应力 - 应变曲线形状可以分为三个阶

段。第一阶段,压缩初始阶段的流变应力瞬态激增阶段,流变应力随应变增加迅速增加并达到峰值应力;第二阶段,流变应力随着应变的增加缓慢下降;第三阶段,流变应力随应变增加趋于稳定并达到稳态应力。

(2)半固态 ZCuSn10 铜合金单向压缩真应力 - 应变曲线中,随着预变形量的增加,流变应力降低;随着温度的增加,流变应力降低;随着应变速率的增加,流变应力升高。较小应变量的真应力 - 应变曲线与大应变量的真应力 - 应变曲线基本重合。

(3)半固态 ZCuSn10 铜合金单向压缩真应力 - 应变曲线中,大部分在应变量为 0.1 左右时达到峰值应力。半固态 ZCuSn10 铜合金的单向压缩力学行为可以用包含流变应力 σ、应变速率 $\dot{\varepsilon}$、应变量 ε、温度 T 和液相率 f_L 等参数的本构关系模型来描述。其分阶段的本构关系模型为:

$$\sigma = \exp(55.08 - 55294.367/T)\dot{\varepsilon}^{0.079}\varepsilon^{0.18}[1 - \beta f_L]^{13.085} \ , \quad f_L = (\frac{t_M - t_L}{t_M - t})^{\frac{1}{1-K}} \ , \quad \varepsilon < 0.1$$

$$\sigma = \exp(-10.878 + 16036.273/T)\dot{\varepsilon}^{0.136}\varepsilon^{-0.262} \ , \quad \varepsilon > 0.1$$

参考文献:

[1] R. Canyook, J. Wannasin, S. Wisuthmethangkul, M. C. Flemings. Characterization of the microstructure evolution of a semi–solid metal slurry during the early stages [J]. Acta Materialia, 2012, (20): 3501–3510.

[2] 余小鲁, 李付国, 李淼泉. 半固态材料触变成形通用本构方程及优化 [J]. 机械工程学报, 2007, 43 (10): 72–76.

[3] 孙国强. 半固态加工技术及其应用 [J]. 稀有金属, 2003, 27(3): 382–384.

[4] 罗守靖, 姜巨福, 杜之明. 半固态金属成形研究的新进展、工业应用及其思考 [J]. 机械工程学报, 2003, 39(11): 52–60.

[5] 罗守靖, 田文彤, 谢水生, 毛卫民. 半固态加工技术及应用 [J]. 中国有色金属学报, 2000, 10(6): 765–773.

[6] L A Lalli. A model for deformation and segregation of solid liquid mixtures [J]. Metal Trans. 1985, 16A: 1393–1403.

[7] K Muammer, V Victor, W Thomas. Application of the finite element method to predict material flow and defects in the semid –solid forging of A356 Aluminum alloys [J]. Journal of Materials Processing Technology, 1996, 59(5): 106–112.

[8] 康永林, 毛卫民, 胡壮麒. 金属材料半固态加工理论与技术 [M]. 学技术出版社, 2004.

[9] C G Kang, B S Kang, J Kim. An investigation of the mushy state forging process by the finite element method [J]. Journal of Materials Processing Technology, 1998, 80: 444–449.

[10] L Daniel, L Josep, W L Wu. A constitutive model for the tensile deformation of a binary aluminum alloy at high fractions of solid [J]. Metall Mater Trans, 2006, 37B: 4431–4439.

[11] GB/T73142005. 金属材料室温压缩试验方法 [S]. 中国国家标准化管理委员会, 2005.

[12] 王 磊. 材料的力学性能 [M]. 沈阳: 东北大学出版社, 2005.

基于果蝇优化算法的铸钢件冒口优化的研究

王　瞳，周建新，殷亚军，沈　旭，周　琴

(华中科技大学　材料成形与模具技术国家重点实验室)

摘　要：冒口设计是铸造工艺设计中的重要环节。为节约金属材料和降低生产成本,本文深入研究了铸钢件的冒口工艺,根据铸钢件的补缩原理和果蝇优化算法的特点,建立了冒口优化数学模型。运用果蝇优化算法对冒口尺寸进行了优化设计,通过模拟结果和实际结果对比,验证了果蝇优化算法的可行性。

关键词：果蝇优化算法；铸钢件；冒口优化

The Research on Riser Optimization in Steel Casting Process Based on Fruit Fly Optimization Algorithm

WANG Tong, ZHOU Jian-xin, YIN Ya-jun, SHEN Xu, ZHOU Qin

(State Key Laboratory of Material Processing and Die & Mould Technology, Huazhong University of Science and Technology)

Abstract: Riser design is an important part of casting process design. In order to save metal materials and reduce production cost, the process of steel casting riser is studied in this paper. According to the feeding principles of steel casting and the characteristics of fruit fly optimization algorithm, the riser optimization model is established. The riser size is optimized by using fruit fly optimization algorithm. Compared the simulation result with the actual result, the feasibility of fruit fly optimization algorithm is verified.

Key words: fruit fly optimization algorithm; steel casting; riser optimization

基金项目：国家数控重大专项(2012ZX04012-011),材料成形与模具技术国家重点实验室2015年重点自主研究项目(数字化智能化铸造技术及应用)。

第一作者：王　瞳,男。铸造工艺优化。wangtongfly@foxmail.com。

通讯作者：周建新,男。材料成形过程模拟仿真及数字化技术。zhoujianxin@hust.edu.cn。

铸钢件补焊过程的热应力及变形行为研究

王　文，周建新，庞盛永

(华中科技大学　材料成形与模具技术国家重点实验室)

摘　要：本研究针对铸钢件补焊过程建立的一个三维瞬态热力耦合模型。在热分析中,采用体积热源模型描述焊接热源输入,考虑了工件内部的导热及工件与环境之间的对流换热作用,以及材料的熔化与凝固过程中释放的潜热。在结构分析中,采用 MISES 准则判断材料的弹塑性状态,认为材料近似服从双线性等向强化的本构模型,及采用生死单元法处理增材过程。基于该模型,对铸钢补焊过程的热应力及变形行为进行了研究,发现了热应力及变形的分布及演化规律,并基于模拟结果对铸钢件补焊过程的缺陷形成进行了分析。

关键词：补焊；热力耦合模型；数值模拟

Simulation of Thermal Stress and Distortion During Multi-pass Repair Welding of Cast Steel

WANG Wen, ZHOU Jian-xing, PANG Sheng-yong

(State Key Laboratory of Materials Processing and Die & Mould Technology, Huazhong University of Science and Technology (HUST))

Abstract: In this study, a transient three dimensional thermo-mechanical coupling model is established for the repair welding process of cast steel. In the thermal analysis, volumetric heat source is adopted and, the influence of conduct heat transfer within the workpiece and the convection heat transfer between the workpiece and the environment are considered, and the latent heat of melting and solidification are also considered. As to the stress analysis, the Mises yield criterion is used to judge the plastic or elasticity state, the constitutive relation of the material is treated as double-linear, and the addition of the material is treated with the birth-to-death element method. Based on the model, the distribution and evolution regulation of the dynamic thermal stress and distortion are obtained. Based on the simulation results, several important characteristics, closely relating to the formation of several welding defects, are also analyzed.

Key words: repair welding; thermo-mechanical model; numerical simulation

基金项目：国家数控重大专项(2012ZX04012-011),材料成形与模具技术国家重点实验室 2015 年重点自主研究项目(数字化智能化铸造技术及应用)。

第一作者：王　文,男。材料成形过程计算机数值仿真技术。wenwang_phd_stu@qq.com。

通讯作者：周建新,男。材料成形过程模拟仿真及数字化技术。zhoujianxin@hust.edu.cn。

复合材料结构件热压罐内多物理场模拟

董长春[1], **周建新**[1], **廖敦明**[1], **孙 飞**[1], **朱大雷**[2]

(1.华中科技大学 材料成形与模具技术国家重点实验室,武汉 430074　2.北京卫星制造厂,北京,100094)

摘　要: 热压罐成型是生产航空航天用树脂基复合材料的最重要的成型方法。热压罐提供了复合材料成型所需要的温度和压力条件。温度使树脂发生交联固化成形,压力使预浸料中多余的树脂排出铺层体系并起到紧实铺层的作用。本文建立了热压罐物理模型,对罐内流动场及温度场进行数值模拟。通过本文的模拟,得到了罐内空气流动场、温度场及模具温度场,再现复合材料制造过程中环境物理场的变化,为实际工艺的改进提供了一定的参考。

关键词: 热压罐;复合材料;多物理场

Multi-physicals Simulations inside the Autoclave in Manufacturing Composite Structure

DONG Chang-chun[1], ZHOU Jian-xin[1], LIAO Dun-ming[1], SUN Fei[1], ZHU Da-lei[2]

(1. State Key Laboratory of Materials Processing and Die & Mould Technology, Huazhong University of Science and Technology, Wuhan 430074, China　2. Beijing Spacecraft Manufacturing Factory, Beijing 100094, China)

Abstract: Autoclave molding is the most important forming method in manufacturing of resin matrix composites in aeronautics and astronautics. Autoclave supplies temperature and pressure conditions for the manufacturing of composites. Resin crosslinking and curing in the effect of temperature while pressure makes excess resin drained from the preform system and compacts the preform. Autoclave physical model is presented in this paper, the flow field and temperature field for autoclave is numerically simulated. Through the simulation, air flow field, temperature field and mold temperature is obtained inside the autoclave. The change of environmental physical fields in the manufacturing of composite materials is reappeared, and a certain reference for the actual process improvement is provided.

Key words: autoclave; composite materials; multi-physical fields

第一作者:董长春,男。材料成形过程模拟仿真及数字化技术,dongchangchun@hust.edu.cn。
通讯作者:周建新,男。材料成形过程模拟仿真及数字化技术,zhoujianxin@hust.edu.cn。

双级固溶处理对超高强铝合金7A04
力学性能的影响及工艺优化

何　欢，张　婷，孙　祥，符跃春，沈晓明，吴忠艺，曾建民

(广西大学 材料科学与工程学院,有色金属及特色材料加工重点实验室)

摘　要：本文研究了在相同的时效工艺下，双级固溶处理对超高强铝合金 7A04 力学性能的影响。首先在不同参数条件下对 7A04 合金进行双级固溶处理,接着进行 120℃/24h/空冷时效处理,然后测试其抗拉强度 σ_b、屈服强度 $\sigma_{0.2}$、伸长率 δ 和冲击吸收功 A_k 等力学性能数据,并结合扫描电镜显微组织观察,对固溶处理工艺参数对该合金力学性能的影响进行了分析讨论。

关键词：固溶处理；超高强铝合金；力学性能

Effect of Double−stage Solid Solution Treatment on Mechanical Properties of 7A04 Ultra−high Strength Aluminum Alloy

HE Huan, ZHANG Ting, SUN Xiang, FU Yue−chun, SHEN Xiao−ming,
WU Zhong−yi, ZENG Jian−min

(Key Laboratory of New Processing Technology for Materials and Nonferrous Metal,
College of Materials Science and Engineering, Guangxi University, Nanning)

Abstract: The aim of this paper is to study the influence of double-stage solid solution treatment to the mechanical properties of 7A04 ultra-high strength aluminum alloy under the same aging process. Firstly, double-stage solid solution treatments with different parameters were carried out. Then aging treatments were made at 120℃ for 24h. And then mechanical properties such as tensile strength σ_b, yield strength $\sigma_{0.2}$, elongation δ and impact absorbing energy A_k were tested. Combining SEM microstructure observation, the effect mechanism of solid solution treatment on the mechanical behavior of the alloy is analyzed and discussed.

Key words: ultra-high strength aluminum alloy; solution treatment; mechanical properties

基金项目:广西教育厅基金项目(No. YB2014013);广西理工科学实验中心项目(合同编号:20150042);广西自然科学基金项目(No. GUIKENENG 14-045-04)；有色金属及材料加工新技术教育部重点实验室开放基金项目(GXKFZ-04, GXKFJ09-25, GXKFJ11-09);国家自然科学基金项目(61474030)。

第一作者:何　欢,男。材料表面改性。noblehe@gxu.edu.cn。

低能离子注入对石墨烯结构和性能的影响

吴　鑫[1]，赵海燕[1]，裴家云[1]，严　冬[1]，雷永平[2]

(1.清华大学 机械工程系,清华大学摩擦学国家重点实验室；2.北京工业大学 材料科学与工程学院)

摘　要：掺杂可以打开石墨烯零带隙的能带结构,对其电学性能进行控制,采用低能离子注入的方法可以对石墨烯薄膜进行定位定量掺杂。本文采用实验和分子动力学模拟结合的方法研究了低能氮离子注入下石墨烯结构的掺杂行为和掺杂后的结构性能。结果表明:低能离子束注入的方法可以对石墨烯结构进行有效掺杂,氮离子主要以置换的方式进行掺杂;掺杂的同时会带来石墨烯结构的局部破坏,低能量离子作用下石墨烯结构主要以吸附原子为主,较高能量离子束作用下石墨烯结构以空位缺陷为主;石墨烯掺杂位置会出现应力集中,拉伸过程中易于发生裂纹萌生;掺杂后的电学输运特性受掺杂离子的置换位置影响,石墨烯薄膜呈现明显的半导体性能。

关键词：石墨烯；低能离子；掺杂；分子动力学模拟

Transformation of the Structure and Properties of Graphene under Low Energy Ion Beam Irradiation

WU Xin[1], ZHAO Hai-yan[1], PEI Jia-Yun[1], YAN Dong[1], LEI Yong-ping[2]

(1.State Key Laboratory of Tribology, Department of Mechanical Engineering, Tsinghua University;
2. College of Materials Science & Engineering, Beijing University of Technology)

Abstract: The zero bandgap of graphene can be effectively opened by doping method, among which, low energy ion beam irradiation may be used to dope graphene with appointed site and doping dose. In this study, experimental and molecular dynamics studies are combined to research the doping possibility of graphene by low energy ion irradiation. The results show that the low energy ion beam irradiation is an effective method to dope graphene, in which the nitrogen atom is presented mainly as substitution. During graphene doping, there are also defects, mainly vacancies, introduced into graphene sheet. Graphene sheets is heavily damaged if the energy of ion beam is high. The doping behavior should generate stress concentration, where the crack is easy to be initiated under uniaxial tensile loading. The electrical transport properties of graphene is influenced by the doping site and doping dose. And graphene sheet displays a properties of semiconductor after doping.

Key words: graphene; low energy ion beam irradiation; doping; molecular dynamics simulation

基金项目：北京市自然科学基金资助项目(3142010);高等学校博士学科点专项科研基金资助项目(20130002110088);
　　　　　清华大学摩擦学国家重点实验室自主研究课题资助项目(SKLT2014A03)。

第一作者：吴　鑫,男。石墨烯的加工。xwu10@hotmail.com。

通讯作者：赵海燕,男。焊接及连接。hyzhao@tsinghua.edu.cn。

阳极氧化时间对阳极氧化铝膜的影响

温静娴 [1,2]，黄祖江 [1]，周　敏 [3]，杨　阳 [1]，李伟洲 [1]

(1.广西大学 材料科学与工程学院；2.广西理工科学实验中心；3.富士康科技集团)

摘　要： 利用阳极氧化技术在铝片上制备多孔氧化铝膜，研究阳极氧化时间对多孔氧化铝膜的表面形貌、硬度和厚度、耐蚀性与抗热裂性的影响。结果表明：阳极氧化铝(AAO)膜的孔面积和孔径随阳极氧化时间的增加而增大，孔密度在阳极氧化时间为50min后变小。阳极氧化铝(AAO)膜的厚度与硬度随阳极氧化时间的延长明显增加，其中厚度随阳极氧化时间的增加呈显著线性变化。当阳极氧化时间超过50min后，氧化膜硬度下降。阳极氧化铝(AAO)膜的耐蚀性随阳极氧化时间增加而增强，其中阳极氧化时间为60min的耐蚀性最好，20min最差。不同阳极氧化时间制备的阳极氧化铝(AAO)膜表面都有不同程度的开裂，且多数裂纹横贯整个表面。其中20min时的裂纹密度最多，随着时间的增加，裂纹密度出现减少，但30min和40min的裂纹密度变化不大。

关键词： 阳极氧化；处理时间；耐蚀性；抗热裂性

Effect of Anodic Oxidation Time for Anodic Alumina Film

WEN Jing-xian[1,2], HUANG Zu-jiang[1], ZHOU Min[3], YANG Yang[1], LI Wei-zhou[1]

(1. School of Materials Science and Engineering, Guangxi University;
2. Guangxi Experiment Centre of Science and Technology; 3. Foxconn Technology Group)

Abstract: In order to study the effect of anodic oxidation time for anodic alumina film's surface morphologies, hardness and thickness, corrosion resistance, and thermal cracking resistance, anodic aluminum oxide(AAO) film was obtained by anodizing on aluminum. The results indicated that, with the increase of anodizing time, pore area and diameter increased. Pore density decreased after 50 min anodizing time. Anodic aluminum oxide (AAO) film's hardness and thickness increased with anodizing time. Thickness changed in a linear fashion, and hardness decreased after 50 min anodizing time. Corrosion resistance of anodic aluminum oxide (AAO) film increased as anodizing time increased. Corrosion resistance after 60 min anodizing time was the best and after 20 min anodizing time was the worse. Anodic aluminum oxide (AAO) film obtained by different anodizing time emerged different extent cracks, which was through the whole surface. Among them, crack density after 20 min anodizing time was the maximum. Crack density reduced with anodizing time, however, it showed little change after 20 min and 30 min anodizing time.

Key words: anodizing; anodizing time; corrosion resistance; thermal cracking resistance

基金项目： 广西理工科学实验中心开放基金(YXKT2014026)；广西自然科学基金(2014GXNSFCA118013,2010GXNSFD013006)；
广西高等学校高水平创新团队项目(第二批)；广西自然科学基金创新研究团队项目(2011GXNSFF018001)。
第一作者： 温静娴，女。电子显微分析。wjxfh@126.com。
通讯作者： 李伟洲，男。金属材料表面处理。liwz2008@hotmail.com。

0 引言

铝及铝合金具有良好的导热和导电性、比强度高、塑性好和易加工的特点,是有色金属中使用量最大,应用面积最广的材料[1]。铝及铝合金在空气下会自然生成 $Al_2O_3 \cdot H_2O$ 或 Al_2O_3 氧化膜,可以起到一定的保护作用[2]。但这层氧化膜薄且疏松、硬度小、机械强度低、耐蚀性差,在高温下极易开裂,无法满足使用要求。因此需要在使用前对铝及铝合金进行表面处理,其中最常用的方法是阳极氧化[3—5]。阳极氧化后的阳极氧化铝(AAO)膜具有优良的耐蚀性、耐磨性、隔热性、绝缘性以及其特殊的孔结构具有许多功能性应用[6]。阳极氧化铝(AAO)膜的热膨胀系数和热传导性与铝基体相差较大(氧化膜的热膨胀系数和热传导系数分别约为铝基体的 0.2 和 0.1 倍[7])。阳极氧化铝(AAO)膜在较高温度服役下,内部会产生很大的热应力,致使膜开裂甚至脱落,从而导致氧化膜的性能下降[8],因此阳极氧化铝(AAO)膜的抗热裂性决定铝材表面的质量和使用寿命。

目前,许多研究认为氧化膜内应力是影响氧化膜开裂的主要原因。内应力分为生长应力和热应力[9—12]。氧化膜恒温生长时产生生长应力,温度变化时基体与氧化膜热膨胀系数差异引起热应力。而氧化膜内用力释放的三种方式分别为[13]:(1)氧化膜塑性变形;(2)金属基体塑性变形;(3)氧化膜开裂和脱落。氧化膜的塑性变形主要是通过高温蠕变进行[14]。A.M.Huntz[9]等通过对 NiO,Cr_2O_3 和 Al_2O_3 氧化膜的应力测试,结果表明氧化膜和基体的热膨胀系数不同而引起的热应力造成氧化膜开裂甚至剥落;氧化膜内主要存在生长应力和热应力,如果应力没有通过蠕变释放,在其表面将会出现氧化膜的破裂。刘伟华[15]对比研究了纯铝、ZL201 和 LY12 铝合金阳极氧化在 300℃ 范围内的抗热裂性,发现氧化膜孔隙率的降低和残余应力增大都可以阻碍裂纹的扩展;纯铝在封闭后情况较好,不容易产生裂纹,而合金中的金属间颗粒、氧气泡的存在以及残余应力导致缺陷形成,这些缺陷会使铝合金阳极氧化后沸水封孔的过程中容易产生裂纹,导致氧化膜力学性能降低。

虽然在阳极氧化铝(AAO)膜制备、封孔等工艺上已经有了深入的研究,但目前关于阳极氧化铝(AAO)膜热裂性的研究还比较少。本文通过改变阳极氧化时间参数,研究了不同阳极氧化时间对阳极氧化铝(AAO)膜的表面形貌、硬度与厚度、耐蚀性和抗热裂性的影响。

1 实验材料和方法

1.1 实验流程

铝阳极氧化实验所采用的材料为纯度 99.5% 的工业纯铝,其化学成分列于表 1,试样尺寸为 40mm×30mm×1mm。采用一次阳极氧化法制备 AAO 膜,阴极为铝电极,采用双阴极板,阳极为预处理(预处理参数如表 2 所示)后的工业纯铝片,阳极置于双阴极板中间进行阳极氧化;电解槽为聚丙烯塑料材质,电源为 RXN-605D 型数显稳压直流电源,电流表量程为 0.6A~3A,电解液为 11wt.% 硫酸溶液[16],电流密度为 2.5A/dm²,溶液温度为 30℃。实验采用恒流阳极氧化法,实验中使用电动搅拌器对电解液进行搅拌。实验结束后,用自来水冲洗再放入去离子水中浸泡 1 分钟,最后吹干待用。

表 1 铝片的成分及含量

Tab.1 Chemical composition of aluminum

元素	Al	Cu	Si	Fe
含量(wt%)	余量	0.015	0.20	0.25

表 2 预处理配方及参数

Tab.2 Formulation and parameter of pretreatment

步骤	配方	温度	时间
酸洗	16wt.% H2SO4	室温	5min
碱洗	50g/L NaOH	60℃	1min
除灰	58wt.%HNO3	室温	2min

1.2 测试方法

利用 X 射线衍射仪(XRD)对 AAO 膜进行物相分析,测试条件如下:D/max 2500V 型衍射仪,Cu 靶的 Kα 射线进行扫描,管压为 40kV,管流为 200mA,扫描范围为 20~80o,扫描速度为 10o/min。采用配有能谱仪(EDS)的扫描电镜(SEM)观察 AAO 膜表面形貌和测量截面厚度。AAO 膜的硬度测试采用数显显微硬度仪,所加载荷为 100g,加载时间为 20s;测试时在 AAO 膜表面随机取六个点进行测量,然后取硬度平均值。AAO 膜的极化曲线和电化学阻抗谱(EIS)[17]用 CHI-660e 电化学工作站,实验介质为 3.5wt.%的 NaCl 溶液,试样测试面积为 1cm²,频率范围 0.1Hz~100kHz,振幅为 0.005V。阳极氧化铝膜热裂处理:将阳极氧化铝试样置于马弗炉内,升温速率为 10℃/min,升温至 400℃后保温 1 小时,随炉冷却至室温。

结合 AAO 膜表面形貌 SEM 图对其孔面积进行计算。孔面积 S 孔计算公式如下:$S_孔 = K\pi r^2$,其中 K 为孔密度,r 为平均孔径。在表面孔洞较均匀处选取三张边长为 a(本次实验中 a 为 140nm)的正方形图片,在每一张图中随机选取 10 个孔洞,测量直径并计算出平均孔径 r。通过测量三张图片中孔的数量,计算出平均孔密度,其中将图片中不全的孔拼凑成一个完整的孔[18],如图 1 所示。

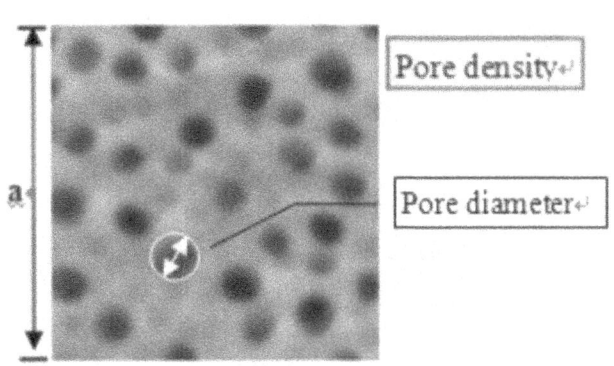

图 1 AAO 膜表面孔面积测量示意 图

Fig.1 Schematic illustration of hole area on AAO film

2 试验结果与讨论

2.1 阳极氧化时间对阳极氧化铝膜表面形貌的影响

图 2 为不同阳极氧化时间制备的阳极氧化铝膜表面形貌。有图可见,孔洞分布规则有序,随着处理的时间增加,孔径增大。结合表 3 分析可知孔面积和孔径随处理时间延长而增大。孔密度在处理时间 50min 后变小,推测是阳极氧化在 50min 时,AAO 膜的生长速率有所下降,表面发生局部溶解,且溶解的速率开始大于生长速率。孔洞不断溶解,导致孔壁和孔间距缩小,部分孔洞开始连通,此时膜表面的孔径显著增大,而孔密度减小。

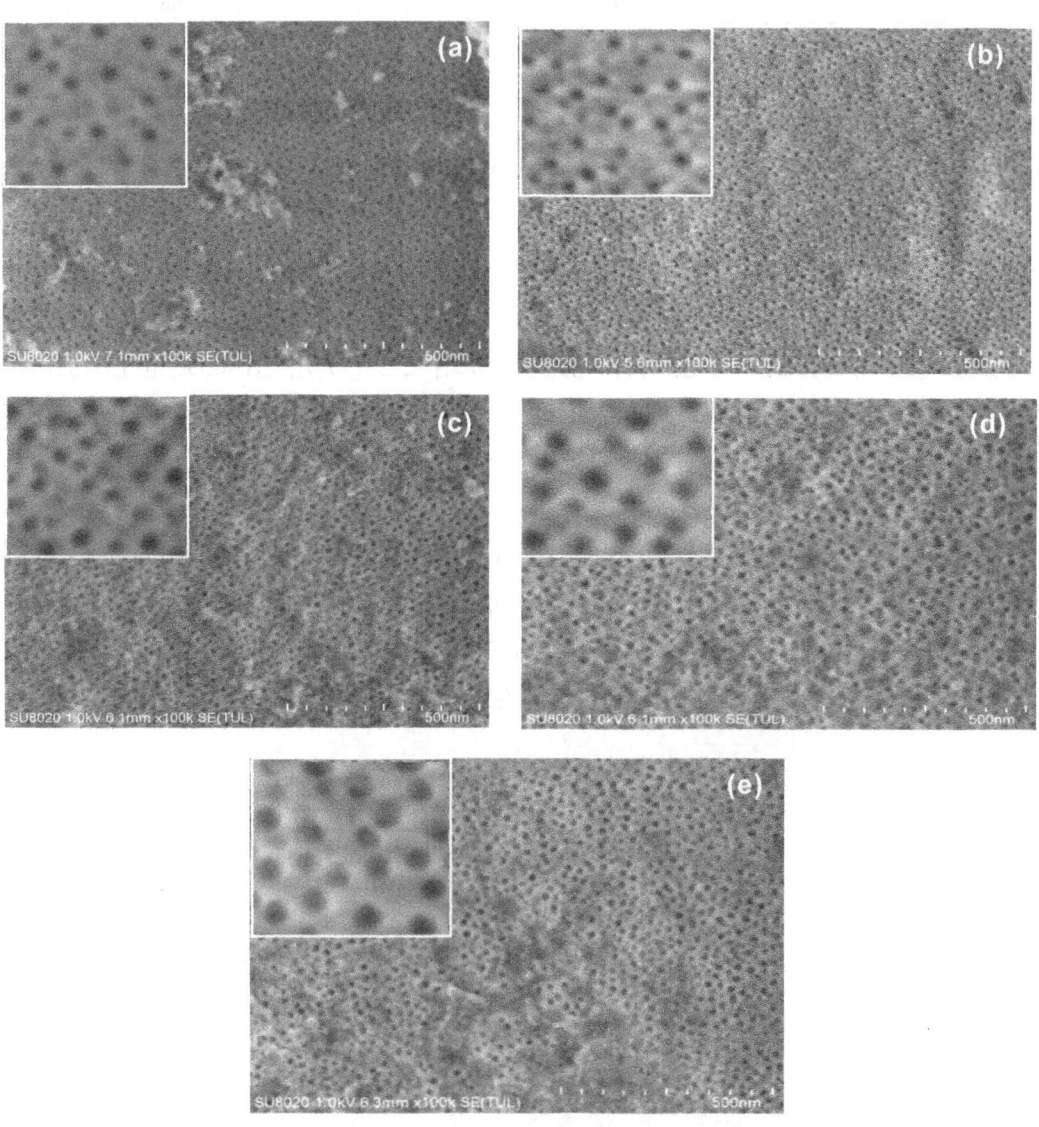

图 2　不同阳极氧化时间制备的 AAO 表面形貌

Fig.2　Surface morphologies of AAO film after different anodizing time

(a)20min, (b)30min, (c)40min, (d)50min, (e)60min

表3　不同氧化时间制备的 AAO 膜孔密度、孔径值和孔面积

Tab.3　Pore density, diameter and area of AAO film after different anodizing time

处理时间(min)	孔密度 K(个/a²)	平均孔径(nm)	孔面积 S_孔(nm²)
20	21	10.9	2413.6
30	23	12.0	2560.0
40	24	13.6	3484.6
50	17	15.8	3531.5
60	20	17.3	4698.9

2.2 阳极氧化时间对阳极氧化铝膜硬度与厚度的影响

阳极氧化铝膜由阻挡层和多孔层构成,阻挡层的厚度只与氧化电压有关[16,19],而多孔层的厚度则取决于膜的生长速度与溶解速度,膜的生长速度大于溶解速度时,AAO 膜厚度才会增加,反之则厚度降低。图 3 为阳极氧化铝膜厚度与硬度随处理时间的变化,随着处理时间的增加,AAO 膜厚度和硬度都明显增大,其中厚度随处理时间变化呈显著线性变化;当处理时间超过 50min 后,AAO 膜硬度下降。

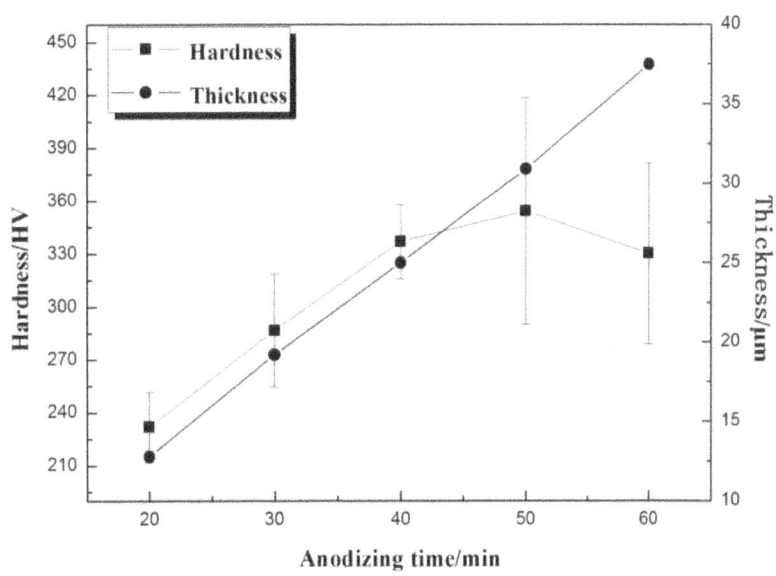

图 3　阳极氧化时间对阳极氧化膜硬度和厚度的影响

Fig.3　Effect of different anodizing time on hardness and thickness of anodic oxide film

2.3 阳极氧化时间间对阳极氧化铝膜耐蚀性的影响

图 4 是不同阳极氧化时间制备的阳极氧化铝膜在 3.5%NaCl 溶液中的极化曲线,其中 20min 制备的 AAO 膜腐蚀电流密度为 $1.26\mu A \cdot cm^{-2}$,40min 的腐蚀电流密度为 $0.44\mu A \cdot cm^{-2}$,60min 时腐蚀电流密度为 $0.33\mu A \cdot cm^{-2}$。腐蚀电流密度越小,腐蚀速率越慢,说明耐蚀性越好[20,21]。因此耐蚀性依次为:60min>40min>20min,AAO 膜耐蚀性随处理时间增加而增强,由图 3 知 60min 的膜层厚度最大,膜层厚度增加能够提高氧化膜耐蚀性。

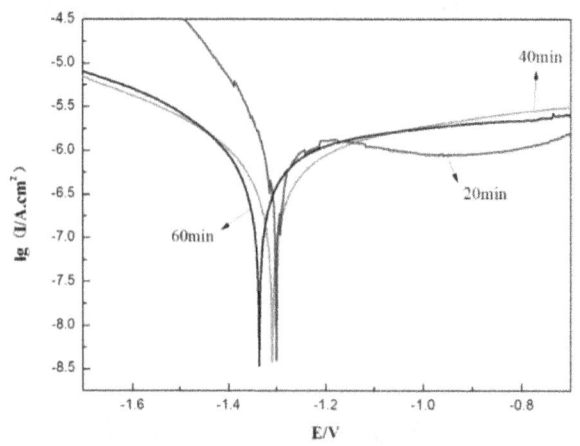

图4 不同阳极氧化时间制备的AAO膜在3.5%NaCl溶液中的极化曲线
Fig.4 Polarization curves of AAO film after different anodizing time in 3.5% NaCl solution

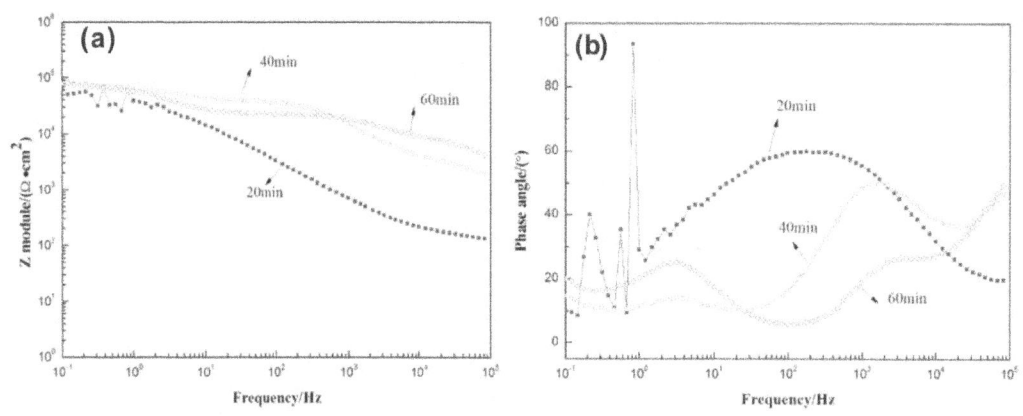

图5 不同氧化时间制备的AAO膜在3.5%NaCl溶液中的EIS图
Fig.5 EIS plots of AAO film after different anodizing time in 3.5% NaCl solution
(a) Amplitude-frequence curves, (b) Phase-frequence curves

图5为不同阳极氧化时间制备的AAO膜在3.5%NaCl溶液中的EIS图。图5(a)中三种不同处理时间后的AAO膜在低频段阻抗总体上基本相似,40min和60min的相似度更高,说明膜层厚度在阳极氧化起始阶段就已经确定,再增加处理时间对膜层厚度几乎无影响。图5(b)中高频段阻抗区别明显,60min制备的AAO膜高频阻抗最大,40min次之,20min最小;阻抗值越大则耐蚀性越好[17,22]。因此耐蚀性依次为:60min>40min>20min,与极化曲线的描述相同。EIS高频阻抗图谱中所包含的半圆形峰的个数就是时间常数的个数[17],由此可知20min制备的AAO膜只有1个时间常数,而40min和60min的AAO膜有2个时间常数。

2.4 阳极氧化时间对阳极氧化铝膜抗热裂性的影响

不同阳极氧化时间制备的阳极氧化铝膜会产生不同的残余生长应力,这些会导致AAO膜热裂性的不同。通过光照射样品表面,利用裂纹反光的特性,对其进行拍照,结合裂纹密度对其进行分析。裂纹密度是指单位面积内的裂纹条数,即$\rho=m/s$,其中ρ为裂纹密度,单位为条/cm²,m为样品表面裂纹数量,s为样品面积。

　　图 6 为在不同阳极氧化时间制备的 AAO 膜热处理后的表面图片，从图中可看到 AAO 膜表面都有不同程度的开裂，且多数裂纹横贯整个表面。在相同温度下硬度越大，AAO 膜热应力就会越大，抵抗开裂的能力越弱；AAO 膜越软，蠕变越容易发生，其抵抗开裂的能力越强。当 AAO 膜硬度在 200HV 以上时，蠕变变得非常困难，此时膜会以开裂的方式释放内应力。由图 3 知，AAO 膜的硬度大于等于 230HV，硬度大，抵抗开裂的能力弱，所以 AAO 膜表面都有裂纹。

　　结合表 4，可发现 20min 时的裂纹密度最大，随着处理时间增加，裂纹密度出现减少，而 30min 和 40min 的裂纹密度变化不大。AAO 膜厚度越小，抵抗开裂的能力越差。从图 3 中可见处理时间为 20min 时的 AAO 膜厚度最薄，只有 12.8μm，抵抗开裂的能力差，因此裂纹密度最大，高达 18 条 /cm²。

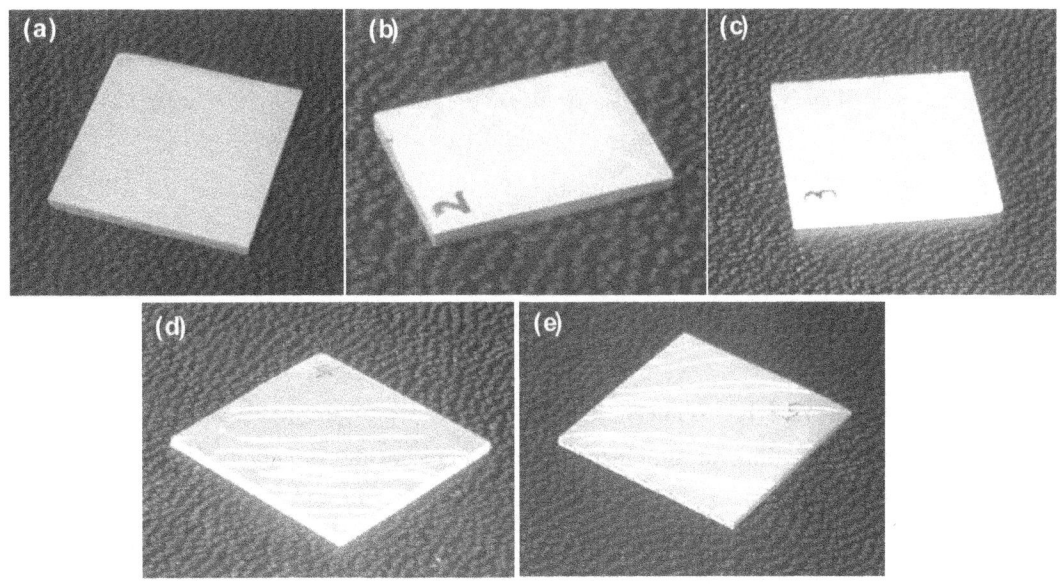

图 6　不同阳极氧化时间下制备阳极氧化铝膜热处理后的表面

Fig.6　Morphologies of AAO film formed under different anodizing time after thermal exposure
(a)20min，(b)30min，(c)40min，(d)50min，(e)60min

表 4　不同阳极氧化时间制备的阳极氧化铝膜热处理后裂纹密度

Tab.4　Crack density of AAO film under different anodizing time after thermal exposure

氧化时间(min)	20	30	40	50	60
裂纹密度(条/cm²)	18.0	14.3	14.2	13	14

3　结论

　　(1)阳极氧化铝膜的孔面积和孔径随阳极氧化时间增加而增大，孔密度在处理时间 50min 后变小。

　　(2)阳极氧化铝膜厚度与硬度随阳极氧化时间的延长都明显增加，其中厚度随处理时间变化呈显著线性变化；当处理时间超过 50min 后，氧化膜硬度下降。

　　(3)阳极氧化铝膜的耐蚀性随阳极氧化时间增加而增强，60min 制备的阳极氧化铝膜耐蚀性最好，20min 最差。

　　(4) 不同阳极氧化时间制备的阳极氧化铝膜表面都有不同程度的开裂，且多数裂纹横贯整个表面。其中 20min 时的裂纹密度最多,随着处理时间增加,裂纹密度出现减少,而 30min 和 40min 的裂纹密度变化不大。

参考文献:

[1] 何思宇. 铝合金表面微弧氧化膜的组织结构与性能 [D]. 山东大学, 2012.

[2] 崔昌军, 彭乔. 铝及铝合金的阳极氧化研究综述 [J]. 全面腐蚀控制, 2002, 16(6): 12–17.

[3] 陈艳杰, 范洪远, 陈道琪. 6063 铝合金两种阳极氧化工艺的氧化膜性能研究 [J]. 热加工工艺, 2011, 40(6): 128–130.

[4] 王学华, 马连姣, 许震,等. 高孔密度阳极氧化铝模板的制备及结构表征 [J]. 武汉工程大学学报, 2008, 30(1): 48–50.

[5] Chaure N B, Stamenov P, Rhen F M F, et al. Oriented cobalt nanowires prepared by electrodeposition in a porous membrane [J]. Journal of Magnetism & Magnetic Materials, 2005, 290–291: 1210–1213.

[6] 马骏. 铝合金阳极氧化耐磨防腐涂层制备与性能研究 [D]. 河海大学, 2007.

[7] Sheasby P G, Pinner R. The surface treatment and finishing of aluminum and its Alloys [J]. Nature, 2001, 39(8): 233–234.

[8] 陈爽. 提高铝合金阳极氧化膜在高温下的抗开裂性能研究 [D]. 北京化工大学, 2009.

[9] Huntz A M. Stresses in NiO, Cr2O3 and Al2O3 oxide scales [J]. Materials Science & Engineering A, 1995, 201(1–2): 211–228.

[10] Vrublevsky I, Parkoun V, Schreckenbach J, et al. Effect of the current density on the volume expansion of the deposited thin films of aluminum during porous oxide formation [J]. Applied Surface Science, 2003, 220: 51–59.

[11] 钱余海, 李美栓, 张亚明. 氧化膜开裂和剥落行为 [J]. 腐蚀科学与防护技术, 2003, 15(02): 90–93.

[12] 李美栓, 辛丽, 钱余海,等. 氧化膜应力研究进展 [J]. 腐蚀科学与防护技术, 1999, 11(5): 300–305.

[13] Delaunay D, Huntz A M. Mechanisms of adherence of alumina scale developed during high–temperature oxidation of Fe–Ni–Cr–Al–Y alloys [J]. Journal of Materials Science, 1982, 17(7): 2027–2035.

[14] Hancock P, Hurst R C, Advances in Corrosion Science and Technology [M], New York: Plenum Press, 1974, 4: 1.

[15] 刘伟华. 热循环作用下铝合金阳极氧化膜的开裂行为与机理研究 [D]. 北京化工大学, 2008.

[16] Poinern, Ali, Fawcett. Progress in Nano–Engineered Anodic Aluminum Oxide Membrane Development [J]. Materials (1996–1944), 2011, 4(3): 487–526.

[17] 曹楚南, 张鉴清. 电化学阻抗谱导论 [M]. 北京: 科学出版社, 2002.

[18] Belwalkar A, Grasing E, Vangeertruyden W, et al. Effect of processing para- meters on pore structure and thickness of anodic aluminum oxide (AAO) tubular membranes [J]. Journal of Membrane Science, 2008, 319(1–2): 192–198.

[19] 朱祖芳. 铝合金阳极氧化膜与表面处理技术 [M]. 北京: 化学工业出版社, 2004.

[20] 陈飞, 解利昕, 孙晨,等. 动电位极化曲线法研究铝合金材料在海水中腐蚀现象[J]. 天津化工, 2014, 28(5): 12–14.

[21] 杨振海, 徐宁, 邱竹贤. 铝的电位—pH 图及铝腐蚀曲线的测定 [J]. 东北大学学报:自然科学版, 2000, 21(4): 401–403.

[22] 贺格平, 梁燕萍, 刘男. 多孔阳极氧化铝膜 EIS 阻抗谱研究 [J]. 表面技术, 2007, 36(6): 9–11.

固态铝中氢扩散的物理模拟

黄　蓓，曾建民

(广西大学 材料科学与工程学院；广西大学 有色金属及加工新技术教育部重点实验室)

摘　要：氢是铝及其铝合金中唯一能够大量溶解的有害元素。由于氢在液态和固态巨大的溶解度差别，在凝固过程中的析出会形成气孔并导致铸件产生缩松，降低材料的力学性能、导热和导电性能以及抗腐蚀性能。因此，控制铝中的氢含量是控制铝产品质量的一个重要任务。本工作根据电桥对于不同含氢量惰性气体导热性的敏感性，通过桥路输出电压的变化测量固态铝中氢扩散过程。测定了不同凝固速度对样品含氢量的影响。实验表明：固态铝中的含氢量随着凝固速度的增加而增加。通过快速凝固的手段，可以测定铝熔体中的含氢量。该技术可以作为铝合金熔炼质量的检测手段。

关键词：氢；固态铝；凝固

Physical Simulation for Hydrogen Diffusion in Solid Aluminium

HUANG Bei, ZENG Jian-min

(Key Laboratory of Education Ministry in Non-ferrous Metals and New Processing Technology;
School of Materials Science and Engineering, Guangxi University)

Abstract: Hydrogen is a soluble detrimental element in aluminum alloys in large quantity. The shrinkage and micro-porosity will result during solidification of casting from the difference of solubility between liquid and solid states, which leads to the decrease of mechanical properties, heat conductivity, electro-conductivity and corrosion resistant. Therefore, it is an important task to control the hydrogen in aluminum alloys. The present work proposes an innovative method to measure hydrogen diffusion in solid aluminum by the change in electricity voltage output through the bridge circuit, based on the sensitivity of inertial gas to the heat conductivity. The effect of solidification rate on hydrogen content of the sample was measured. The tests have shown that the hydrogen in molten aluminum can be detected by means of fast solidification.

Key word: hydrogen; solid aluminum; solidification

基金项目：国家自然科学基金（51064003）；广西自然科学基金(2011GXNSD018009)。
第一作者：黄　蓓，女。铝熔体净化研究。869550255@qq.com。
通讯作者：曾建民，男。铝熔体净化研究。zjmg@gxu.edu.cn。

Al–Zn–Mg–Cu 铝合金非等温时效研究

冯呈庠，曾淼霞，金 曼

(上海大学 材料科学与工程学院)

摘 要：文章研究了两种 Cu 含量的 Al–Zn–Mg–Cu 合金在非等温时效工艺条件下，析出相的变化规律和力学性能特征。通过示差扫描量热分析析出相的析出情况。通过常规力学性能试验检测合金的强度、伸长率、硬度以及耐蚀性。实验结果表明经过非等温时效后，Al–Zn–Mg 合金的抗拉强度达到了 472.25MPa、延伸率为 16.67%；Al–Zn–Mg 合金的抗拉强度为 543.01MPa、延伸率为 18.33%。此外，Cu 含量的增大有利于提高 G.P.区在时效初期的稳定性，并能推迟 η' 相向 η 相的转变。

关键词：非等温时效；Al–Zn–Mg–Cu 合金；Cu 元素；析出相

The Non–isothermal Linear Cooling Treatment of Al–Zn–Mg–Cu Aluminum Alloy of Two Different Cu Content

FENG Cheng–xiang, ZENG Miao–xia, JIN Man

(College of Material Science and Technology, Shanghai University)

Abstract: The precipitates and mechanical properties of Al-Zn-Mg-Cu aluminum alloy with two different Cu contents by the non-isothermal linear cooling treatment have been studied. The change regulation of precipitates has been observed by differential scanning calorimetry (DSC). The strength, elongation, hardness and corrosion resistance of the aluminum alloy are detected by the conventional mechanical property tests. The result shows that the tensile strength of Al-Zn-Mg is up to 472.25MPa and elongation is up to16.67%, while the tensile strength of Al-Zn-Mg-Cu is up to 543.01MPa and elongation is up to 18.33%. Furthermore, Cu could improve the stability of G.P. zone at the early stage of aging treatment, and delay the translation of η' into η during the non-isothermal line cooling treatment.

Key words: the non-isothermal linear cooling treatment; Al-Zn-Mg-Cu alloy; Cu; precipitates

0 引言

Al-Zn-Mg-Cu 合金具有比强度高、密度小、塑性好等特点，因而被认为是一种优秀的轻质材料，广泛应用于交通运输和航空航天领域。一些大型构件在加热时会出现温度梯度，即非等温环境。国内外

第一作者：冯呈庠，男。新型铝合金研发。ryansampson@163.com。

通讯作者：张 梅，女，博士。轻量化金属材料开发和应用。zhangmei3721@i.shu.edu.cn。

学者利用这一特性对铝合金进行时效研究,并得到了较好的综合性能[1]。据报道,非等温时效方式所得到的材料具有更好的综合性能[2]。此外,非等温时效的时间短,生产效率高,具有良好的经济效益。本文主要研究了非等温时效对 Al-Zn-Mg-Cu 系合金性能的影响。

1 试样制备和试验方法

1.1 试样制备

试验用的两种合金化学成分如表 1。以纯铝(99.98wt.%)、纯锌(99.98wt.%)、纯镁(99.75wt.%)和 Al-42.3wt.%Cu、Al-4.21wt.%Zr 为原料配成 Al-7.8 Zn-1.6 Mg-x Cu-0.14 Zr (x=0,0.8,1.6) 的三种成分,经过石墨坩埚熔炼、除气、精炼、打渣后,加电磁搅拌连铸成直径为 Φ72mm 的圆柱型棒料。将浇铸好的棒料加热至 420℃,在轧机上轧制变形 90%。

表 1 试验材料的化学成分(wt.%)
Tab.1 Chemical composition of the test steels(wt.%)

Alloy	Zn	Mg	Cu	Zr	Si	Fe	Al
7000(1#)	7.82	1.57	0.003	0.139	0.031	0.02	Bal
7085(2#)	7.8	1.59	1.64	0.14	0.026	0.07	Bal

1.2 试验方法

在对实验板材 Al-Zn-Mg 合金和 Al-Zn-Mg-Cu 合金进行时效之前,需要先进行固溶处理。实验中对同一批次的试样进行相同的固溶处理,以提供相同的基体组织。将实验板材加热至 470℃,并保温两小时,结束时水冷至室温。在对 Al-Zn-Mg 合金和 Al-Zn-Mg-Cu 合金板材固溶处理后,在箱式炉中进行时效处理。线性升温时效是将经过固溶处理的板材放入箱式炉中, 以 20℃ /h 的速度从室温升至 220℃,随后立即水冷却至室温。

2 试验结果与分析

2.1 硬化特性

两种 Cu 含量的 Al-Zn-Mg-Cu 合金的硬化曲线如图 1 曲线所示。根据试验结果可得,随时效温度的提高,两种合金的硬度都呈增加趋势,并在某一温度达到峰值后下降。Al-Zn-Mg-Cu 合金硬度始终大于 Al-Zn-Mg 合金硬度。其原因在于 Cu 元素在基体内固溶,晶格发生畸变使硬度增加,起到了固溶强化的作用。在合金的时效转变期内,Al-Zn-Mg 合金中不同类型析出相的转变速度相对较快,表现为曲线上升斜率较大。

此外,Al-Zn-Mg-Cu 合金硬度开始下降的温度相对 Al-Zn-Mg 合金滞后。这说明 Cu 元素可以延缓亚稳相 η' 向稳定相 η 的转变。Cu 元素对 Al-Zn-Mg-Cu 合金线性升温时效过程中不同析出相之间的转换速率和转换方式有影响,并且可能形成具有新的晶体学特征的析出相[3]。

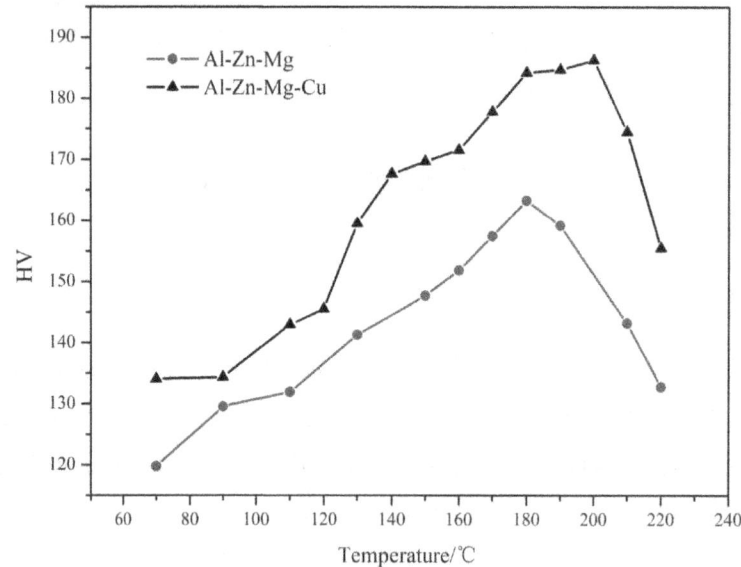

图1 Al-Zn-Mg 合金和 Al-Zn-Mg-Cu 合金以 20℃/h 的速度线性升温时效过程中硬度随温度的变化曲线

Fig.1 The hardness curve of Al-Zn-Mg and Al-Zn-Mg-Cu during the non-isothermal line cooling treatment at 20℃/h

2.2 电导率分析

本实验中,对两种 Cu 含量的 Al-Zn-Mg-Cu 系合金进行非等温时效,升温速率为 20℃/h。图 2 为两种合金线性升温时效过程中电导率随温度的变化曲线。

比较图 2 的两条曲线可以看出,电导率都随着时效温度的升高而上升,且在时效初期,电导率值较低,增长速率缓慢。Al-Zn-Mg 合金在 180℃ 后电导率快速增加。Al-Zn-Mg-Cu 合金在 200℃ 后电导率快速上升。但是在整个时效过程中,Al-Zn-Mg-Cu 合金的电导率小于 Al-Zn-Mg 合金。此外,Al-Zn-Mg-Cu 合金的电导率曲线转变点落后于 Al-Zn-Mg 合金。

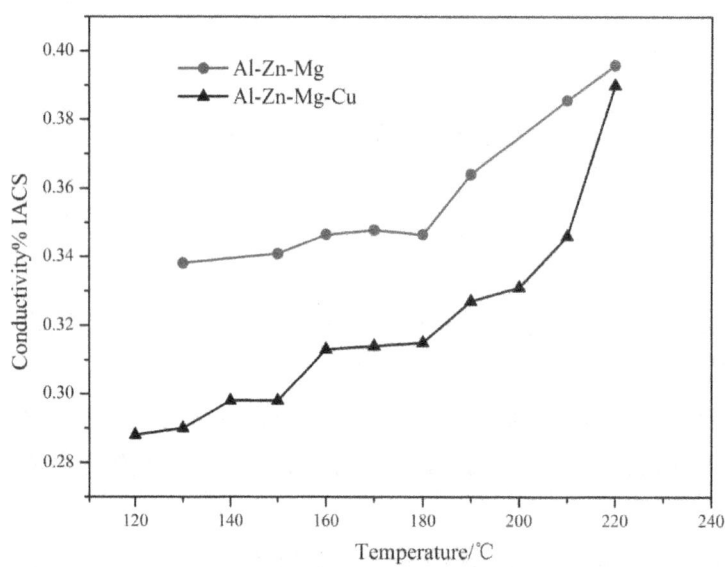

图2 固溶态的 Al-Zn-Mg 合金和 Al-Zn-Mg-Cu 合金以 20℃/h 的速度线性升温时效过程中电导率随温度的变化曲线

Fig.2 The hardness curve of Al-Zn-Mg and Al-Zn-Mg-Cu during the non-isothermal line cooling treatment at 20℃/h

电导率试验结果表明,Cu 在时效过程中能够置换析出相中的 Mg、Zn 原子。由于 Mg、Zn 原子和 Cu 原子尺寸不同,Cu 元素的加入会破坏原有晶体结构的对称性,造成晶格畸变,使得晶格对电子的散射作用增强,因此 Al-Zn-Mg-Cu 合金的电导率更低。另一方面,随着 Cu 含量的增加,固溶体中合金元素的过饱和程度增大,合金的高能区增多,提高了合金元素脱溶析出的驱动力,使得合金元素更容易偏聚形核并且长大。此外,本实验中所用的 Al-Zn-Mg-Cu 合金是经过轧制后空冷处理的试样。合金中粗大的析出相尺寸约为 1-4μm,当粗大的第二相粒子尺寸大于 1μm 时会诱发形核,引起再结晶的发生。合金组织中再结晶的出现,会使晶格对电子散射的作用增强,电导率降低[4]。

2.3 力学性能

经过升温时效的 Al-Zn-Mg 合金和 Al-Zn-Mg-Cu 合金的拉伸性能如图 3 所示。曲线大致可分为两个阶段。直线部分是材料的弹性变形阶段,材料延伸率呈线性增加。随后在屈服点附近进入塑性阶段,合金的延伸率随着载荷的增加而增加,直至试样断裂。拉伸过程中,Al-Zn-Mg 应力应变曲线有微小波动,这可能是因为在 Al 基体中固溶原子受到位错的作用逐渐脱离基体所致。

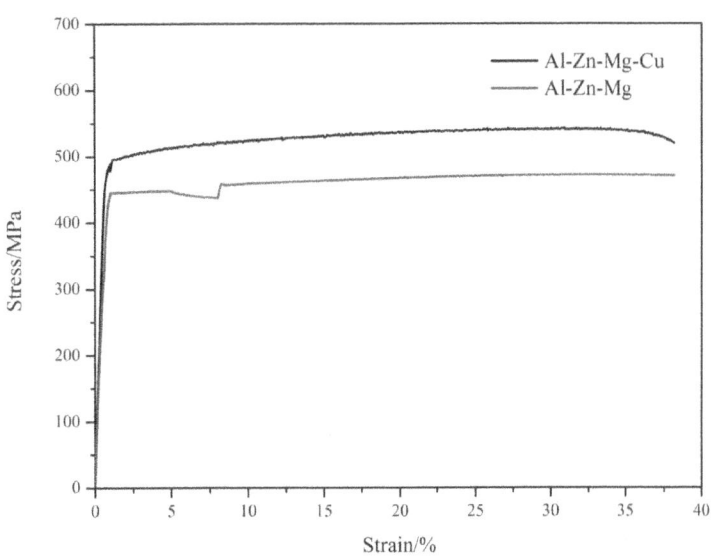

图 3　Al-Zn-Mg 合金及 Al-Zn-Mg-Cu 合金的应力应变曲线
Fig.3　The stress-strain curve of Al-Zn-Mg and Al-Zn-Mg-Cu

两种 Cu 含量的 Al-Zn-Mg-Cu 合金非等温时效后力学性能如表 2 所示。从中可以看出,Al-Zn-Mg-Cu 合金的力学性能明显好于 Al-Zn-Mg 合金。屈服强度增加 7.22%,抗拉强度提升 14.98%,延伸率提高 9.06%。可见,Cu 含量的增加有利于提高 Al-Zn-Mg-Cu 合金的性能。其原因与诸多因素有关,一方面,Cu 固溶于 Al 基体内,使得晶格畸变程度增大,这种畸变使得强度得到提高。另一方面,Cu 在时效过程中析出,形成一些稳定相,这些相的存在阻碍形变时位错的运动。从而提高了合金的屈服强度和抗拉强度。此外,Cu 与 Al 基体中的元素发生反应形成金属间化合物。这些化合物十分细小,在铝液凝固时成为非均质形核核心。当单位体积内非均质形核数较多时,形成的晶粒越细,因而起到了细晶强化作用。细晶不仅可以提高强度,同时也能提高材料的塑性,使材料具有良好的力学性能。此外,拉伸试验结果也表明同一牌号的合金,采用非等温时效可以得到更高的强度。采用过时效态的 Al-Zn-Mg-Cu 合金的抗拉强度为 475MPa[5],而试验中 Al-Zn-Mg-Cu 合金的平均抗拉强度为 543.01MPa,强度提高 14.3%。

表2 两种 Cu 含量的 Al-Zn-Mg-Cu 合金非等温时效后力学性能

Tab.2 The Mechanical properties of Al-Zn-Mg-Cu aluminum alloy with two different Cu contents by the non-isothermal linear cooling treatment

热处理状态	试样	屈服强度/MPa	抗拉强度/MPa	断后伸长率/%
非等温时效	Al-Zn-Mg	443.95	472.25	16.67
	Al-Zn-Mg-Cu	476.01	543.01	18.33

2.4 DSC 分析

图4是两种 Cu 含量的 Al-Zn-Mg-Cu 合金的 DSC 曲线。表3、表4是 Al-Zn-Mg 合金以及 Al-Zn-Mg-Cu 合金不同相的转变温度区间。

DSC 试验结果表明,Cu 的加入对 Al-Zn-Mg-Cu 合金析出的最初阶段会产生很大影响,表现为在图4的曲线上出现了单独的 G.P.区放热峰和吸热峰。不含 Cu 元素的 Al-Zn-Mg 合金,在 DSC 曲线中则观察不到明显的 G.P.区析出和溶解峰。由此可得,Cu 的存在可以有效提高 G.P.区在时效初期的稳定性。有研究表明,Cu 原子可以溶入 G.P.区,提高 G.P.区的稳定温度范围,延缓非平衡相 η' 向平衡相 η 的转变,而本文实验也验证了这一观点。此外,在时效后期,Al-Zn-Mg 合金中 η' 相的长大和回溶,出现在 160~185℃ 温度范围内,而在 Al-Zn-Mg-Cu 合金中则相对滞后,具体温度范围是 180~204℃。这说明 Cu 不仅会对 G.P.区的稳定性产生影响,同时也会提高非平衡相 η' 在时效过程中的相变稳定性。

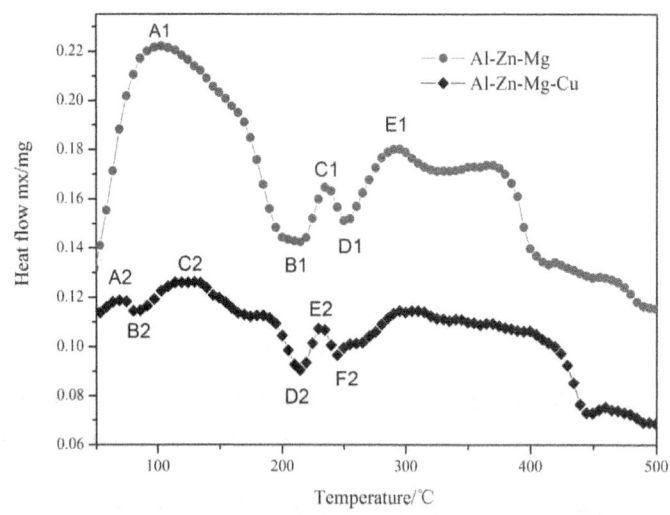

图4 固溶态 Al-Zn-Mg 合金和 Al-Zn-Mg-Cu 合金的 DSC 曲线(扫描速度为 10K/min)

Fig.4 The DSC curve of Al-Zn-Mg and Al-Zn-Mg-Cu at solid solution state(Scan speed: 10K/min)

表3 Al-Zn-Mg 合金不同相的转变温度区间

Tab.3 The transition temperature range of different phases of Al-Zn-Mg

温度区间/℃	反映
61~185(放热峰 A1)	η' 相形核
185~225(吸热峰 B1)	η' 相溶解
225~245(放热峰 C1)	η 相形核
245~265(吸热峰 D1)	η 相溶解

表 4 Al-Zn-Mg-Cu 合金不同相的转变温度区间

Tab.4 The transition temperature range of different phases of Al-Zn-Mg-Cu

温度区间/℃	反映
59~77(放热峰 A2)	G.P.区析出
77~100(吸热峰 B2)	G.P.区溶解
100~180(放热峰 C2)	η' 相形核
180~225(吸热峰 D2)	η' 相溶解
225~240(放热峰 E2)	η 相形核
240~276(吸热峰 F2)	η 相溶解

3 结论

1. 两种 Al-Zn-Mg-Cu 合金的硬度随时效温度的升高而升高，达到峰值硬度后迅速下降，且 Al-Zn-Mg 合金的硬度始终小于 Al-Zn-Mg-Cu 合金的硬度。Al-Zn-Mg 合金的最高硬度为 163.2HV；Al-Zn-Mg-Cu 合金则为 186.33HV。

2. 两种合金的电导率随着温度的升高呈上升趋势。Al-Zn-Mg-Cu 合金电导率始终小于 Al-Zn-Mg 合金的电导率。

3. 非等温时效可以使两种合金得到较好的力学性能。Al-Zn-Mg 合金的抗拉强度为 472.25MPa、延伸率为 16.67%；Al-Zn-Mg 合金的抗拉强度为 543.01MPa、延伸率为 18.33%。

4. Cu 元素可以增加在时效初期 G.P.区的稳定性，并促进 G.P.区的形核。在时效后期，使得 η' 相长大粗化，并且能够推迟 η' 相向 η 相的转化。

参考文献：

[1] J. T. Staley, Durham. Method and process of non-isothermal for aluminum alloys [J]. US Patent. 0237113Al. 2007.

[2] 李科. 7A85 铝合金非等温时效工艺及组织性能的研究 [D]. 哈尔滨：哈尔滨工业大学硕士学位论文. 2008.

[3] 宁爱林，蒋寿山，彭北山. 铝合金的力学性能及其电导率 [J]. 轻金属，2005，6: 34-36.

[4] 熊柏青，李锡武，张永安等. Al-Zn-Mg-Cu 合金的淬火敏感性 [J]. 中国有色金属学报，2011，21(10)：2631-2638.

[5] 杨海虹，胡兴华，王少华，等. 7085 铝合金锻件强度、硬度和电导率关系研究 [J]. 铝加工，2012，4: 37-40.

数值模拟在先进高强钢极限拉深和回弹预测中的应用

汤 潇[1]，宋佳男[1]，张 梅[1*]，王 韬[2]

(1.上海大学 材料科学与工程学院；2.上海汇众汽车制造有限公司)

摘 要：运用 Dynaform 软件对先进高强钢的冲压成形、回弹和极限拉深进行数值模拟。分析压边力、摩擦系数对回弹结果的影响，通过软件探索板材的极限拉深值。实验结果与数值模拟结果基本吻合，当压边力从 150KN 增加到 300KN 时，B 钢的回弹角从 12.45° 降低到了 1.2°，F 钢的回弹角从 12.45°降低到 2.75°。摩擦系数对模拟结果有很大影响，对于 B 钢，当压边力为 200KN 时，随着摩擦系数从 0.03 增加到 0.06，回弹角从 13.769°降低到 0.676°。

关键词：先进高强钢；Dynaform；回弹；极限拉深

The Application of Numerical Simulation in Limit Drawing and Springback Experiments of Advanced High Strength Steels

TANG Xiao[1], SONG Jia-nan[1], ZHANG Mei[1], WANG Tao[2]

(1.Shanghai University, Materials Science and Engineering Institute；
2. Shanghai HuiZhong Automotive Manufacturing Corporation)

Abstract：The forming, springback and limit darwing processes of advanced high strength steel were simulated respectively using the commercial FE software Dynaform. The effects of blank holding force and friction factor on the springback results were analyzing and predicted the limit drawing by Dynaform. The simulation results and the experimental results were matched well. With the increase of blank holding force form 150KN to 300KN the springback angle becomes smaller form 12.45° to 1.2° for steel B and from 12.45° to 2.75° for steel F. The coefficient of friction had a great effect on the simulation results, When the blank holding force is 200KN, with the increase of friction coefficient from 0.03 to 0.06, the springback angle reduces form 1.769° to 0.676° .

Keywords: advanced high strength steel; dynaform; springback; limit darwing

---

第一作者：汤 潇，男。汽车用钢冲压特性研究。253968834@qq.com。
通讯作者：张 梅，女，博士。轻量化金属材料开发和应用。zhangmei3721@i.shu.edu.cn。

22MnB5 后桥横梁热成形工艺仿真分析

薛甬申[1]，金晓春[2]，张　梅[1*]，宋佳男[1]，武　海[2]，万　紫[2]

(1.上海大学 材料科学与工程学院；2.上海汇众汽车制造有限公司 技术中心)

摘　要：热成形作为一项将板料加工和淬火工艺相结合的成形技术，可以使成形零件获得高强度以及高精度。典型的淬火硬化硼钢板 22MnB5 可通过热冲压成形使性能超高强化。本文提出了该钢种针对后桥横梁零件的热成形工艺流程，通过有限元数值模拟仿真软件 Deform 对零件热成形工艺进行仿真分析，分析了零件在成形和随模保压淬火过程中模具和零件温度场的变化，并检测了实际热冲压件的性能和组织，确认了热冲压零件可以达到超高强化的目标，最终组织为低碳马氏体，显微硬度达到 480HV 以上，可为该工艺在后桥横梁零件产业化方面提供依据。

关键词： 22MnB5；热成形；数值模拟；CAE

CAE on Hot Forming Procedure of Rear Axle Component

XUE Yong-shen[1], JIN Xiao-chun[2], ZHANG Mei[1*], SONG Jia-nan[1], WU Hai[2], WAN Zi[2]

(1.School of Materials Science and Engineering, Shanghai University;

2.Technical Center, Shanghai Huizhong Automotive Manufacturing Co.,Ltd,)

Abstract: Hot stamping of ultra high-strength steels (UHSS) is a new and complex forming technology integration metal hot forming and the quenching process, the shape accurancy and strength of parts formed are very high. 22MnB5 is a typical hardened ultra-high strength boron steel, which can obtain better properties through hot stamping. This paper takes 22MnB5 as research object, the hot stamping process is simulated by using DEFORMTM code, analysis the temperature field of 22MnB5 during forming and quenching, offer the basis for the industrial production line.

Key words: 22MnB5; hot forming; numerical simulation; CAE

基金项目: 973 计划项目(编号:2010CB-630802)；国家自然科学基金项目(50934011 和 50971137)。

第一作者: 薛甬申(1990—)，男，硕士研究生。

通讯作者: 张　梅,女,博士。轻量化金属材料开发和应用。zhangmei3721@i.shu.edu.cn。

TRIP 钢的热变形行为研究

甘 斌[1], 张 梅[1*], 赵 雪[1], 朱 妍[1], 钟 勇[2], 李 麟[1]

(1.上海大学 材料科学与工程学院; 2.宝钢集团汽车用钢开发与应用技术国家重点实验室)

摘 要：本文通过 Gleeble-3500 物理模拟试验机，研究 TRIP 钢在温度 900℃~1150℃、应变速率为 $0.1s^{-1}$,$1s^{-1}$,$10s^{-1}$、变形量为 60%下的热变形行为，分析了变形温度和变形速率对流变应力的影响，发现在高温 1150℃及低应变速率 $0.1s^{-1}$ 下材料更容易发生动态再结晶行为。通过回归分析，求得实验钢在 900℃~1150℃下的热变形激活能 Q=264.86 KJ/mol 和应力指数 n=3.7113，建立了实验钢的流变应力本构方程。

关键词：TRIP 钢；动态再结晶；热变形激活能；本构方程

Research on Hot Deformation Behavior of TRIP Steel

GAN Bin[1], ZHANG Mei[1*], ZHAO Xue[1], ZHU Yan[1], ZHONG Yong[2], LI Lin[1]

(1.School of Materials Science and Engineering, Shanghai University;
2.State Key Laboratory of Development and Application Technology of Automotive Steels)

Abstract: The hot deformation behavior of TRIP high strength steel is studied on Gleeble-3500 thermo-mechanical simulator under the temperature from 900~1150℃, strain rate of $0.1s^{-1}$,$1s^{-1}$,$10s^{-1}$ and deformation of 60%. The result show that dynamic recrystallization occurs easily under higher deformation temperature of 1150℃ and lower strain rate of $0.1s^{-1}$. The deformation activation energy and stress exponent n of the tested steel at 900-1150℃ are respectively calculated of 264.86kJ/mol and 3.7113 by regression analysis. Meanwhile, the constitutive equation of the tested steel is also established.

Key words: TRIP steel; dynamic recrystallization; deformation activation energy; constitutive equation

0 引言

当前车身部件需满足安全性和低油耗的要求,低油耗可降低 CO_2、NO_x 的排放量,减少对环境的污染。应用高强度钢材可达到既提高汽车安全性又可减轻汽车自重的目的。具有相变诱发塑性(Transformation Induced Plasticity)效应的 TRIP 钢板能满足上述要求[1]。在汽车用双相钢的发展过程中,有不少用廉价的硅、锰元素取代铬、镍及钼元素的经验,同样在双相钢中发现有残余奥氏体存在,并且具

基金项目："973"计划项目(编号:2010CB-630802);国家自然科学基金项目(编号:50934011 和 50971137)。

作者简介:甘 斌,男,硕士研究生。

通讯作者:张 梅,女,博士。轻量化金属材料开发和应用。zhangmei3721@i.shu.edu.cn。

有 TRIP 效应[2]。随着微合金钢冶炼技术和控轧控冷技术的发展,出现了采用微合金成分,利用控轧控冷技术,以热连轧的形式生产高强度钢板,具有更高的平整度和尺寸精度、更均匀的性能和更好的表面质量[3]。近 20 多年,这种钢受到国外钢铁和汽车行业的高度关注,对其进行了大量的开发研究工作,并且已经商业化,用于制造汽车的冲压构件以及经冲压后的抗撞构件。

本文通过 Gleeble-3500 热力模拟试验机进行新型超高强度 TRIP 钢的热压缩实验,研究轧制工艺参数(变形温度、变形速率)对实验钢流变应力的影响,确定实验钢的热变形激活能和流变应力方程,旨在为实验钢的工业生产提供重要理论依据。

1 实验材料及方法

本实验选用某钢厂提供的代号为 D 的实验 TRIP 钢,其主要元素有 C-Mn-Si-Al,以及微合金元素 Nb,主要元素如表 1 所示。

表 1   D 钢成分
Tab.1   Chemical composition of the studied steel D

	C	Si	Mn	B	Al
D	0.28–0.36	<0.5	<1.5	0.002–0.005	>1.5

试样尺寸加工为 Φ10×15mm(如图 1 所示),通过 Gleeble-3500 热模拟机,进行单道次热压缩实验,同时试样测试应力 - 应变曲线。实验工艺图如图 2 所示,试样加热到 1200℃奥氏体化温度并保温 5 分钟,然后以 5℃/s 冷却到变形温度,在变形温度保温 10s 使试样温度均匀,随后以恒定应变速率进行压缩,变形温度为 900℃~1150℃,每隔 50℃记录应力应变曲线,应变速率分别为 0.1s⁻¹,1s⁻¹,10s⁻¹,变形量为 60%,压缩后试样立即进行水淬以固定组织。

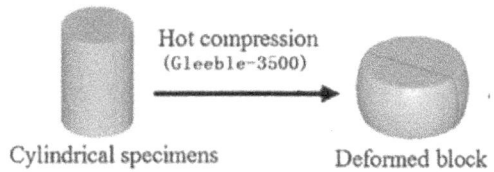

图 1   试样压后示意图
Fig.1   The shape of specimen before and after test

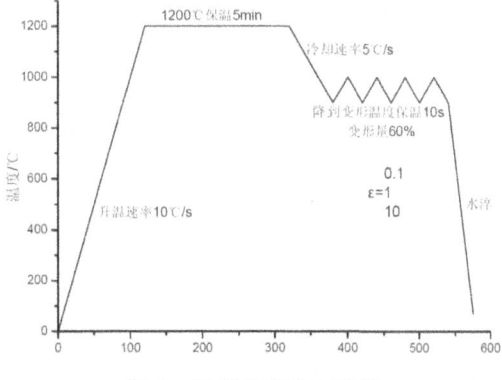

图 2   热压缩实验工艺图
Fig.2   Process of hot compression test of steel D

2 实验结果与分析

2.1 真应力-真应变曲线

热变形流变应力是材料在高温下的塑性指标之一。温度对流动应力影响很大,温度升高,在变形过程中出现动态回复和动态再结晶过程,动态再结晶是通过形核和长大来完成,其机理是大角度晶界(或亚晶界)向高位错密度区域的迁移。这些软化作用使加工硬化减弱。实验证明在一定温度范围内,温度较高时,热激活作用增强位错的活动性能,可能出现新的滑移系统,可能使扩散塑性变形机理同时起作用,使变形容易进行,流动应力下降。同时温度升高有利于回复和再结晶软化过程的发展,可使变形过程造成的变形和缺陷修复,从而塑性提高,流动应力下降。

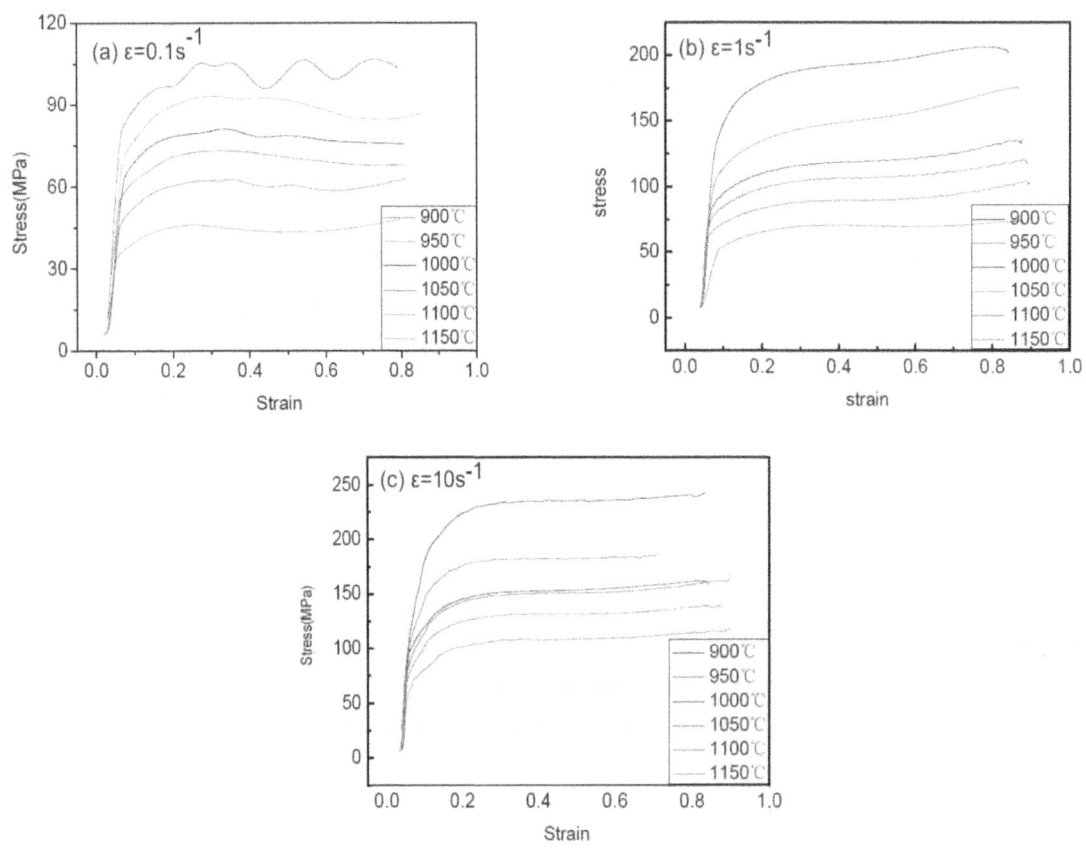

图3 D钢不同变形温度下真应力-应变曲线

Fig.3 True strain-stress curves of D steel under different deformation temperature

图3为实验钢在应变速率为0.1s⁻¹、1s⁻¹和10s⁻¹下不同变形温度(950℃~1150℃)下的真应力真应变曲线。图中可看出,应力随变形温度上升而下降。当应变速率较低时(图3-a),应力应变曲线中,应力随应变的增大而增加,随后出现极大值,接着应力随应变增加而减小,最后达到稳定阶段,此类曲线为动态再结晶型曲线;而当应变速率升高时(图3-c),各温度下的真应力-真应变曲线并未呈现出明显的动态再结晶特征,应力随应变的增大不变,达到稳定阶段,此类为动态回复型曲线;还有一类为加工

硬化型曲线,其应力随着应变一直增大(图 3-b)。由此表明,在高温以及低应变速率下,实验钢容易发生动态再结晶行为,在低变形温度和高应变速率下,动态再结晶不易启动。

当应变速率一定时,实验钢的峰值应力(变形抗力)和峰值应变均随着变形温度的升高而减小(图 3-a),表明变形温度越高,动态再结晶越容易发生。动态再结晶是一个热激活过程,变形温度的升高导致空位原子扩散、位错进行攀移以及交滑移驱动力的增大,由材料热变形激活能控制的动态再结晶的形核率和晶核长大的驱动力不断增大,动态再结晶进行更加充分,从而部分消除加工硬化现象,使材料热变形过程中的峰值应力降低;另外,随着温度升高,原子热运动加剧,原子间结合力减弱,临界切应力降低,也降低了材料的变形抗力[4]。

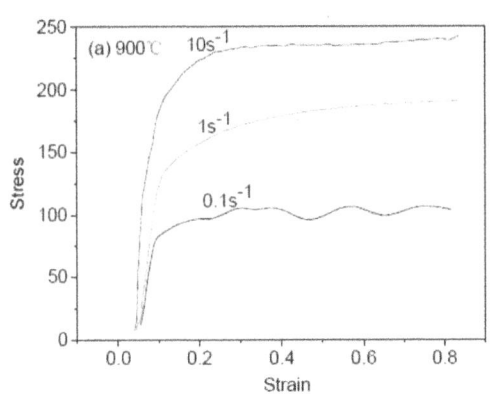 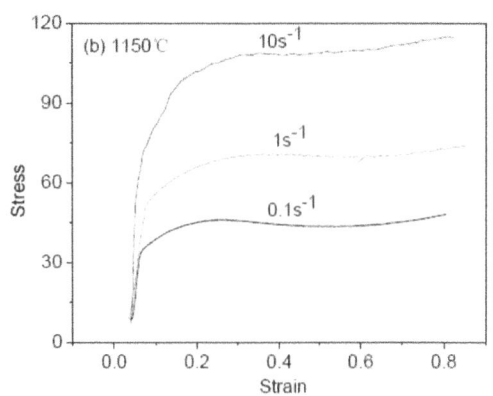

图 4 D 钢 900℃及 1100℃时,不同应变速率下的真应力–真应变曲线
Fig.4 Strain–stress curves of D steel under different strain rates

变形温度一定时,峰值应力和峰值应变都随应变速率的增大而增大,随着应变速率提高,加工硬化导致的位错密度增大显著,而由动态回复和动态再结晶导致的位错密度降低难以同步抵消形变产生的位错积累,硬化与软化作用间差距拉大导致流变应力增加,从而峰值应力和峰值应变均增大。图 4(a)中,实验钢 900℃下变形只有在 0.1s⁻¹ 的应变速率下其曲线才为动态再结晶型,而 4(b)中,1150℃下变形,无论何种应变速率下实验钢的应力应变曲线都出现应力峰值,全部为动态再结晶型。由此可见:在相同温度下,应变速率较低时,动态再结晶容易启动,温度的增加也有利于动态再结晶行为的发生。

2.2 热变形方程

动态再结晶是由热激活控制的过程。金属的热变形过程用涵盖了变形温度和应变速率的 Zener-Hollomon 参数方程[5]来描述。根据 Z 参数的概念,热变形过程中流变应力 σ 和变形条件之间存在如下关系:

$$Z = \dot{\varepsilon} \exp(\frac{Q_{def}}{RT}) = A[\sinh(\alpha\sigma)]^n \tag{1}$$

其中,σ_p 为峰值应力,$\dot{\varepsilon}$ 为变形速度,Q 为激活能,R 为气体常数,T 为绝对温度,A,a,n 是与钢种有关的材料常数,a 变化很小。根据 Uvira 和 Jonas 等[6]的研究,本文所用实验钢 a 最佳取值一般为 0.012MPa⁻¹。该 Z 参数等于变形速度 $\dot{\varepsilon}$ 与一个温度函数 exp(Q/RT)的乘积,故又称为经过温度补偿的变形速度。塑性加工金属学的许多现象与此参数有关,如高温变形时的变形抗力、动态回复后的亚晶粒尺寸、动态再结晶成核机构、动态再结晶后的晶粒大小等,都是 Zener-Hollomon 参数的函数。

为依据热压缩实验结果计算得出实验钢的热变形激活能，对式(1)经变形得

$$\dot{\varepsilon} = A[\sinh(\alpha\sigma_p)]^n \cdot \exp(\frac{-Q_{def}}{RT}) \qquad (2)$$

对式(2)两边取对数整理得：

$$\ln\dot{\varepsilon} + \frac{Q_{def}}{RT} - \ln A = n.\ln\left\{\sinh\left(\alpha\sigma_p\right)\right\} \qquad (3)$$

当 $\dot{\varepsilon}$ 一定时及当 T 一定时，式(3)两边分别对 $\frac{1}{T}$ 及对 $\ln\dot{\varepsilon}$ 求偏导，整理得

$$Q_{def} = Rnb \qquad (4)$$

其中

$$b = \frac{\left[\partial\ln\sinh\left(\alpha\sigma_p\right)\right]}{\partial\frac{1}{T}}\bigg|_{\dot{\varepsilon}} \qquad (5)$$

$$n = \frac{\partial\ln\dot{\varepsilon}}{\partial\left[\ln\sinh\left(a\sigma_p\right)\right]}\bigg|_T \qquad (6)$$

图 5、图 6 分别为 lnsinh($\alpha\sigma_p$)与 $\frac{1}{T}$、ln$\dot{\varepsilon}$与 lnsinh($\alpha\sigma_p$)的关系，经线性拟合得到直线斜率的平均值即 b、n，从而得出 Q$_{def}$，计算结果如表 2 所示。

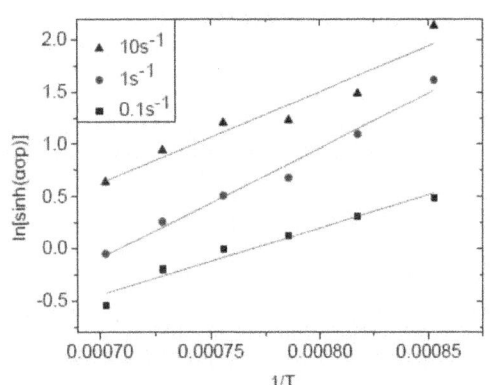

图 5 lnsinh(ασp)与 的关系
Fig.5 Relationship between lnsinh(ασp) and

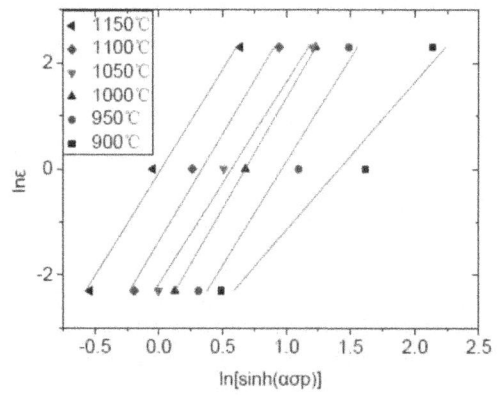

图 6 ln 与 lnsinh(ασpp)的关系
Fig.6 Relationship between ln and lnsinh(ασp)

表2 D 钢激活能及相关参数

Tab.2 Deformation activation energy and related parameters

钢种	b	n	Q$_{def}$(kJ/mol)
D	8582.805	3.711	264.86

将各变形条件下的 n 值、b 值、α 值以及 Qdef 代回式(1)可得到 A 值为 4.45×109，再将所得 A 值代回式(1)，可得到 Z 参数与应变速率、变形温度、峰值应力的函数：

$$Z_{DTRIP} = \dot{\varepsilon}\exp\left(\frac{264860}{RT}\right) = 4.45\times10^9\left[\sinh\left(\alpha\sigma_p\right)\right]^{3.7113}$$

3 结论

（1）本实验采用代号为 D 的 TRIP 钢热变形实验过程中，流变应力受变形温度和应变速率的影响显著，且随变形温度的升高而减小，随应变速率的增大而增大。由真应力 - 真应变曲线可知，实验钢在高变形温度和低应变速率下易发生动态再结晶。

（2）通过线性拟合以及回归分析，得到 D 钢在 900℃ 到 1150℃ 下的热变形激活能 Q 为 264.86KJ/mol。

（3）用 Zener-Hollomon 参数来描述实验钢的热变形行为，,,通过回归分析求得 Z 参数为：

$$Z_{DTRIP} = \dot{\varepsilon} \exp\left(\frac{264860}{RT}\right) = 4.45 \times 10^9 \left[\sinh\left(\alpha\sigma_p\right)\right]^{3.7113}$$

参考文献

[1] 唐代明. TRIP 钢中合金元素的作用和处理工艺的研究进展 [J]. 钢铁研究学报,2008,01: 1-5.

[2] 马鸣图,吴宝榕. 双相钢——物理和力学冶金 [M]. 北京: 冶金工业出版社,1988.

[3] 张浩. 纳米级析出强化高强钢的工艺研究 [D]. 辽宁科技大学,2008.

[4] 鲍思前，赵刚，余驰斌，叶传龙. 含 Nb 微合金钢动态再结晶行为研究 [J]. 武汉科技大学学报, 2008,05: 543-546.

[5] C. Zener, J. H. Hollomon. Effect of Strain Rate Upon Plastic Flow of Steel [J]. Journal of AppliedPhysics, 1944, (15): 22-32.

[6] Laasraoui, A,Jonas J J. Pridiction of steel flow stresses at high temperatures and strain rates [J]. Metallrgical Transactions, 1991, A22: 1545-1577.

10Mn 钢的高温拉伸行为

李振宇[1]，张 梅[1*]，钟 勇[2]，王 利[2*]，李 麟[1*]

(1.上海大学 材料科学与工程学院；2.宝钢集团汽车用钢开发与应用技术国家重点实验室)

摘 要：采用 Gleeble−3500 型热模拟试验机进行实验，研究 M10 钢在变形温度为 600℃~1300℃、应变速率为 0.004~40s^{-1} 条件下的高温拉伸变形行为，探究其热塑性、热强度随温度的变化规律以及应变速率对其热塑性的影响。结果表明：M10 钢在 800℃~1200℃时有着良好的塑性，塑性指标断面收缩率 RA 最高可达近 90%，其脆性区间为 600℃~800℃，但塑性指标 RA 也均高于 50%。随着温度的降低，热强度呈线性上升趋势。总之，钢种的高温塑性良好。

关键词：热塑性；变形温度；应变速率；脆性区间；断面收缩率

Hot Ductility Behavior of Steel 10Mn

LI Zheng-yu[1], ZHANG Mei[1*], ZHONG Yong[2], WANG Li[2], LI Lin[1]

(1.School of Materials Science and Engineering, Shanghai University;
2.State Key Laboratory of Development and Application Technology of Automotive Steels, Baosteel Group)

Abstract: Hot tensile experiment of steel 10Mn was carried out on a Gleeble-3500 thermal mechanical simulator. The effects of deformation temperature(600~1300℃) and strain rate(0.004~40s^{-1}) on hot ductility behavior of the steel were researched. The results reveal the hot ductility and tension strength changing rules with the temperature, and the influence of strain rate on the hot ductility characteristics. The results shows that steel 10Mn has a good hot ductility between 800~1200℃, with a reduction of area (RA) of up to nearly 90%. And the III brittle zone of the steel is 600 ~ 800℃, with RA of not less than 50%. With temperature decreasing, tensile strength of the steel has a linearly rising trend. In conclusion, the studied steel shows quite good high temperature ductility.

Key words: hot ductility; deformation temperature; strain rate; brittle zone; percentage reduction of area (RA).

0 引言

奥氏体中锰钢在中、低冲击功下有良好的加工硬化性能，表面有高硬度而心部仍有良好的韧度，比当前我国广泛应用的高锰钢(Mn13)的耐冲击磨料磨损性能高[1−4]，钢的高温力学性能反映了钢在高

基金项目：国家重点基础研究发展计划资助项目(2007CB936803)。

第一作者：李振宇(1992—)，男，硕士研究生。

通讯作者：张 梅，女，博士。轻量化金属材料研发和应用。zhangmei3721@i.shu.edu.cn。

温下的行为变化规律。一般以断面收缩率 RA 和强度极限 R_m 作为钢的高温力学性能的指标,可从中总结出热塑性和热强度的变化规律。研究钢的高温力学性能及其变化机理,是订制和完善制造工艺的基础。钢的高温力学性能受很多因素的影响,如变形温度、应变速率和奥氏体晶粒度等。这些可变因素增加了研究、理解钢的高力学性能的复杂性,同时也为钢的高温力学性能的改善提供了条件[5—6]。

随着汽车对于安全性和自身轻量化的要求越来越高,第一代汽车用钢已不能满足人们对未来汽车发展趋势的需求,因为其在安全性和轻量化方面还是不够出色。而第二代汽车用钢由于合金含量较高,导致其成本昂贵,在工艺性能和冶金生产方面也都存在很多弊端。最新一代汽车用钢的代表中锰钢,其成本和性能均介于第一代和第二代汽车用钢之间,不失为未来汽车用钢发展的方向。所以,近年来中锰钢因其优异的性能吸引广大学者进行研究,但对其高温变形特性的研究有待深入。

合金元素对材料的微观组织和力学性能有重要的影响。锰(Mn)元素是一种奥氏体稳定元素,它不但能扩大奥氏体相区,而且还能抑制珠光体和贝氏体转变,冷却至室温后,能获得室温稳定的残余奥氏体;同时,锰的扩散和配分还影响着退火过程中奥氏体的形成。本实验研究了 Mn10 钢高温塑性随温度和应变速率的变化规律,为实际生产提供依据。

1 试样制备和试验方法

1.1 试样

试验用 Mn10 钢取自轧制过程中间坯,加工成热拉伸的标准试样,如图 1 所示。然后在 Gleeble-3500 热模拟试验机上进行热拉伸实验。

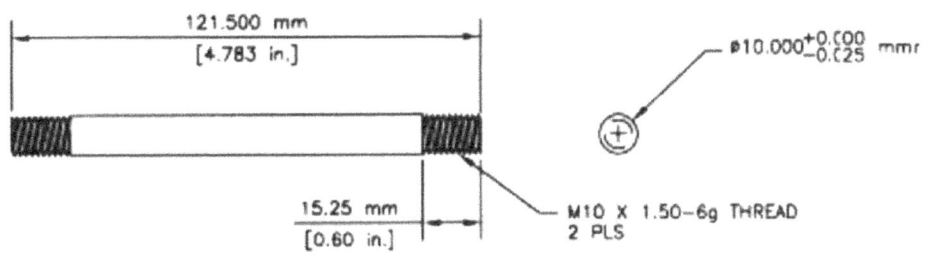

图 1 热塑性试验标准试样

Fig.1 Hot ductility test standard sample

1.2 试验方法

实验分两部分进行:① 变形温度的影响:将试样以 10℃/s 的速率加热至 1300℃,保温 3min,然后以 1.5℃/s 的冷却速率冷却至变形温度,分别为:600℃、650℃、700℃、750℃、800℃、850℃、900℃、950℃、1000℃、1100℃、1200℃、1300℃。保温 30s 后以 $4\times10^{-3}s^{-1}$ 的应变速率进行拉伸实验,直到拉断为止。试样拉断后,采用水淬冷却以保留高温下的断口形貌。② 应变速率的影响:将试样以 10℃/s 的速率加热至 1300℃,保温 3min,然后以 1.5℃/s 的冷却速率冷却至 1100℃,保温 30s 后分别以 0.04、0.4、4、40s⁻¹ 的应变速率进行拉伸实验,直到拉断为止。

试样全部拉断后按顺序贴好标签放入试样袋,进行线切割,每根试样只留断口后的 8mm 长,然后采用超声波清洗仪清洗 15min,最后进行 SEM 扫描,观察其断口的组织形貌以及裂纹形态和夹杂物,分析变形温度及应变速率对 10Mn 钢高温力学性能及断裂机理的影响。

2 试验结果与讨论

2.1 变形温度对热塑性和热强度的影响

图 2 和图 3 分别是材料的高温拉伸热塑性曲线和高温强度曲线。

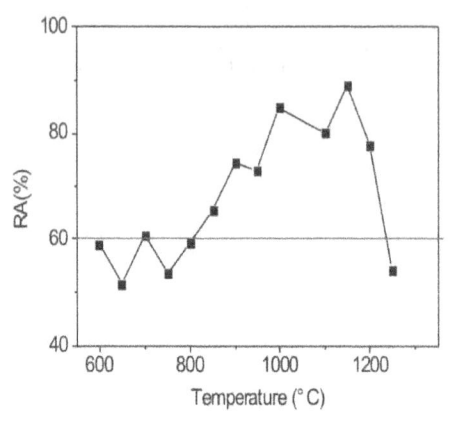
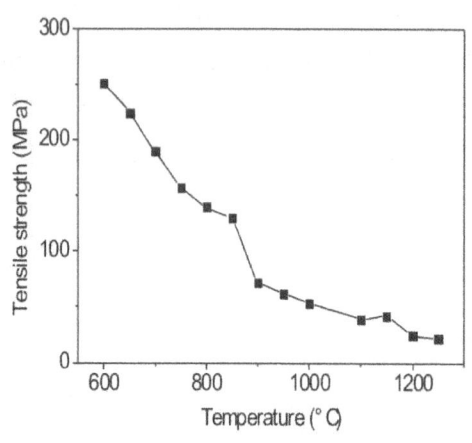

图 2　热塑性曲线

Fig.2　Hot ductility curves

图 3　热强度曲线

Fig.3　Peak stress-temperature curve

如图 2 可见,在 1250℃ 温度附近为第 I 脆性区间,在 800℃～1200℃ 拥有优异的热塑性,最高可达将近 90%,塑性区间宽而深。在 600℃～800℃ 温度范围内为第 III 脆性区间,但是脆性不算大,断面收缩率最小值均为 50% 以上。在 700℃ 时断面收缩率也达到 60%。

由图 3 可见,随着变形温度的下降,热强度存在明显上升的变化趋势。在 900℃ 及以上温度时 M10 钢的热强度在 50MPa 以下;而当温度降至 800℃ 时,材料的热强度就达到 100MPa 以上。随着温度进一步下降,热强度增加更快。

2.2 应变速率对高温性能的影响

不同应变速率下的高温拉伸热塑性曲线如图 4 所示,变形温度 1050℃,应变速率在 $0.004\sim40s^{-1}$ 时,M10 钢具有优异的塑性,断面收缩率在 78%～89% 之间。因此,应变速率的变化对其热塑性的影响不大。

随着应变速率的增加,材料中螺位错的交滑移和刃位错的攀移不能充分进行,交滑移和攀移所提供的软化程度较小,故出现峰值应力相对滞后所需的应变量应当明显增加[7,8]。可是在图 5 中,不同应变速率所对应的应变量不但没有明显增加,反而还有微弱减小的趋势,出现这种情况的原因可能是 M10 钢内部位错比较有序,缺陷较少,在增大应变速率的时候反而可以在小应变范围内快速拉断。与热塑性不同的是,随着应变速率的增加,峰值应力呈上升趋势,如图 6 所示。

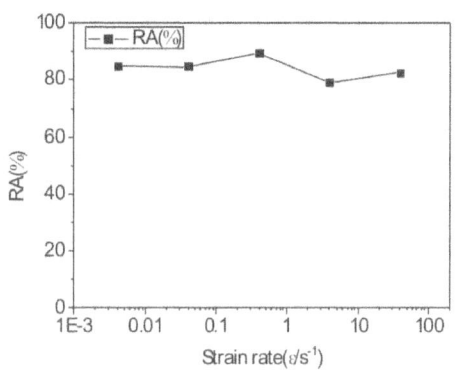

图 4　热塑性曲线
Fig.4　Hot　ductility　curve

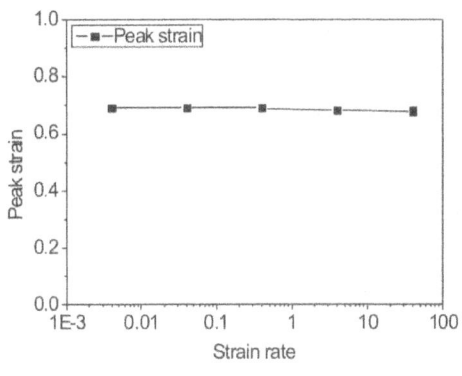

图 5　峰值应变–应变速率曲线图
Fig.5　Peak strain–strain rate curve

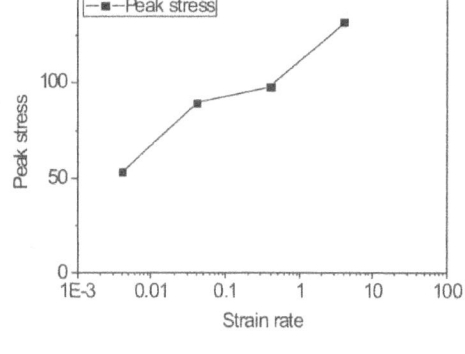

图 6　峰值应力–应变速率曲线图
Fig.6　Peak stress–strain rate curve

2.3 断裂机理的讨论

在 0.004s⁻¹ 应变速率下，不同变形温度的试样热拉伸断口 SEM 形貌见图 7 所示,650℃～1000℃ 温度范围内拉伸断口整体呈塑性断裂形貌,在断口处均可以观察到等轴韧窝,并且随着变形温度的升高,韧窝的直径减小但是深度增加,在 500 倍放大倍数下该规律更加明显。Crowther 和 Mintz[9]研究表明: 高温拉伸过程中,随着温度的升高,在宏观缩颈开始之前韧窝或微孔的长大不明显。在韧窝或微孔形成初期,由于晶界滑动变形的晶间裂纹。随后,裂缝被扭曲成细长的空隙,这些空隙之间发生缩颈,直至最终断裂。

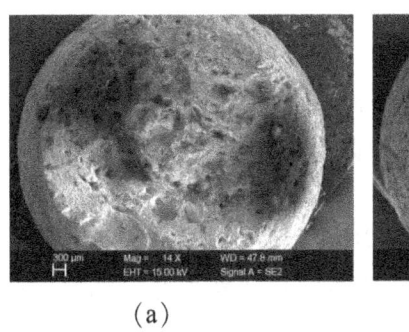
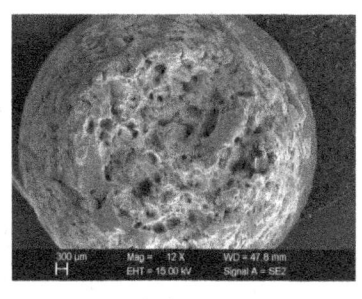
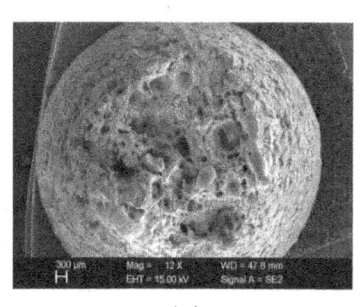

（a）　　　　　　　（b）　　　　　　　（c）

(d)　　　　　　　　(e)　　　　　　　　(f)

图7　应变速率 0.004s-1 下不同变形温度试样断口形貌
Fig.7　SEM fractographs for the specimens under strain rate of 0.004s⁻¹ at different temperatures
(a) and (d) 650℃;　(b) and (e)950℃;　(c) and (f) 1000℃

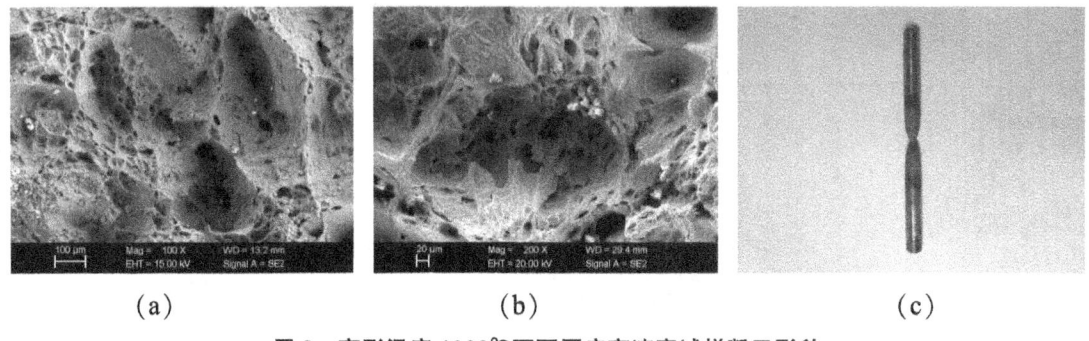

(a)　　　　　　　　(b)　　　　　　　　(c)

图8　变形温度 1000℃下不同应变速率试样断口形貌
Fig.8　SEM fractographs for the specimens at 1000℃ with strain rate of (a)0.004s⁻¹;　(b)0.04s⁻¹;　(c)0.4s⁻¹

在变形温度 1000℃ 时,不同应变速率下的拉伸断口 SEM 形貌见图8。当在高应变速率($\geq 0.4s^{-1}$)进行拉伸时,试样均没有拉断,但是出现明显的颈缩。出现这种情况的原因可能是应变速率太大,导致在拉伸结束时还没达到最大力,所以没有拉断。另外也可能是试样本身塑性过于良好。

当在低应变速率($\leq 0.04s^{-1}$)下进行拉伸时,试样均能拉断,但是在 $0.04s^{-1}$ 应变速率时其 SEM 断口图(图 8b)中可以看出,断口处并没有十分密的韧窝,甚至在高倍下根本观察不到韧窝,但是却表现出沿晶断裂的形貌,随着应变速率的降低,变形过程中的温度变得越来越高,使得断口局部的温度急剧增加。有研究结果表明[10,11]:在高温条件下,晶界强度弱化,变形过程中裂纹会在晶界处形核,从而呈现沿晶断裂形貌[12]。

3 结论

(1)在应变速率 $0.004s^{-1}$,变形温度为 800℃～1200℃ 和变形温度为 1000℃,应变速率为 0.004～$40s^{-1}$ 时,M10 钢均具有良好的塑性,断面收缩率均在 60% 以上,最高可达 89%。不同的变形温度下,峰值应力随温度线性降低,而随应变速率的增加峰值应力变化不大。

(2)在 $0.004s^{-1}$ 应变速率下,热拉伸的断口均呈韧窝形貌,随着变形温度的增加,韧窝变得大而深;变形温度 1000℃:在应变速率$\geq 0.4s^{-1}$ 时,试样均没有拉断,原因可能是塑性太好;当应变速率低于 $0.4s^{-1}$ 时,热拉伸断口呈沿晶断裂形貌。

参考文献:

[1] 姜启川. 金属学报,1990,26(1): B71

[2] 陈希杰. 高锰钢[M]. 北京:机械工业出版社,1989,344.

[3] 王明胜. 机械工程材料,1993,17(1): 17.

[4] Zhang F Cet al. Chinese Journal of M echanicalEngineering, 1992, 5(4): 284.

[5] 张 晨,岳尔斌,仇圣桃. 钢的高温力学性能及影响因素分析 [J].连铸,2008(6):6—10.

[6] 郭学锋, 杨文朋, 宋佩维. 往复挤压 Mg-Al-Si 合金的高温拉伸性能 [J]. 材料热处理学报,2012,33 (2): 7—11.

[7] 哈宽富. 金属力学性质的微观理论[M]. 北京: 科学出版社,1983.

[8] Chen X M,Song S H,Sun Z C,et al.Effect of microstructural features on the hot ductility of2.25Cr-1Mo steel[J].Materials Science and Engineering A, 2010, 527: 2725-2732.

[9] Crowther D N,Mintz B. Influence of carbon on hot ductility of steels[J].Journal of Materials Science and Technology, 1986(2) : 671-676.

[10] Hamada A S,Karjalainen L P. Hot ductility behaviour of high-Mn TWIP steels [J]. Materials Science and Engineering A, 2011, 528:1819-1827.

[11] 吴希俊.晶界结构及其对力学性质的影响(II) [J].力学进展,1990,20(2) : 159-173.

[12] 吴希俊.晶界结构及其对力学性质的影响(I) [J].力学进展,1989,19(4) : 433-441.

高强钢焊接研究现状

孙玉良[1], 张 梅[1]*, 孙保良[2], 侯宝辉[1]

(1.上海大学 材料科学与工程学院；2.上海汇众汽车制造有限公司 技术中心)

摘 要：论述了高强钢在焊接过程的特点以及焊接过程中容易出现的问题。高强钢因具有较高的强度和硬度,焊接稳定性较差,易出现焊接裂纹和热影响区性能的变化,降低构件的使用性能。对高强钢焊接性的研究主要集中在焊缝强韧性匹配、裂纹敏感性和热影响区组织性能三个方面。高强钢在焊接前应进行材料焊接性,焊材以及焊接工艺的综合考量,在保证强度的前提下,增强塑韧性,将焊接缺陷降到最低。

关键词：高强钢；焊接接头；性能

Status of High Strength Steel Welding

SUN Yu-Liang[1], ZHANG Mei[1]*, SUN Bao-liang[2], HOU Bao-hui[1]

(1. School of Materials Science and Engineering, Shanghai University;
2. Technical Center, Shanghai Huizhong Automotive Manufacturing Co.,Ltd)

Abstract: It discusses the characteristics of high strength steel in the welding process and the welding process prone to problems.High-strength steel for high strength and hardness, welding stability is poor, prone to cracks and weld heat-affected zone change performance, reduce performance components.Welding of high strength steel research focuses on the weld strength and toughness, crack sensitivity, microstructure and mechanical performance of the heat affected zone. High-strength steel welding should be carried out before a comprehensive consideration of material weldability, welding and soldering processes, under the premise of ensuring the strength, enhancing the plasticity and toughness, and reducing welding defects.

Keywords: high-strength steel; welded joints; performance

第一作者:孙玉良,男,硕士研究生。

通讯作者:张 梅,女,博士。轻量化金属材料开发和应用。zhangmei3721@i.shu.edu.cn。

HR550 钢的焊接模拟研究

马庆楠[1]，孙玉良[1]，张　梅[1*]，万　紫[2]，李　麟[1]

(1.上海大学 材料科学与工程学院；2.上海汇众汽车制造有限公司 技术中心)

摘　要：本文用 Gleeble-3500 热模拟机对 HR550 钢的焊接性能进行焊接模拟实验，分析了不同峰值温度和冷却速度对 HR550 钢的金相组织、显微硬度、冲击性能和冲击断口形貌的影响。结果表明，虽然峰值温度增加、冷却速度增大对金相组织有一些影响，但常温冲击性能和低温冲击性能比较稳定，都能得到与母材相对匹配的组织以及硬度。绝大多数冲击断口均为韧性断口，只有峰值温度 1320℃、1260℃结合 $t_{8/5}$ 为 100s 时样品才发生低温脆性断裂。总体看，HR550 钢的焊后综合力学性能良好。试验结果可为 HR550 钢焊接工艺的合理制定提供依据。

关键词：Gleeble-3500；HR550 钢；焊接模拟；组织；冲击性能

HAZ Thermo Cycle Simulation of HR550 Steel

MA Qing-nan[1], SUN Yu-liang[1], ZHANG Mei[1*], WAN Zi[2], LI Lin[1]

(1.School of Materials Science and Engineering, Shanghai University;
2.Shanghai Huizhong Automotive Manufacturing Co.,Ltd)

Abstract：Welding simulation tests of steel HR550 were conducted using Gleeble-3500 thermal simulator under different peak temperature (T_p) and cooling rate ($t_{8/5}$). Microstructure, hardness profile, and impact toughness tests, as well as observation of impact fracture morphology were taken with simulation samples. The results show that HR550 steel has good mechanical properties and toughness. With the increase of Tp and $t_{8/5}$, the microstructure of the samples changes a lot. However, the impact performance at room temperature and -20℃ is stable and the fracture surface is ductile. The microstructure and mechanical properties match with the base metal. Brittle fracture occurs only at -20℃ for T_p=1320℃ and T_p=1320℃, combined with $t_{8/5}$ =100s conditions. The research results provide a reasonable basis for welding process development for steel HR550.

Key words: Gleeble-3500; steel HR550; welding simulation; microstructure; impact toughness

基金项目：973 计划项目(编号：2010CB-630802)；国家自然科学基金项目(50934011 和 50971137)。
第一作者：马庆楠，男，硕士研究生。
通讯作者：张　梅，女，博士。轻量化金属材料开发和应用。zhangmei3721@i.shu.edu.cn。

FB780 钢的扩孔性能与组织

赵 雪[1]，汤 潇[1]，宋佳男[1]，万 紫[2]，张 梅[1*]

(1.上海大学 材料科学与工程学院；2.上海汇众汽车制造有限公司 技术中心)

摘 要：研究了铁素体贝氏体双相钢 FB780 的扩孔性能及显微组织。通过扩孔实验，应用光学显微镜、扫描电镜等对钢中的基体组织和析出相形貌、尺寸、成分等演变规律进行系统研究，并着重研究显微组织、裂纹形成及扩展机制等对钢材扩孔性能的影响。

关键词：FB780；扩孔率；微观组织；裂纹扩展

Hole Expansion Property and Microstructure of FB780 Steel

ZHAO Xue[1], TANG Xiao[1], SONG Jia-nan[1], WAN Zi[2], ZHANG Mei[1*]

(1.School of Materials Science and Engineering, Shanghai University;
2.Shanghai Huizhong Automotive Manufacturing Co.,Ltd)

Abstract: The hole expansion property and microstructure of ferrite bainite dual-phase steel FB780 was investigated. We carried out the hole expansion test in the first place. The steel's organization, morphology, size and composition of precipitates and inclusions, were studied with optical microscopy, scanning electron microscopy, etc. It focuses on the influence of microstructure, crack formation and extension mechanism on reaming performance.

Keywords: FB780; Hole expansion rate; Microstructure; Crack propagation

基金项目："973"计划项目(编号：2010CB-630802)；国家自然科学基金项目(编号：50971137)。

作者简介：赵 雪(1990—)，女，硕士生。shuzhaoxue@126.com。

通讯作者：张 梅，女，博士。轻量化金属材料开发和应用。zhangmei3721@i.shu.edu.cn。

含 Nb 超高强度钢的扩孔性分析及数值模拟

宋佳男[1]，张　梅[1*]，赵　雪[1]，汤　潇[1]，万　紫[2]，王　韬[2]

(1.上海大学 材料科学与工程学院；2.上海汇众汽车制造有限公司,技术中心)

摘　要： 在最近的研究和开发中，钢铁企业为开发高的强度—塑性—扩孔率均衡性的钢种付出了大量的努力。本文研究对象为新开发的 800MPa 级超高强度含 Nb 铁素体贝氏体双相钢 FB780,测试并分析了材料的扩孔特性。同时结合板料成形非线性有限元分析软件 DYNAFORM,对板料的扩孔过程进行数值模拟,最终为应用中最优化的工艺选择做好技术储备。

关键词： 高强度钢；扩孔性；显微组织；数值模拟

Test and Numerical Simulation on Hole-expansion Ratio of Ultra High Strength Steels Containing Nb

SONG Jia-nan[1], ZHANG Mei[1*], ZHAO Xue[1], TANG Xiao[1], WAN Zi[2], WANG Tao[2]

(1. School of Materials Science and Engineering, Shanghai University;
2. Technical Center, Shanghai Huizhong Automotive Manufacturing Co.,Ltd)

Abstract: The steel enterprises have devote huge efforts to developing high strength- ductility-hole expansion ratio balance steels. In this paper, ferrite bainite dual-phase steel named FB780 which contains Nb has been investigated. Hole-expansion ratio of studied steel were tested and analysed. The hole-expansion process was simulated numerically by using the dynamic explicit non-linear finite element software eta/DYNAFORM, to provide an optimized process for hot rolled dual-Phase steels.

Key words: high strength steel(HSS); hole-expansion property; microstructure; numerical simulation

---

基金项目: 973 计划项目(编号:2010CB-630802);国家自然科学基金项目(编号:50934011 和 50971137)。
第一作者: 宋佳男,女,硕士研究生。
通讯作者: 张　梅,女,博士。轻量化金属材料开发和应用。zhangmei3721@i.shu.edu.cn。

数值模拟与物理模拟在汽车板件成形中的应用

李怡然[1]，宋佳男[1]，汤 潇[1]，张 梅[1*]，王 韬[2]

(1.上海大学 材料科学与工程学院；2.上海汇众汽车制造有限公司 技术中心)

摘 要：板件冲压成形是材料成形中重要手段之一，汽车、航空、航天、船舶和家电等均有涉及。据统计，汽车上有60%~70%的零件是采用冲压工艺生产出来的。如何高效地得到高品质冲压件是一个永恒的话题。研究冲压成型的方法有数值模拟和物理模拟。传统的研究方法主要是物理模拟，主要是应用各种试验方法获得板料的拉深成形性能。物理模拟方法对合理地选取拉深成形原材料、分析和判断生产中出现的与板材性能有关的质量问题、设计合理的拉深成形工艺和产品结构具有重要的实际意义。然而，物理模拟技术的缺点在于费用高、周期长、不能模拟动态成形过程和复杂拉深成形(如汽车覆盖件成形)等。近年来，随着计算机软硬件技术的飞速发展以及数值算法的不断完善，利用计算机进行的研究拉深成形的数值模拟技术已开始进入应用阶段。通过对拉深成形过程进行数值模拟，可以大大缩短模具设计周期，成倍地降低产品成本。本文主要介绍物理模拟和数值模拟技术的基本原理和相关应用以及二者的各自利弊，从而形成正确的研究方法体系。

关键词：板件冲压成形；数值模拟；物理模拟

Application of Numerical and Physical Simulation in Automotive Sheet Forming Process

LI Yi-ran[1], SONG Jia-nan[1], TANG Xiao[1], ZHANG Mei[1*], WANG Tao[2]

(1.School of Materials Science and Engineering, Shanghai University;
2.Shanghai Huizhong Automotive Manufacturing Co.,Ltd)

Abstract：Sheet stamping is one of the important means during material forming, automotive, aviation, aerospace, shipbuilding and home appliances are involved. According to statistics, 60% to 70% of the parts of the car are produced using stamping process . How to get high-quality products and efficiently stamp is an eternal topic. Research methods on stamping include numerical and physical modeling. Traditional research methods are mainly physical simulation ones, it is mainly the application of various test methods to obtain sheet metal deep drawing performance. Physical simulation of deep drawing has important practical significance to reasonable selecting raw materials, analyzing and judging quality issues related to forming performance during stamping, and rational design of deep drawing process and product structure. However, the disadvantage of physical simulation technology are the high cost, long cycle, and can not simulate the dynamic forming process and

第一作者：李怡然，男，硕士研究生。

通讯作者：张 梅，女，博士。轻量化金属材料开发和应用。zhangmei3721@i.shu.edu.cn。

simulate complex deep drawing （such as automotive panel forming）, etc. In recent years, with the rapid development of computer hardware and software technology and continuous improvement of numerical algorithms, using computer drawing forming numerical simulation technology has entered the application stage. Through the deep drawing process simulation can greatly reduce the mold de sign cycle, reduce product costs exponentially. This paper describes the basic principles of physical modeling and numerical simulation technology, related applications as well as their respective advantages and disadvantages of both, and establishes proper research methods for readers.

Keywords: sheet stamping forming; numerical simulation; physical simulation

铀铌合金凝固过程微观组织演变的数值模拟研究

苏 斌，王震宏，邬 军，罗 超，沙 萌

(中国工程物理研究院材料研究所)

摘 要：铀铌合金是一种重要的核材料,在核工程及军工领域得到了普遍的应用。通过铸造方式制备铀铌合金时,凝固组织形貌和成分分布对凝固后的产品性能影响很大。在过去的研究中,国内外学者对铀铌合金的凝固过程进行了一些实验,但是实验方法周期较长,研究费用较高,而利用数值模拟手段可以大大缩短实验周期和节约成本。本文采用 Cellular Automaton(CA)方法对铀铌合金铸件凝固过程的微观组织演化进行了数值模拟,模型考虑了固/液界面的溶质再分配、微观固相分数、曲率以及各向异性等因素。为验证数值计算结果,进行了浇注与测温实验,将数值模拟结果和实验结果进行了对比,验证了模拟结果的准确性。研究结果对提高铀合金铸件质量具有重要的理论意义和工程应用价值。

关键词：铀铌合金；凝固过程；组织演变；数值模拟

Numerical Simulation of Solidification Microstructure of U–Nb Alloy Using Cellular Automaton Method

SU Bin, WANG Zhen–hong, WU Jun, LUO Chao, SHA Meng

(China Academy of Engineering Physics)

Abstract: In recent years, numerical simulation has become a powerful tool for simulating the microstructure evolution. In this paper, the microstructure evolution of U-Nb alloy during solidification process was simulated by using cellular automaton method. The preferential growth orientation, solute redistribution in both liquid and solid, solid/liquid interface solute conservation, interface curvature and the growth anisotropy were all considered in the model. By using the model, the microstructure evolution at different locations in U-Nb alloy casting can be simulated, and the grain size and solute distribution can be predicted. It is of great significance to improve the performance and quality of U-Nb alloy casting.

Key words: U-Nb alloy; solidification process; microstructure evolution; numerical simulation

---

第一作者:苏 斌,男。核材料制备及材料加工过程模拟仿真。subin@caep.cn。

通讯作者:王震宏,男。核材料制备及成型技术。wzh703@126.com。

GCr15 钢离异共析转变工艺研究

尹志新，梁均全，陈俊霖

(广西大学 材料科学与工程学院)

摘　要： 本文利用 Gleebe3500 型热模拟实验机，测定了 GCr15 钢在升温和降温过程中相变开始点和相变结束点，采用扫描电镜和硬度实验，考察、分析了奥氏体化温度和时间对碳化物颗粒形态及分布的影响，以所测冷却过程中的共析转变开始点和结束点为依据，分析了多种热处理工艺，研究结果如下：(1)GCr15 钢以大约 7℃/分钟的速度加热到 805℃保温 30 分钟后，在缓慢冷却过程中，从共析转变开始点到大约 715℃之间，共析转变的主要方式为离异共析转变，直接形成球状碳化物和铁素体组织。从 715℃到共析转变结束点之间，共析转变过程中出现片状珠光体，在此区间内，随温度的下降，片状珠光体转变趋于主导。(2)在上述奥氏体化条件下，炉冷至 715℃等温 150 分钟后，再随炉冷却到 650℃时出炉空冷至室温，可获得完全的球化珠光体组织，组织中碳化物平均尺寸为 1.35μm，布氏硬度达到 181HBW 左右，球化时间总共约为 6.5 小时。

关键词： 轴承钢；离异共析转变；球化退火

The Research of Process on Divorced Eutectoid Transformation of GCr15 Steel

YIN Zhi-xin, LIANG Jun-quan, CHEN Jun-lin

(School of Materials Science and Engineering)

Abstract: In this paper, the starting and ending eutectoid transformation temperature of GCr15 steel were measured by using of Gleeble-3500 thermal simulation testing machine. The influence of austenitizing time, remained carbide morphology, as well as isothermal temperature and time on the microstructure and hardness of GCr15 are examined by means of analysis of scanning electron microscopy and hardness testing after heat treatment through box heat treatment furnace. The results are as follow:(1) While heated to 805℃ at a speed of about 7℃/min and held for 30 minutes, the main mode of eutectoid transformation of GCr15 steel is divorced eutectoid one from starting temperature of eutectoid transformation to about 715℃ in the process of slowly cooling. And the microstructure of spheroidal carbides and ferrites are formed directly from two phase region (austenite+θ). Lamellar pearlite appeared and tends to dominate with temperature drop from 715℃ to end temperature of eutectoid transformation.(2) Under the condition of above austenitizing, the complete spheroidizing pearlite could be obtained if the sample of GCr15 is cooled to 715℃ and is held for 150 minutes, then again cooled to 650℃ all with the furnace, and finally taken out from furnace cooling in air to room temperature. The average size of carbide particles in the microstructure is about 1.35μm. Brinell hardness is around 181 HBW. The time of spheroidizing is shorten to about 6.5 hours.

Key words: bearing steels；divorced eutectoid transformation；spheroidizing annealing

---

第一作者：尹志新，男。金属材料。holly@gxu.edu.cn。

ZTC4 铸造钛合金铸件微观组织实验研究

郄喜望 [1,2]，**王红红** [2]，**吴国清** [3]，**南　海** [1]

(1.北京航空材料研究院；2.北京百慕航材高科技股份有限公司；3.北京航空航天大学)

摘　要：铸件结构、铸造工艺等因素对铸件微观组织的影响很大。为研究不同工艺下某 ZTC4 大型铸件各部位组织的差异，本文通过铸件结构分析，选取三种不同工艺下典型铸件的厚区(A 区)、关键区(B 区)、薄区(C 区)和厚薄转接区(D 区)为研究对象，定量分析该铸造钛合金不同位置试样的 β 晶粒尺寸及 α 片层间距。结果表明，铸件各部分的 β 晶粒尺寸、α 片层间距与铸件壁厚相关，随着试样厚度的增加，其 β 晶粒尺寸、α 片层间距呈线性增大的趋势；铸件中心和表层组织存在明显的对应关系，可以考虑通过铸件的表面组织评估，推测铸件内部组织状况；对比 1#、2# 和 3# 铸件在相同位置处试样的 β 晶粒尺寸，发现 2# 铸件试样的 β 晶粒尺寸均存在明显的晶粒粗大现象。

关键词：铸造钛合金；微观组织；晶粒尺寸；片层间距；定量分析

Experiment Research for Casting Microstructure of ZTC4 Titanium Alloy

QIE Xi-wang[1,2], WANG Hong-hong[2], WU Guo-qing[3], NAN Hai[1]

(1.Beijing Institute of Aeronautical Materials; 2.BAIMTEC Material CO.,LTD; 3.Beihang University)

Abstract: Due to the effects of casting structure and process, the difference microstructure obiviously in ZTC4 (Ti-6Al-4V) casting. To investigate the casting microstructure between difference casting parts in three different processes of ZTC4, the four typical sections in ZTC4 casting, including the thick area (area A), the key area (area B), the thickness of thin area (area C) and the transformation area (area D), have been selected. And a quantitative analysis of the different location of the casting titanium alloy specimens for β grain size and α lamella spacing has been conducted through the whole analysis of the casting. The results show that the β grain size and α lamella spacing of casting are related with the increase of the thickness of the specimen. In this case, the β grain size and α lamella spacing of casting increase linearly; the microstructure in core section and surface section of casting have obvious corresponding relation, and the casting internal microstructure can be speculated by considering the nondestructive surface evaluation of the casting; Compared with the β grain size and α lamella spacing in the same location of 1#, 2# and 3# castings, the the β grain size in 2# sample casting present the phenomenon of coarse grains obviously.

Key words: cast titanium alloy; microstructure; grain size; lamella spacing; quantitative analysis

基金项目:科工局某发动机材料专项课题(大尺寸钛合金中介机匣整体铸件研制,项目编号:JPPT-F2008-9-1)。

第一作者:郄喜望,男。铸造钛合金工艺、组织及性能研究。qiexiwang@139.com。

热模拟机用于优化搪玻璃钢生产工艺参数研究

李桂艳，刘凤莲，赵宝纯，刘志伟

（鞍钢股份有限公司技术中心）

摘　要：利用 Gleeble－3800 热模拟试验机，研究搪玻璃钢轧后连续冷却转变曲线，相变组织形成规律，并以此为基础进行多道次控轧控冷工艺模拟试验，优化出精轧的开轧温度、终轧温度、穿水后冷却速率等工艺参数。在优化后的控轧控冷工艺参数控制下，搪玻璃钢得到理想组织：铁素体+少量珠光体，晶粒度 9 级，铁素体基体上分布均匀弥散的 TiC 第二相粒子，符合焊接性、搪烧密着性、抗鳞爆性要求。经多次搪烧后力学性能稳定，具有良好的综合性能。

关键词：热模拟；工艺参数；搪玻璃钢

中图分类号：TG142　　**文献标志码**：A

Process Parameter Optimizing for Glass－lined Pressure Vessel Steel Production by Using Simulation Test

LI Gui－yan, LIU Feng－lian, ZHAO Bao－chun, LIU Zhi－wei

（Technology Center of Angang Steel Co.,Ltd）

Abstract: The controlled rolling and controlled cooling simulation tests were performed on a Gleeble-3800 thermal mechanical simulator to study the continuous cooling transformation and microstructure evolution of the glass-lined pressure vessel steel. The process parameters such as start rolling temperature, finish rolling temperature and the cooling rate were analyzed to find out a suitable process. With the customized parameters, the ideal microstructure is mainly composed of ferrite and pearlite. From Transmission electronic microscopy observation, there are second phases of TiC homogeneously distributed on the ferrite, which contributes to good weldability, enameled adherence and fish-scale resistant. It is proved by enameling baking test that the developed steel has high comprehensive mechanical properties.

Key words: simulation; process parameter; glass-lined pressure vessel steel

---

第一作者：李桂艳，女，高工。

GCr15 锻钢密封盘材质抗磨性的研究

许云龙 [1], 崔晓明 [2], 王 宁 [2]

(1.内蒙古兰太实业股份有限公司；2.内蒙古工业大学 材料科学与工程学院)

摘 要： 针对本企业自备电厂除尘系统中吹灰密封盘材料抗磨性差、使用寿命短、更换频繁等问题，提出用 GCr15 锻钢代替原来材料的研究课题。本文着重研究热处理对 GCr15 的硬度和抗磨性能的影响，并将 GCr15 的试验材料与厂家提供样件的硬度、耐磨性进行比较。结果表明，GCr15 的抗磨性能远远高于厂家提供样件的抗磨性能，达到了密封盘的使用要求。

关键词： GCr15；抗磨性；球化退火；淬火

Study on Abrasion Resistance of GCr15 Forging Steel Disc Material Sealing

XU Yun-long[1], CUI Xiao-ming[2], WANG ning[2]

(1.Inner Mongolia Lantai industrial Co.,Ltd.Inner Mongolia Alxa League;
2.School of Materials Science and Engineering, Inner Mongolia University of Technology)

Abstract: In view of the poor abrasion resistance, short service life and replacing frequent problems of the dust removal system of dust removal system in the dust removal system of the enterprise, this paper presents a research subject of GCr15 forging steel instead of raw materials. This paper focuses on the effects of heat treatment on the hardness and abrasion resistance of GCr15, and the test materials and manufacturers to provide hardness, abrasion resistance of the GCr15 samples are compared. The results showed that the abrasion resistance of GCr15, much higher than the manufacturers to provide the abrasion resistance of samples, the performance meets the requirements of the sealing disk.

Keywords: GCr15; abrasion resistance; spheroidizing annealing; hardening

0 引言

本项研究的目的是提高发电厂流化床锅炉除尘系统中的吹灰密封盘的抗磨性，延长使用寿命。该工件安装在除尘系统中，当吹灰时，将其打开，由压缩空气带动煤灰通过密封盘口排出，吹灰完毕关合，每天开关达 800 余次。煤灰的强烈磨擦作用，对其产生严重磨损(见图 1)，其寿命一般只有 20 天左右。

第一作者：许云龙,内蒙古兰太实业股份有限公司热动力分厂工程师。18648313299。

GCr15 钢是一种被广泛应用的滚动轴承钢,其合金含量较少、价格比较便宜并具有较高的硬度和良好的抗磨性,所以选择其作为研究的材质。

图 1　吹灰密封盘
Fig.1　Blowing ash seal plate

1　试验内容

1.1 化学成分分析

GCr15 锻件的化学成分见表 1[1],厂家样件化学成分见表 2。

表 1　GCr15 化学成分 (%)

元素	C	Mn	Si	S	P	Cr	Mo	Ni	Cu	Ni+Cu
含量	0.95-1.05	0.20-0.40	0.15-0.35	≤0.020	≤0.027	1.30-1.65	≤0.10	≤0.30	≤0.25	≤0.50

表 2　厂家样件的化学成分 (%)

样品	C	S	Si	Cr	Mn	P	Mo	Ni
1 号	0.5824	0.0035	0.2121	0.2712	0.7160	0.0322	0.0441	0.3314
2 号	0.5609	0.0042	0.1836	0.2642	0.5881	0.0378	0.0377	0.2649
平均	0.5711	0.0038	0.1979	0.2677	0.6521	0.0350	0.0409	0.2982

1.2 热处理

主要检验热处理工艺对 GCr15 组织、性能的影响。进一步做磨料磨损性能试验,以比较二者的优劣。

1.2.1 球化退火工艺

本试验工艺是:将试样放入箱式电阻炉中随炉升温至 800℃ (一般 GCr15 为 780℃ ~ 810℃),保温60 分钟,随炉冷却至 650℃后,出炉空冷。

1.2.2 淬火和回火工艺

淬火温度应严格控制在 840+10℃ 范围内，实施油淬，淬火后应立即回火，回火温度为 150℃～160℃，保温 60 分钟，出炉空冷。

1.3 检测硬度

本试验采用洛氏硬度计测量硬度值，每组试样测 5 个点，最后计算平均硬度值。GCr15 和厂家材料测量结果见表 3、表 4。

表 3　GCr15 的硬度值

	第一点	第二点	第三点	第四点	第五点	平均硬度
球化退火态	25	24	25	25	24	24.6
淬火、回火态	62	63	63	63	62	62.6

表 4　厂家试样的硬度值

	第一点	第二点	第三点	第四点	第五点	平均硬度
厂家材料	27	29	27	30	28	28.2

1.4 磨损试验

磨损试样取自厂家提供的样件和自制的 GCr15 锻件。用线切割机在厂家材料和 GCr15 的样件上切出直径为 20mm、厚度为 5mm 的试样备用。

本试验在型号为 WW-W1 的立式万能摩擦磨损试验机上进行，具体工艺参数是：施加作用力 100N，转速 100r/min，磨损时间 3min，磨料为 100 目砂纸。试验前将试样放到实验装置上进行预磨，待试样表面光滑后取下，将试样与放样装置紧紧粘牢，再用型号为 AL104 的电子天平称量，并以此作为原始重量，3min 更换砂纸并称重，记录每次测得的数据。磨损试验数据如表 5、表 6、表 7 所示

2　实验结果分析

2.1 热处理的显微金相组织分析

2.1.1 球化退火热处理组织

按照上述工艺进行球化退火热处理，组织如图 2 所示。图中基体组织为均匀分布的球状珠光体，碳化物颗粒较小，呈点、球状分布。

2.1.2 淬火、低温回火热处理组织

按照上述工艺进行淬火、低温回火后，组织如图 3、图 4 所示。在图 4 中有回火马氏体及细小的颗粒状碳化物及少量的残余奥氏体，马氏体的亮区及黑区仍比较明显，属正常淬火、低温回火的显微组织。在正常温度淬火后，将出现隐针状(黑色)及细小针状(亮区)马氏体组织，这是轴承钢淬火后特有的显微组织。亮区一般作网状分布，在奥氏体晶界处，该处的碳化物首先发生溶解，故含碳量及含铬量要

比晶内多,Ms 点则较低,淬火冷却时形成的马氏体不易发生自回火,故不易受浸蚀,呈白色。而奥氏体晶内的碳化物溶解较少,所以该处未溶解的碳化物颗粒较多,奥氏体中碳和铬溶入较少,Ms 点较高,在淬火冷却时易发生自回火,故浸蚀后呈黑色。

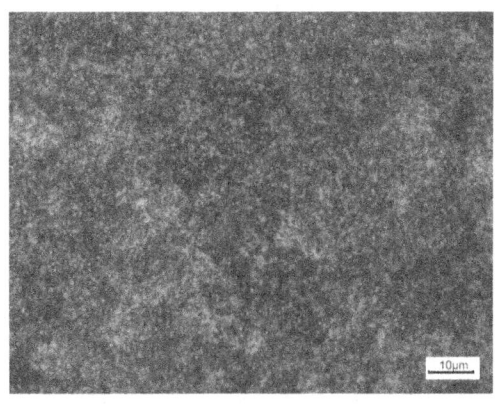

图 2 GCr15 球化退火态 1000×
Fig.2 Nodular annealing state 1000×

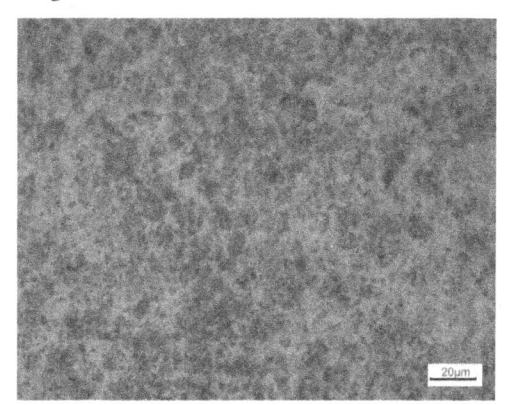

图 3 GCr15 淬火态 200×　　　　　　图 4 GCr15 淬火态 500×
Fig.3 GCr15 Quenching state 200×　　　Fig.3 GCr15 Quenching state 500×

随淬火温度的提高,亮区面积也相应增加,黑区则相应减少。同时随着加热温度的升高,钢中碳化物不断溶入基体,致使淬火后的残留奥氏体数量增多,由于马氏体及残留奥氏体都是不稳定的相,所以淬火后必须及时进行回火处理,一方面可以消除淬火造成的内应力,另一方面可以使组织趋于稳定,同时还能稳定工件的尺寸。

2.2 硬度结果分析

由表 3 可见,试验材料 GCr15 球化退火后的硬度为 HRC25,这是适宜切削加工的硬度,经淬火、回火处理的硬度为 HRC62~63,而厂家提供样件的硬度为 HRC28.2,明显低于 GCr15 的淬火硬度。材料达到 HRC62~63 这样的硬度,必然会具有良好的抗磨性能。这说明所制定的热处理工艺是合理的,达到了预期的效果。

2.3 磨损试验结果分析

磨损试验数据如表 5、表 6、表 7 所示:

表 5   GCr15 淬火、回火热处理磨损情况(g)

	原始质量	3min 后质量	6min 后质量	9min 后质量	总磨损量
1 号试样	96.2816	96.2739	96.2695	96.2617	0.0199
2 号试样	95.8148	95.8104	95.8044	95.8014	0.0134
3 号试样	95.9827	95.9786	95.9739	95.9652	0.0175

注:表中质量系试验机固定试样辅助装置质量与磨损试样质量之和

表 6   厂家试样磨损情况(g)

	原始质量	3min 后质量	6min 后质量	9min 后质量	总磨损量
1 号试样	96.5321	96.5198	96.5085	96.4960	0.036
2 号试样	95.6702	95.6561	95.6452	95.6352	0.0350
3 号试样	96.3700	96.3573	96.3468	96.3379	0.0321

注:表中质量系试验机固定试样辅助装置质量与磨损试样质量之和

表 7   厂家试样、GCr15 磨损量对比(g)

	1 号试样	2 号试样	3 号试样	平均磨损量
GCr15 试样磨损量	0.0199	0.0134	0.0175	0.0170
厂家试样磨损量	0.0361	0.0350	0.0321	0.0344

由表 7 可以看出,厂家试样磨损量约为 GCr15 试样的两倍,说明 GCr15 材料具有良好的抗磨性能。再对照表 3 的硬度值进行分析,GCr15 的硬度为 HRC62～63,远远高于厂家样件的硬度 HRC28.2,可见硬度直接影响磨损量,在粉料磨损的情况下,硬度越高,则磨损量越小。

2.4 碳含量与硬度关系分析

从表 1、表 2 中可见,厂家的样件碳含量为 0.57%,而 GCr15 的碳含量为 0.95%～1.05%,相差约为 1 倍,含碳量高,组织中的强化相—渗碳体多,则硬度高,抗磨性也高。

3 结论

通过对试验结果进行全面分析,得出以下结论:

(1)GCr15 的磨损量远低于厂家材料的磨损量,而 GCr15 的硬度远远高于厂家样品的硬度,可见,在粉料磨损条件下,硬度高,则抗磨性也高。

(2)GCr15 必须进行球化退火预处理和淬火、低温回火的最终处理两个环节,球化退火获得良好的切削加工性能,淬火、回火获得的组织为回火马氏体、均匀细小的碳化物及少量的残余奥氏体,文中所述处理工艺是可行的。

(3)碳元素是提高材质硬度的主要元素,碳含量高,硬度高;硬度高,则抗磨性也高。

参考文献:

[1] 崔占全. 工程材料 [M]. 北京:机械工业出版社, 2013.

齿轮热处理过程数值模拟分析

徐广晨 [1,2]

(1.沈阳工业大学；2.营口理工学院)

摘　要：齿轮是机械传动与变速的重要零件，在各类机械中的运用广泛。以锻造技术生产齿轮具有节省材料、成本低、效率高、机械性质佳等优点，故齿轮锻压方法与技术的发展已越来越受重视。本研究结合齿轮几何模型、有限元法模拟技术，进行正齿轮热精锻，针对其微观组织之晶粒细化效果做深入探讨。研究结果显示，锻压速率对晶粒细化有明显影响，动态再结晶分布与应变分布成正比、与静态再结晶分布成反比；由本研究建立的微观组织变化分析的模型，可有效用于预测正齿轮在热处理变化行为。

关键词：热处理；Deform-3D；晶粒细化

Numerical Simulation Analysis in Thermal Processes of Spur Gears

XU Guang-chen[1,2]

(1.Shenyang University of Technology；2.Yingkou University of Technology)

Abstract: Being an important part of mechanical transmission, gear has been extensively applied in various machines.Gear produced by forging possesses the advantages of material reduction, low cost, higher efficiency, and better mechanical properties. In this study, the geometric model and finite element simulation technique are combined to analyze the thermal process of spur gears, the grain-refinement effect in microstructure during hot precision forging of spur gears is discussed. The results show that grain-refinement is obviously influenced by forging velocity, and the dynamic recrystallization distribution is in direct proportion to strain distribution, but in inverse proportion to static recrystallization distribution. The numerical modeling developed to calculate the carburized layer and to analyze the phase transformation, can be used effectively to predict the thermo-mechanical behavior of spur gear during heat treatment.

Key words: thermal processes; Deform-3D; grain-refinement

基金项目：院青年自然科学基金资助项目(QN-L-201402)。

第一作者：徐广晨，男。机械热处理。kanu012@163.com。

基于加工硬化率的法兰铸坯动态再结晶及组织演变

秦芳诚，李永堂，巨　丽，齐会萍，雷步芳

(太原科技大学 材料科学与工程学院)

摘　要： 在 Gleeble-3500 热力模拟机上研究法兰 Q235B 钢铸坯材料的高温压缩行为，采用加工硬化率方法，识别出动态再结晶临界条件，引入无量纲参数 Zener-Hollomn，表征该材料动态再结晶演变的临界应变、临界应力峰值应变以及稳态应变模型并对其再结晶显微组织演化进行分析。在高温、低应变速率下，动态再结晶更容易发生，晶粒长大更迅速；随着应变速率增大，动态再结晶难于启动，晶粒直径减小。在 1050℃ 和 $1s^{-1}$ 下变形到达稳态时的晶粒细化效果最为显著。

关键词： 加工硬化率；法兰铸坯；动态再结晶；临界应变；晶粒细化

Microstructure Evolution and Dynamic Recrystallization of Flange Cast Blank Based on Working Hardening Rate

QIN Fang-cheng, LI Yong-tang, JU Li, QI Hui-ping, LEI Bu-fang

(School of Materials Science and Engineering, Taiyuan University of Science and Technology)

Abstract: The compression behavior at elevated temperature of Q235B flange cast blank was investigated by Gleeble-3500 thermo-mechanical simulator. The critical conditions for dynamic recrystallization were identified based on working hardening rate method, and then the critical models of dynamic recrystallization such as model of critical strain, model of critical stress, model of peak strain and model of steady stress were represented by considering Zener-Hollomn. The microstructure development during recrystallization was analyzed in detail. At a higher temperature and low strain rate, the dynamic recrystallization is easy and grain growth is rapid. The occurrence of dynamic recrystallization is difficult, and the grain size decreases with the increase of strain rate. The grains refining effect is significant when deformed to steady-state at the conditions of 1050℃ and $1s^{-1}$.

Key words: working hardening rate; flange cast blank; dynamic recrystallization; critical strain; grain refinement

---

基金项目： 国家自然科学基金重点(51135007)、国家自然科学基金(51174140、51205270)、高等学校博士点优先发展计划(20111415130001)、山西省回国留学人员科研(2011-084)和山西省自然科学基金(2014011026-2)资助项目。

第一作者： 秦芳诚(1988—)，男，博士研究生。精确塑性成形过程微观组织演变研究，。qfcqfc1988@163.com。

加热速度对 22MnB5 超高强钢奥氏体化温度影响的研究

靳永明，梁卫抗，刘　壮，陶文杰，董正达，李　阳

(华中科技大学 材料成形与模具技术国家重点实验室)

摘　要：采用热膨胀法，在 Gleeble3500 热力模拟实验机上对 22MnB5 钢板材进行不同速度加热 (10℃~100℃/S)试验，研究不同加热速度下奥氏体化开始温度 Ac1 与结束温度 Ac3 的变化规律。根据得到的膨胀曲线，建立加热速度与与奥氏体化温度之间的理论模型。结果表明：加热速度对奥氏体化过程有很大影响，Ac1 与 Ac3 都随着加热速度的增加而相应提高，同时对 Ac1 的影响比 Ac3 要大，并且随着加热速度的提高，奥氏体化温度区间变窄。该研究对于超高强钢板料热成形电阻加热过程工艺应用起到指导作用。

关键词：22MnB5 热成形钢；热膨胀法；加热速度；奥氏体化温度；电阻加热

Investigation on Effect of Heating Rates on Austenization Temperature of 22MnB5 Ultra-high Strength Steel

JIN Yong-ming, LIANG Wei-kang, LIU Zhuang, TAO Wen-jie, DONG Zheng-da, LI Yang*

(State key Laboratory of Materials Processing and Die & Mould Technology ,Hua Zhong University of Science and Technology)

Abstract: The influences of heating rate on austenization start temperature Ac1 and the end temperature Ac3 of 22MnB5 hot stamping steel were scientifically investigated by Gleeble 3500 with expansion methods. Then theoretical model between heating rate and austenization temperature was built up by the expansion curve.The results show that heating rate has a great effect on austenization temperature and Ac1,Ac3 increases with the increase of heating rate ,meanwhile,it has a more obvious influence on Ac1 than Ac3,the temperature range for austenization dropped .The research will play a guiding role In the resistance heating of 22MnB5 hot stamping steel.

Key words: 22MnB5 hot stamping steel; expansion methods; heating rate; austenization; resistance heating

第一作者:靳永明(1991—)，男，硕士。高强度钢板热冲压。JinYM@hust.edu.cn。
通讯作者:李　阳(1979—)，男，博士。高强度钢板热冲压工艺优化。liyang@hust.edu.cn。

Gleeble3500 热／力模拟试验机实验误差与程序设计

蔡晓文，杨雄飞

(攀钢集团研究院有限公司 钒钛资源综合利用国家重点实验室)

摘 要：QuikSim 软件是 Gleeble3500 的用户操作界面，它提供了控制系统和数据分析软件之间的转换，通过此软件的程序设计可实现不同语言间的转换，实现材料的热/力试验模拟。QuikSim 程序的微调可使实验结果更精准,从而更真实地反映模拟对象。

关键词：热/力模拟试验；QuikSim 程序；零点；应力应变

Experimental Error and Programming of Gleeble 3500 Thermal /Froce Simulation Machine

CAI Xiao-wen, YANG Xiong-fei

(Pangang Group Research Institute Co., Ltd., State Key Laboratory of Vanadium and Titanium Resources Comprehensive Utilization)

Abstract：QuikSim software is a graphical user-computer interface for Gleeble 3500 that is designed to make program and provide a link between the control system and a data analysis software tool, which achieves material thermal / force simulation. The result was exacter when the QuikSim program was adjusted properly, and simulate objects would be more realistically reflected through the thermal/force simulation.

Key words: thermal/mechanical simulation; QuikSim program; zero; Stress-Strain

第一作者:蔡晓文(1982—),女,工程师,硕士。金属材料、热模拟工艺。greatcxw@126.com。

热模拟在钢种研发过程中的应用

屈朝霞，夏立乾，宋国斌

（宝钢中央研究院）

摘 要： 热模拟给钢种的开发过程提供了很大的便捷性，在研发初期，通过较短时间、较少试样即可获得材料的相关性能特征。本文结合宝钢在产品研发过程中的实际工作经验，对常用的一些热模拟试验方法进行归纳。主要介绍了运用热模拟试验机完成高温拉伸、高温压缩等试验内容。获得材料的高温塑性和变形抗力，通过试验所得数据，反馈到产品开发当中，指导优化钢种成分和制造过程工艺。

关键词： 热模拟；高温塑性；变形抗力

The Application of Thermal Simulation in Steel Development

QU Zhao-xia, XIA Li-qian, SONG Guo-bin

（Research Institute （R&D Center）, Baosteel Group Corporation）

Abstract: Thermal simulation provides a great convenience in the development of new type steels. In the initial period thermal simulation could be used to obtain related characteristics with less samples and short time. This paper summarized some usual thermal-simulating methods based on the experience of the steel development in Baosteel. The tests completed by thermal-simulating tester, as high-temperature tensile and compression, to analyze the high temperature plasticity and deformation resistance, were introduced. This experimental data could be fed back to the steel development as a guidance to optimize the composition and manufacturing process.

Key words: thermal simulation; high temperature plasticity; deformation resistance

---

第一作者：屈朝霞，女。材料焊接性。quzx@baosteel.com。

热模拟在钢种焊接性研究中的应用

夏立乾，屈朝霞，许　磊

（宝钢中央研究院）

摘　要：新钢种在开发过程中需考虑后续使用中的焊接特性。借助热模拟方法，可在研发前期快速地初步判断材料的焊接性。本文介绍了宝钢在新产品研发过程中通过 Gleeble 热模拟试验机，进行 SH-CCT、焊接热影响区热模拟等试验。结合试验数据和实际需求，分析钢种的焊接性，反馈到钢种开发过程，优化钢种成分。

关键词：焊接；热模拟；焊接性

The Application of Thermal Simulation in the Research of Steel Weldability

XIA Li-qian, QU Zhao-xia, XU Lei

（Research Institute (R&D Center), Baosteel Group Corporation）

Abstract: It is necessary to consider the weldability in the development of new type steels. A quick judge could be made by using thermal simulation in the early development. This paper introduced the tests, such as SH-CCT and HAZ simulation, completed with Gleeble thermal-simulating tester, in the new product development in Baosteel. The experimental data could be analyzed in terms of the practical demands, and guide the optimization of steel components.

Key words: welding; thermal simulation; weldability

第一作者：夏立乾，男。材料焊接性。xialiqian@baosteel.com。
通讯作者：屈朝霞，女。材料焊接性。quzx@baosteel.com。

镁合金熔体超声处理过程的数值模拟

赵福泽，杨院生

（中国科学院 金属研究所）

摘　要：利用有限元方法对镁合金熔体的超声处理过程中的声场分布、流场分布以及空化效应进行了数值模拟。结果显示，声压呈辐射状分布，距离超声探头端面越远，其声压幅值越低。超声作用下，熔体中存在沿探头端面-中轴线-容器底面-容器内壁面-探头端面的环流。熔体空化效应表现出如下特性：声压较小时，熔体中存在的空化气泡为微小气泡以及巨大气泡能存活，而处于中间大小的气泡发生溃灭；声压较大时，能存活的微小气泡更小，同时巨大气泡不能存活。综合考虑了超声下镁合金熔体的各种变化，针对镁合金熔体处理的不同应用目的，分别给出了不同的实际操作建议。

关键词：超声处理；数值模拟；镁合金；声场；声流场；空化效应

Numerical Simulation for the Ultrasonic Melt Treatment Process of Magnesium Alloys

ZHAO Fu-ze, YANG Yuan-sheng

（Institute of Metal Research, Chinese Academy of Sciences）

Abstract: The sound field, acoustic streaming and cavitation of the magnesium alloy melt during ultrasonic melt treatment were investigated using numerical simulation method. The simulation results show that the sound pressure distributes in a radial form and the farther the positon is from the ultrasonic probe end face, the lower the sound pressure is. There is a ring current in the melt following a route of the probe end face—center line—bottom surface—wall—probe end face. The cavitation in magnesium alloy expresses the following characteristics: when sound pressure is low, the bubbles exist in the melt have a very small or big scale, and the middle-sized bubbles would collapse immediately; while in high sound pressure condition, big bubbles would not exist and the small bubbles survived are smaller than before. After making a comprehensively analyses about the above effects happening in the magnesium melt under ultrasonic, we present some suggestions for several different applications of ultrasonic melt treatment.

Key words: ultrasonic treatment; numerical simulation; magnesium alloys; sound field; acoustic streaming; cavitation

---

基金项目：国家自然科学基金项目(51274184)。

第一作者：赵福泽，男。镁基复合材料。fzzhao11s@imr.ac.cn。

通讯作者：杨院生，男。先进材料凝固与制备。ysyang@imr.ac.cn。

ZA81M 镁合金的热变形行为和热加工图研究

朱绍珍 [1,2]，罗天骄 [2]，张廷安 [1]，杨院生 [2]

(1.东北大学 材料与冶金学院；2.中国科学院 金属研究所)

摘　要： ZA81M 是一种具有优良强韧性的多元微合金化新型镁合金。利用 Gleeble 3800 热模拟试验机研究了 ZA81M 镁合金在温度为 200℃~350℃ 和应变速率为 $0.001s^{-1}$—$1s^{-1}$ 范围内的热变形行为，获得了该合金在上述变形条件下的流变应力-应变曲线，以动态材料模型为基础建立了 ZA81M 镁合金的热加工图。研究表明，合金的流变应力-应变曲线具有明显的动态再结晶特征，且流变应力随着应变速率的增加而增大，随着变形温度的增加而减小。运用回归分析方法建立了 ZA81M 镁合金流变应力预测模型，该模型与实验结果能较好的吻合。从加工图中可以看出，随着应变增加合金的非稳态区域变大，ZA81M 合金在高温和低应变速率下具有良好的加工性。

关键词： 镁合金；热变形行为；流变应力；热加工图；动态再结晶

Hot Deformation Behavior and Processing Maps of As-cast Magnesium Alloy ZA81M

ZHU Shao-zhen[1,2], LUO Tian-jiao[2], ZHANG Ting-an[1], YANG Yuan-sheng[2]

(1. School of Materials and Metallurgy, Northeastern University；2. Institute of Metal Research, Chinese Academy of Science)

Abstract: The hot deformation behavior of as-cast magnesium alloy ZA81M was studied by hot compression tests at temperatures of 200oC-350oC and strain rates of $0.001s^{-1}$—$1s^{-1}$ with a Gleeble 3800 thermal simulator testing machine. True stress-strain curves of the alloy at different temperatures and strain rates were obtained which show that the flow stress increases significantly with increasing strain rate and decreases as the temperature increases. A flow stress model based on the regression analysis was developed to predict the flow behavior of ZA81M alloy during the hot compression, which shows a well agreement with experimental results. The processing maps established according to dynamic materials model show that the alloy has good hot workability at higher temperature and lower strain rate.

Key words: magnesium alloy; hot deformation; flow stress; processing maps; dynamic recrystallization

基金项目： 国家重点基础研究发展计划资助项目(2013CB632205)。

第一作者： 朱绍珍，男。新型镁合金的开发与研究。szzhu12s@imr.ac.cnl。

通讯作者： 杨院生，男。先进材料的凝固与制备。ysyang@imr.ac.cn。

NixAlTiCrFeCoCu 高熵合金的组织与性能

李安敏[1]，黄宇炜[1]，陈若怀[1]，欧阳志凤[2]，郑奇峰[1]

(1.广西大学 材料科学与工程学院；2.广西玉柴动力机械有限公司)

摘　要： 利用真空电弧炉制备 $Ni_xAlTiCrFeCoCu$ (各元素原子百分比为 x:1:1:1:1:1:1, x=0.5、1.0、1.5、2.0、2.5、3.0)高熵合金,研究铸态合金的显微组织与性能。结果表明,$Ni_xAlTiCrFeCoCu$ 系列高熵合金的显微组织主要由具有面心立方结构(FCC)的固溶体和体心立方结构(BCC)的固溶体组成。合金组织主要有花瓣状组织、菊花状组织、富 Cu 的晶间组织和富 Ni 相;菊花状组织是共晶组织,是由贫 Cr、Fe 的 α 相和富 Cr、Fe 的 β 相组成;花瓣状组织是 α 相。随着 Ni 含量增加,花瓣状组织逐渐减少,随后消失(x=2.0),而富 Ni 相(γ)出现。除了 NiAlTiCrFeCoCu 合金,其他 5 种合金的屈服强度都高于 1.5GPa,断裂强度约 2GPa,合金的压缩断口都为脆性断口。$Ni_{20}AlTiCrFeCoCu$ 高熵合金在 H_2SO_4 溶液中的耐腐蚀性能最好, 在 NaCl 溶液中,$Ni_{15}AlTiCrFeCoCu$、$Ni_{20}AlTiCrFeCoCu$、$Ni_{30}AlTiCrFeCoCu$ 的耐腐蚀性能均比 304 不锈钢的耐腐蚀性能好。

关键词： 高熵合金；显微组织；压缩性能；耐腐蚀性能

Microstructures and Properties of NixCuAlCoFeCrTi High Entropy Alloys

LI An-min[1], HUANG Yu-wei[1], CHEN Ruo-huai[1], OUYANG Zhi-feng[2], ZHENG Qi-feng[1]

(1.School of Material Science and Engineering, Guangxi University; 2.Yuchai Group Co.Ltd.)

Abstract: A series of $Ni_xAlTiCrFeCoCu$ (x:molar ratio, x=0.5, 1.0, 1.5, 2.0, 2.5, 3.0) High Entropy Alloys (HEAs) were prepared in a water-cold copper hearth of vacuum arc furnace. The microstructures and properties of as-cast HEAs were analyzed. The results of experiments revealed that the phases of the as-cast HEAs were mainly composed by simple solid solution of body-centered cubic (BCC) and face-centered cubic (FCC) structure. There were four typical cast structures in HEAs. They were petals shape organization, chrysanthemum shape organization, intergranular organization which was rich of copper and the phase which was rich of Ni. Chrysanthemum shape organization was eutectic organization that was composed of α phase and β phase. The petals shape organization was α phase. With the increase of Ni, petals shape organization reduced and even disappeared when x was 2.0, and the phase which was rich of Ni appeared. The yield strength of all other alloys is higher than 1.5GPa and the breaking strength is about 2GPa except NiAlTiCrFeCoCu alloy, and the fractures of these HEAs belonged to brittle fracture. $Ni_{20}AlTiCrFeCoCu$ had the best performance of corrosion resistance in H_2SO_4 solution, and the corrosion resistance of $Ni_{15}AlTiCrFeCoCu$, $Ni_{20}AlTiCrFeCoCu$ and $Ni_{30}AlTiCrFeCoCu$ alloys were better than that of the 304 stainless steel in 3.5% of NaCl solution.

Key words: high-entropy alloys; microstructure; compressive property; corrosion resistance

---

基金项目： 2013 年度广西高等教育教学改革工程重点资助项目(2013JGZ104)；广西自治区教育厅科研资助项目(2013YB012)；广西有色金属及特色材料加工重点实验室基金项目(13-A-02-03)。

第一作者： 李安敏,女。高熵合金的强韧化机制。lamanny@126.com。

汽车车身用激光拼焊技术和装备研发现状

侯宝辉，孙玉良，张 梅

(上海大学 材料科学与工程学院)

摘 要：激光加工技术的出现是对传统的焊接工艺和焊接方法具有重大影响的技术变革，产生了巨大的经济效益和社会效益，应用前景十分广阔。激光拼焊自动化装备作为成套关键装备与战略技术，不仅影响着汽车行业的发展，而且将对国家的综合国力以及可持续发展有着巨大而深远的影响。本文综述了激光拼焊技术和装备的发展历程和现状。

关键词：激光；拼焊技术；焊接设备

Development Status of Laser Welding Technique and Equipments for BIW

HOU Bao-hui, SUN Yu-liang, ZHANG Mei

(School of Materials Science and Engineering, Shanghai University)

Abstract：The emergence of laser processing technology is a technological change which has a significant impact on the traditional welding processes and welding methods, produced huge economic and social benefits, and has very broad application prospects. Laser welding automation equipment as sets of key equipment and technology strategy, not only affects the development of the automotive industry, but will also has a huge and far-reaching impact on the comprehensive national strength and sustainable development of the country. This paper reviews the history and current development of laser welding technology and equipment.

Keywords: laser welded blanks; weld equipments; welded joints

第一作者：侯宝辉，男，硕士研究生。
通讯作者：张 梅，女，博士。轻量化金属材料开发和应用。zhangmei3721@i.shu.edu.cn。

镁合金薄壁材稳态模拟及模具改进

林　涛 [1]，杨院生 [1,2]，刘云腾 [1]，衣冠玉 [1]，周吉学 [1]

(1.山东省科学院 新材料研究所；2.中国科学院 金属研究所)

摘　要：以薄壁镁合金方管为研究对象，采用 Deform 有限元软件对其挤压过程进行稳态模拟，分析模具内金属的速度场、应变场以及温度场。分析型材出口位置流速的分布不均匀性并改进初始模具结构，在方管中心加强筋位置，增加模具出口尺寸，流速增加，整体的金属流速更为均匀。有效解决了挤压过程中金属分布不均匀的问题，改进后的模具模拟结果符合实际挤压要求。

关键词：薄壁镁合金型材；模具；稳态模拟；Deform

Steady-state Simulation and Die Improvement on the Extrusion of Magnesium Profile with Thin-wall

LIN Tao[1], YANG Yuan-sheng[1,2], LIU Yun-teng[1], YI Guan-yu[1],ZHOU Ji-xue[1]

(1. Shandong Academy of Sciences New Materials；2. Institute of Metal Research Chinese Academy of Sciences,)

Abstract: In this paper, steady-state simulation of the extrusion process of thin-wall magnesium profile was conducted. The velocity field, strain field and temperature field were analyzed. The results show that the non-uniform metal flow at the die exit is caused by the un-proper structure. Then, at the strengthen rib of the die, the exit was expanded, the velocity was increased and the whole velocity distribution is more uniform. Finally, the problem of the non-uniform distribution of the metal flow was solved effectively. The results of simulation for the improved die are in accordance with the practical production requirements.

Key words: magnesium profile with thin-wall; die; steady-state simulation; deform

第一作者：林　涛，男。新材料。tlin@alum.imr.ac.cn。

凝固过程微观组织的界面前沿跟踪模拟

衣冠玉[1]，杨院生[1,2]，刘波祖[1,3]

(1.山东省科学院 新材料研究所；2.中国科学院 金属研究所；3.山东建筑大学)

摘　要：通过对二元合金凝固物理过程的研究,建立凝固过程前沿跟踪微观组织生长模拟模型,对凝固过程的晶粒竞争生长进行模拟分析。采用有限差分方法模拟了二元合金凝固过程的热传导和溶质扩散,并考虑晶粒生长各向异性及固/液界面成分过冷。为了对模型进行验证,将模拟结果和理论模型的预测结果进行了比较,结果非常接近。模型成功模拟了凝固过程溶质再分配对枝晶生长的影响以及枝晶的竞争生长。

关键词：界面跟踪；竞争生长；凝固模拟

Front-Tracking Modelling of Competitive Crystal Growth during Alloy Solidification

YI Guan-yu[1], YANG Yuan-sheng[1,2], LIU Bo-zu[1,3]

(1.New Materials Research Institute, Shandong Academy of Sciences；
2.Institute of Metal Research, Chinese Academy of Sciences; 3.Shandong Jianzhu University)

Abstract: A front-tracking model is developed to simulate the competitive crystal growth during alloy solidification. In this model, crystal growth of binary alloys is governed by heat and mass transport and the evolution of the solid/liquid (S/L) interface is solved as a part of solution process. At the S/L interface, local equilibrium is maintained. Anisotropy of crystal growth and the S/L interface curvature undercooling are included in the model. The model is able to qualitatively reproduce most of the dendrite features observed experimentally, such as branching, overgrowth and coarsening. The model is validated by a comparison of simulated results with theoretical model. Using the developed model, the solute interaction and the competitive nature of crystal growth during alloy solidification is visualised.

Keywords: front-tracking; competitive growth; solidification modelling

第一作者:衣冠玉,男。新材料。yiguanyu@163.com。

热等静压及热处理对铸造钛合金 TG6 组织和拉伸性能的影响

朱郎平，李建崇，莫晓飞，魏战雷，李　岩，罗　倩，南　海

（北京航空材料研究院）

摘　要：研究了熔模精密铸造近 α 高温钛合金 TG6 在铸态、热等静压及退火热处理等状态下的组织及力学性能。结果表明，合金铸态组织中存在缩松，塑性只有 0.8%；热等静压消除了缩松缺陷，塑性提高至 3.7%；退火热处理后合金组织中的 β 板条发生部分溶解并析出条状及颗粒状硅化物，在 α 板条中弥散分布大量微细的粒状析出物；退火后合金的塑性提高到 11% 以上，但在 600℃~700℃出现明显降低；在 700℃~800℃范围内随着退火温度的升高，合金拉伸强度及塑性同时下降。

关键词：TG6；铸造；退火；微观组织；力学性能

Effect of Hot Isostatic Pressing and Heat Treatment on Microstructure and Tensile Property of Cast TG6 Titanium Alloy

ZHU Lang–ping, LI Jian–chong, MO Xiao–fei, WEI Zhang–lei, LI Yan, LUO Qian, NAN Hai

（Beijing Institute of aeronautical matetialsa）

Abstract: TG6 titanium alloy was prepared by investment casting, and its microstructure and tensile propertystudies were performedunder different process states.The results show that minor shrinkage porosity exists in the as casted alloy, and the elongation is 0.8%; shrinkage porosity was diminished after HIP process, andthe elongationincreased to 3.7%; banded and granular silicide phase precipitated in β lathes after annealing process, and β lathes dissolved partially, large amounts offineparticleswere dispersed in the base of α lath; the ductility at room temperature improvedto up to 11% after annealing process, while there was a rapid decrease at 600℃~700℃; the tensile strength and ductility of the alloy drop slightly with the rise of annealing temperature within 700℃~800℃.

Key words: TG6; cast; annealing; microstructure; tensile properties

基金项目：北京市科技计划项目（编号：Z141104004414053）。

第一作者：朱郎平，男。铸造钛合金。Elangping.china@gmail.coml。

通讯作者：李建崇，男。铸造钛合金。jchlee1983@163.com。

奥氏体变形对热轧 DP600 双相钢连续冷却相变的影响

黄绪传，殷　胜，周丽萍

(上海梅山钢铁股份有限公司技术中心)

摘　要：利用 Gleeble3500 热模拟试验机对 DP600 热轧双相钢进行奥氏体未变形和变形两种条件下的连续冷却相变测定实验，获得两种实验条件下的连续冷却相变 CCT 曲线。对比分析了两种实验条件下获得的组织和连续冷却相变曲线，探讨奥氏体变形对 DP600 双相钢连续冷却相变的影响。结果显示：奥氏体区变形明显提升了铁素体相变开始温度，增大了珠光体相变临界冷却速度，使 CCT 曲线向左上方移动，同时促进了贝氏体及马氏体的形成。

关键词：双相钢；奥氏体变形；连续冷却；影响

Effect of Austenite Deformation on Continuous Cooling Transformation of Hot-rolled DP600 Dual Phase Steel

HUANG Xu-chuan, YIN Sheng, ZHOU Li-ping

(Technology Center of Meishan Iron & Steel Co.)

Abstract：The continuous cooling transformation measurement experiments of DP600 hot rolled dual phase steel under the condition of the two kinds of austenite non-deformation and deformation were performed by Gleeble3500 thermal simulation testing machine, Won the continuous cooling transformation CCT diagram under the two kinds of experimental conditions. The organization and the continuous cooling transformation curve under the two kinds of experimental conditions were analyzed。The effect of austenite deformation on continuous cooling transformation of hot-rolled DP600 dual phase steel were discussed. The results showed that, the austenitic deformation enhances the ferritic transformation starting temperature obviously, increases the critical cooling rate of pearlite transformation, makes the CCT curve shift to the left and upper, Promotes the formation of bainite and martensite.

Key words: dual phase steel；austenite deformation；continuous cooling；effect

---

第一作者：黄绪传(1976—)，男，硕士，高级工程师。钢铁材料热成型工艺技术。619040@baosteel.com。

基于多因素正交实验的含铜钢析出行为研究

安小凡 [1,2]，王海燕 [2]，高雪云 [3]，郑梦珠 [2]，陈树明 [2]，吴志峰 [1]

(1.包头钢铁职业技术学院 机械系；2.内蒙古科技大学 材料与冶金学院；3.中冶东方工程技术有限公司长材事业部)

摘　要： 通过对含 Cu 低碳钢进行热模拟实验，利用形变强化和析出强化的整合作用，使含铜钢得以强化。运用正交实验法进行实验设计，在 Gleeble-1500D 热模拟机上完成热压缩实验，利用扫描电子显微镜观察含铜钢的显微组织形貌，并进行能谱分析。借助显微硬度计测试了实验钢的硬度，通过方差分析得出了变形温度、变形量和变形速率对实验钢硬化效果的影响规律。结果表明，含铜钢的强化效果与变形温度、变形量、变形速率有关，三因素影响实验钢强化效果的主次顺序为：变形量、变形温度、变形速率。

关键词： 含铜钢；热变形；正交实验；析出强化

Precipitation Behavior in Cu-containing High Purity Steel Based on Orthogonal Experiment Method

AN Xiao-fan[1,2], WANG Hai-yan[2], GAO Xue-yun[3], ZHENG Meng-zhu[2], CHEN Shu-ming[2], WU Zhi-feng[2]

(1.Department of Mechanical Engineering, Baotou Iron&Steel Vocational Technical College,; 2.School of Material and Metallurgy, Inner Mongolia University of Science and Technology; 3. Beris Engineering and Research Corporation)

Abstract: In this paper, hot simulation experiments were carried out on Cu containing low carbon steel, to obtain strengthening by strain hardening and precipitation. Experiment was designed by orthogonal experiment, and Gleeble-1500D hot simulator was used for hot compression test. Microstructure of high purity steel containing copper was investigated by means of scan electron microscope (SEM), and hardness was characterized by micro-hardness measurements. The effects of deformation temperature, reduction and deformation rate on enhance of the steel was analyzed by variance calculation. The result showed that strength was improved by strain inducing strengthening, which has relation with deformation temperature, reduction and deformation rate. The main order of these three factors affect strengthening of test steel were deformation rate, deformation temperature, strain rate.

Key words: Cu-containing steel; hot deformation; orthogonal experiment method; precipitation strengthening

基金项目： 国家自然科学基金 (51101083)。

第一作者： 安小凡 (1972—)，包头钢铁职业技术学院高级讲师。

通讯作者： 王海燕 (1975—)，内蒙古科技大学副教授。windflower126@163.com。

华铸 CAE 铸造模拟软件在铸钢工艺优化设计中的应用

刘 霄

(昆明广维通机械设备有限公司)

摘　要：随着计算机技术的飞速发展,计算机数值模拟仿真技术能准确、形象地描述铸造充型凝固过程,预测铸造缺陷,优化浇注及补缩系统设计、消除缩孔、热裂等铸造缺陷、提高工艺出品率、从而为判断工艺方案的可行性提供科学依据。为提高铸钢件在铸造生产中的质量,本文基于"华铸CAE"软件在铸钢件工艺优化设计方面的应用,结合实际生产案例,介绍几个生产中典型的铸钢件工艺优化实例,通过对典型铸钢件的浇注系统、冒口等进行铸造工艺优化,获得了较为合理的工艺优化方案,使得材料利用率提高了 15%~20%,工艺出品率最高达到 83%,生产成本降低了 30%~40%。

关键词：华铸CAE；数值模拟；缺陷预测；缩松；裂纹；工艺优化

Optimization Design Application of the Steel Castings Process Based on the Casting Simulation Software of Intecast CAE

LIU Xiao

(Kunming–Dimensional Machinery Equipment Co., Ltd.)

Abstract: With the rapid development of computer technology, the computer numerical simulation technology can accurately and vividly describe the casting filling and solidification process, and it also can predict the location of casting defects and optimize the pouring and feeding system design, eliminate the shrinkage, thermal cracking and other casting defects and improve process yield rate, thus providing a scientific basis for judging the feasibility process plan. To improve the quality of steel castings in the foundry, this paper, based on the optimization design software application of steel casting process in the Intecast CAE, introduces several typical steel casting production process optimization examples by combining with the actual production case. Among them, the typical steel casting gating system and riser system were optimized. And a more reasonable process optimization program was obtained in which the material utilization increase 16.5%, yielding rate increase 83%, production costs decrease by 35%-39%.

Keywords: intecast CAE; numerical simulation; defect prediction; shrinkage cracks; process optimization

第一作者:刘　霄(1965—),女,高级工程师。铸造工艺研究及技术管理。liuxiao6441@163.com。

电子束焊接钛合金表面残余应力的测试和有限元分析

刘晓佳[1]，林　健[1]，雷永平[1]，吴中伟[1]，刘　昕[2]，付鹏飞[2]

(1.北京工业大学 材料科学与工程学院；2.北京航空制造研究所 高能束流加工技术重点实验室)

摘　要：采用电子束焊接 Ti60 钛合金，并利用残余应力测试仪和有限元分析方法测试和模拟焊后残余应力值。通过测试和模拟焊前预热、焊后缓冷和焊前预热焊后缓冷三种焊接工艺下的残余应力值，研究其残余应力分布规律。研究结果表明，在垂直焊缝截面上，纵向残余应力 σx 的数值模拟结果与实验测试结果在变化趋势上基本一致：在焊缝中心线附近，纵向残余应力 σx 的数值较大。随着距焊缝中心线距离的增大，纵向残余应力的数值逐渐减小；数值较大的残余拉应力区域对于三种焊接工艺而言，存在一定的差异，这与接头的焊缝形状尺寸有关。在平行焊缝截面上，实测和模拟纵向残余应力 σx 的分布规律相似，但是具体数值上还存在一定差异。

关键词：有限元分析；Ti60 合金；焊接残余应力

The Titanium Alloy Surface Residual Stress Test and Finite Element Analysis by EBW

LIU Xiao-jia[1], LIN Jian[1], LEI Yong-ping[1], WU Zhong-wei[1], LIU Xin[1], FU Peng-fei[2]

(1.College of Materials Science and Engineering, Beijing University of Technology; 2. National Key Laboratory of Science and Technology on Power Beam Processes, Beijing Aeronautical Manufacturing Technology Research Institute.)

Abstract: The study use electron beam welding connect the Ti60 titanium alloy plate, it is tested and simulated the welding residual stress by the residual stress tester and finite element analysis, which it is welded by three welding process including preheating before welding、slow cooling after welding and both at the same time. And then we can study the residual stress distribution. In the vertical weld section, the results show that the tested and simulated result trends of longitudinal welding residual stress are similar. The longitudinal welding residual stress is larger near the welding center line. But longitudinal welding residual stress decreases along with the increase of the distance from the welding center line. For the three welding process, the regions with larger residual tensile stress exist certain difference. It is related to welding shape and size. In the parallel weld section, the distribution of tested and simulated longitudinal welding residual stress is similar. But the concrete numerical exists certain difference.

Key words: finite element analysis; Ti60 titanium alloy; welding residual stress

基金项目：国家自然科学基金资助项目(51005004, 51275006)；北京市自然科学基金资助项目(3132006)；
　　　　　北京市教委项目科研项目(面上) (KM2012100050010)。
第一作者：刘晓佳(1990—)，男，硕士。焊接残余应力的测试及数值模拟。E-mail: s201409031@emails.bjut.edu.cn。
通讯作者：林　健(1979—)，男，副教授，硕导。汽车车身的轻量化研究。E-mail: linjian@bjut.edu.cn。

大型挖掘机动臂有限元分析

齐敏杰 [1,3]，张学宾 [1,2,3]，宋克兴 [1,2,3]，张彦敏 [1,2,3]

(1.河南科技大学 材料科学与工程学院 ；2.有色金属共性技术河南省协同创新中心；
3.河南省有色金属材料科学与加工技术重点实验室)

摘 要：本文基于SolidWorks和ANSYS对大型挖掘机动臂铸件结构进行静力分析，首先在软件SolidWorks中建立挖掘机动臂的三维实体模型，之后借助大型有限元分析软件ANSYS对动臂进行结构静力分析，得到动臂的静强度和静刚度。ANSYS分析所得到的结果可以为后续动臂的优化设计提供可靠的科学指导。

关键词：SolidWorks；ANSYS；挖掘机动臂；有限元分析

FEA of the Large Moving Arm of the Excavator

QI Min-jie[1,3], ZHANG Xue-bin[1,2,3], SONG Ke-xing[1,2,3], ZHANG Yan-min[1,2,3]

(1.School of Materials and Engineering, Henan University of Science and Technology; 2.Henan Nonferrous Metal Common Technology Collaborative Innovation Center; 3.Henan Key Laboratory of Advanced Non-ferrous Metals)

Abstract: This paper is the static analysis of the large moving arm's casting structure of the excavator based on SolidWorks and ANSYS. First, 3D solid model of the moving arm of an excavator established by using SolidWorks. Afterwards, carrying out the structure static analysis of the moving arm of the excavator by large-scale finite element analysis software ANSYS. At last, we can get the static strength and static stiffness of the moving arm of the excavator. The result of ANSYS analysis can be provided reliable and scientific guidance for the subsequent optimization design of the moving arm of the excavator.

Keywords: moving arm of excavator; SolidWorks; ANSYS; finite element analysis

0 前言

在建筑施工频繁的时代,挖掘机作为一种既省力又方便快捷的施工工具被广泛使用。动臂作为挖掘机机构中最关键的构件之一,在使用过程中不断承受着复杂的冲击载荷。动臂的传统设计方法主要是以经验设计和类比设计为主,在实际工作时常常会出 现强度和刚度不够的情况,这将对挖掘机工作过程中的安全可靠性造成致命性的影响。因此,对挖掘机动臂的强度和刚度分布情况进行分析具有非常重大的工程意义。此外,分析结果还可以为动臂后续的结构优化设计提供可靠依据。

基金项目:河南省科技厅科技攻关项目(132102210122);河南省杰出人才项目(134200510011);河南省高校科技创新团队(14IRTSTHN007)。
第一作者:齐敏杰(1990—),女。材料加工。
通讯作者:宋克兴,男,教授。材料加工。kxsong123@163.com。

1 三维模型的建立与导入

鉴于 ANSYS 软件在三维模型创建方面的功能还不够完善,本文选择运用了 SolidWorks 三维建模软件完成对挖掘机动臂三维模型的创建,并运用 SoildWorks 与 ANSYS 之间的图形接口,从而实现两者之间的无缝连接。这种导入方法在避免数据丢失的同时,也十分有效地弥补了 ANSYS 软件在三维模型创建方面的不完善[1-2]。挖掘机动臂三维模型由 SolidWorks 软件导入 ANSYS 的具体过程为:点击 SolidWorks 软件界面上方工具栏中的 ANSYS 图标,点击 Workbench,即可进入 ANSYS 界面中进行大型挖掘机动臂结构静力分析。创建完成的大型挖掘机动臂三维几何实体模型如图 1 所示:

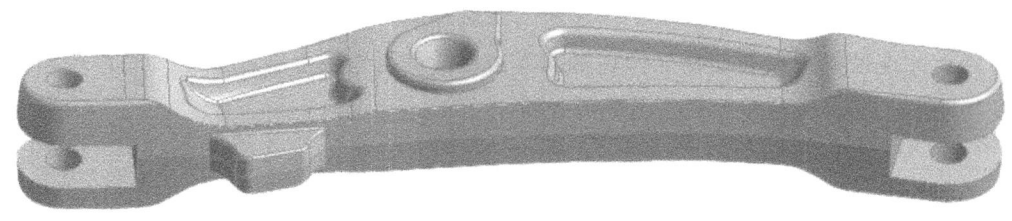

图 1 挖掘机动臂几何模型
Fig.1 Geometrical model of the moving arm of the excavator

2 网格划分

在 ANSYS 软件中对已创建完成的大型挖掘机动臂三维几何实体模型进行网格划分之后,挖掘机动臂三维几何实体模型生成包含有节点和单元的有限元模型。网格划分的过程分为以下三步:一是设置单元属性。在此过程可以设置不同种类的单元,相同种类的单元也可以设置成不同的材料属性、实常数与单元坐标系;二是设置网格控制;三是生成网格。[3]对于单一模型的结构分析可以采用自由网格划分方法。[4]在 ANSYS 中采用自由划分的方法对大型挖掘机动臂的三维几何实体模型进行网格划分后,根据动臂的实际情况,选择动臂的材料参数,确定动臂的密度为 $7.8×10^3$ Kg.m^{-3};弹性模量为 $2.06×10^{11}$Pa;泊松比为 0.3。将网格单元尺寸 (Element Size) 的数值设定为 5mm,动臂模型网格划分的最终结果为 456761 个节点,268388 个单元。动臂三维几何实体模型自由网格划分后的结果如图 2 所示:

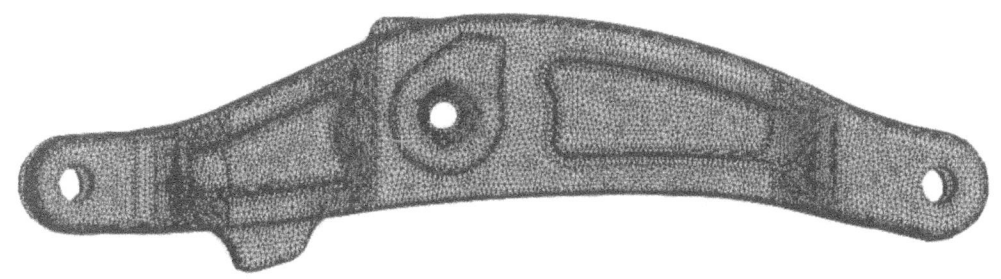

图 2 挖掘机动臂网格模型
Fig.2 Meshed model of the moving arm of the excavator

3 边界条件及加载

为了消除 ANSYS 在有限元计算时总刚度矩阵的奇异性,在设置边界条件时必须消除挖掘机动臂实体模型的刚体位移,可以对某些节点加以约束。其约束条件为:所加约束需要刚好消除全部刚体自由位移,或者把所加约束改为刚度足够大的边界元,作为校核载荷平衡的补充办法。[5]根据边界条件对动臂实体模型各个部位分别施加约束，对动臂工作装置的支点和动臂与液压缸连接的铰点也施加同样的约束;沿三个坐标轴方向的位移。斗齿尖位于动臂与铲斗连接的铰点和动臂工作装置的支点连线的延长线上时,挖掘机工作时动臂和斗杆液压缸作用力臂最大该工况下动臂液压缸作用力臂最大,工作装置可充分发挥斗杆液压缸的挖掘力,故而在动臂上产生最大载荷。[6]经分析,本动臂模型的最大载荷工况是载荷垂直作用在动臂与铲斗连接的铰点和动臂工作装置的支点连线的延长线上时。因此在动臂与铲斗连接的铰点和动臂工作装置的支点连线的延长线上施加外载，此时来分析挖掘机动臂的强度与刚度。挖掘机动臂加载情况如图 3 所示:

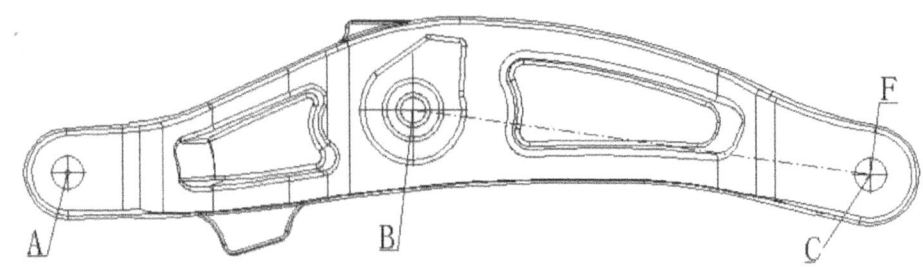

图 3   动臂加载情况

Fig.3   Force status of moving arm

A:动臂和斗杆液压缸连接铰点   B:动臂工作装置的支点   C:动臂与铲斗连接铰点   F:施加载荷

4 静力分析

对大型挖掘机动臂的结构进行有限元分析,全面了解在挖掘机实际工作的过程中动臂受到载荷时的变形和应力分布的情况,为动臂结构设计和动臂后续的优化提供理论依据。通过有限元分析计算可以得出挖掘机动臂受到载荷时的变形云图(图 4)和动臂应力云图(图 5)。

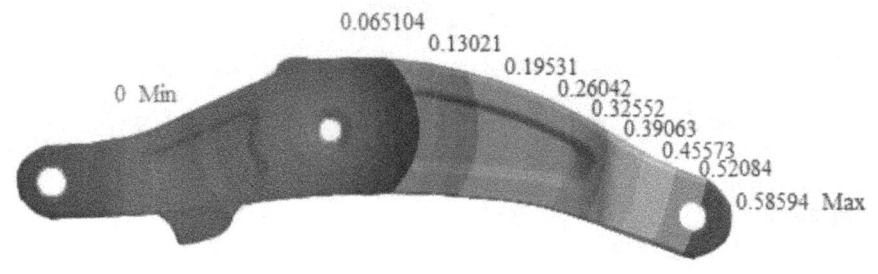

图 4   动臂变形云图

Fig.4   Deformation contours of moving arm

从图 4 中可以看出,动臂的最大变形出现在动臂与铲斗连接的铰点处。动臂与铲斗连接的铰点和动臂工作装置的支点之间前端变形量较大,而后端变形量较小。这可能与动臂受力与动臂工作装置的支点之间的力矩有关,从而导致变形不均匀。由于动臂工作装置的支点和动臂与液压缸连接的铰点被约束,因此动臂工作装置的支点和动臂与液压缸连接的铰点之间没有变形。

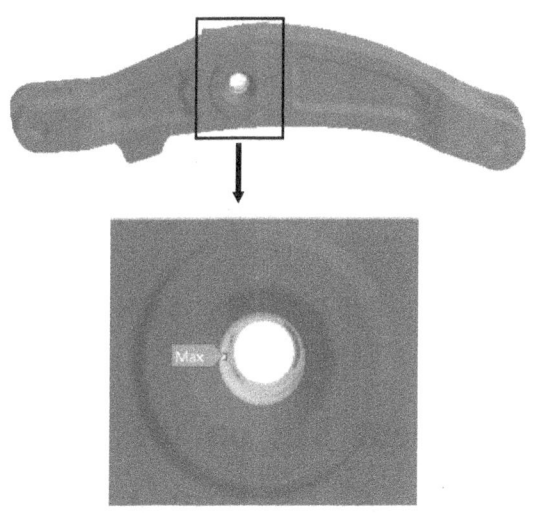

图 5 动臂应力云图

Fig.5 Stress contours of moving arm

从图 5 可以看出,动臂的大部分应力是在较小范围内,而在动臂工作装置的支点处应力比较集中(由图 5 局部放大图可见),出现峰值,其原因主要是因为轴座是整个动臂与平台的联接部件,是整个动臂在工作过程中的支撑点[5],因此在设计时应该给予这个部位以足够的重视。

5 结语

运用 SoildWorks 与 ANSYS 之间良好的图形接口,从而实现两者之间的无缝连接,这种导入的方法在避免数据丢失的同时,也十分有效地弥补了 ANSYS 在三维实体模型创建方面的不完善,从而大大提高了有限元计算的可靠性。通过对大型挖掘机动臂的结构静分析,全面了解在挖掘机工作过程中动臂强度和刚度的分布情况。静力分析结果表明,动臂的最大变形发生在动臂与铲斗连接的铰点处。应力比较集中的地方出现在动臂工作装置的支点处,在设计时这些地方都需要给予足够的重视。挖掘机动臂静力分析的结果为后续的动臂的设计以及优化提供了科学的参考依据。

参考文献:

[1] 徐献忠, 王伟兵. 有限元 AWB 分析及其与 CAD 的接口 [J]. 科技创新导报, 2010(17):32–33.

[2] 郝钟雄. ANSYS 与 CAD 软件的接口问题研究 [J]. 机械设计与制造, 2007(7): 75–76.

[3] 王宇, 卢玲, 李文韬. 基于 Ansys 的有限元网格划分方法应用研究 [J]. 起重运输机械, 2014(3): 53–56.

[4] 王瑞, 陈海霞, 王广峰. ANSYS 有限元网格划分浅析 [J]. 天津工业大学学报, 2002, 21(4): 8–11.

[5] 丁华, 朱茂桃, 赵剡水. 液压挖掘机动臂的有限元分析 [J]. 中国公路学报, 2003, 16(4): 118–120.

[6] 林明智, 邢树鑫. 液压挖掘机动臂有限元分析 [J]. 工程机械, 2010, 41(12): 43–45.

基体掺杂 Mg 对 SiC/Al 界面结合的影响模拟

周贤良，吴开阳，邹爱华，华小珍

(南昌航空大学 材料科学与工程学院)

摘 要：采用基于密度泛函理论的第一性原理平面波超软赝势方法，在优化 SiC/Al 界面结构基础上，探讨了掺杂 Mg 对 SiC/Al 界面作用。研究表明：C 终止界面的结合强度于 Si 终止面，以 Si 为键桥的 bridge-site 结构，是 SiC/Al 界面结合的主要方式。掺杂 Mg 元素时，Mg 原子会和 Si 原子成键，增加了界面结合，界面附近的基体原子和增强体中的原子相互作用也增强，这有利于 SiC/Al 复合材料制备。

关键词：掺杂 Mg；SiC/Al 界面；界面结合；第一性原理

Effect of Alloy Element Mg on the Interface Bonding of SiC/Al by Simulation

ZHOU Xian-liang, WU Kai-yang, ZOU Ai-hua, HUA Xiao-zhen

(School of Material Science and Engineering, Nanchang HangKong University)

Abstract: First-principle ultrasoft pseudopotential approach of the plane wave based on density functional theory was used to study the optimized energy and electronic structure of SiC/Al. The interfaces of SiC/Al substituted by alloy element Mg was investigated. The results show that the bonding strength of C-terminated interface is higher than that of Si terminated interface. Bridge-site model with bridge Si is the main combination way of SiC/Al interface. The introduction of Mg can enhance the interface strengthens and the atomic interactions between the matrix and reinforcements in the interface, which benefits for the preparation of SiC/Al composites.

Key words: addition of Mg; SiC/Al interface; interface bonding; first principles

第一作者及通讯作者：周贤良，男，教授。材料加工。zhouxl209@163.com。

TA2 瞬时液相扩散连接界面特征与相变机理

陈思杰，朱春莉

(河南理工大学 材料科学与工程学院)

摘　要：在开放环境下，使用氩气保护，对 TA2 工业纯钛瞬时液相扩散连接中界面特征和相变规律进行了研究。采用 Cu-Ni-Sn-P 系非晶合金箔作为中间层，设定不同的等温凝固温度，对接头的界面形貌进行了观察，测试了接头的剪切强度。研究表明：在一定压力和保温时间条件下，等温凝固温度对于界面形貌和性能有很大影响，低于 850℃时，等温凝固进行不充分，有残留中间层。等温凝固温度高于 850℃，残留中间层消失，接头结合性能随之提高。

关键词：TA2；瞬时液相扩散焊；焊接温度；界面结构

Interfacial Microstructure and Phase Transitions of TA2 Joints with Transient Liquid Phase Bonded Process

CHEN Si-jie, ZHU Chun-li

(School of Materials Science and Engineering, Henan Polytechnic University)

Abstract：Titanium TA2 sheets were lap joined by transient liquid phase bonding with different bonding temperatures in argon atmosphere. Cu-Ni-Sn-P amorphous foil was used as interlayer. Influence of bonding temperature on joint microstructures was investigated. The shear strength of the joints was tested. The testing results showed that the isothermal solidification of the liquid interlayer was not completed when the bonding temperature lowe r than 850℃, and the remaining interlayer could be observed. Atomic diffusion speed up with the increase of temperature and the remaining interlayer reduce gradually; the joint microstructures were all solid solution at 850℃. The joints shear strength increases with temperature increasing.

Key words: TA2; transient liquid phase bonding; bonding temperature; interfacial microstructure.

基金项目：河南省科技攻关资助项目(142102210434)。
第一作者：陈思杰，男。先进连接技术。chen_sijie@126.com。

TC18钛合金热变形临界动态再结晶条件判别

曲凤盛，王震宏，张新建，龙 波，魏怡芸

(中国工程物理研究院材料所)

摘 要：TC18钛合金是一种近β型高强钛合金，其淬透性好，强度高等原因广泛的应用于航空航天领域。本研究采用Gleeble1500D热模拟试验机在变形温度750~950℃，应变速率$10^{-2}s^{-1}\sim 10s^{-1}$的条件下进行TC18钛合金热模拟压缩试验。研究了TC18钛合金变形时的动态再结晶临界条件。结果表明：在本实验条件下，TC18钛合金在本试验条件下呈现两种曲线特征类型的应力-应变曲线，其θ-σ曲线均呈现拐点及$-d\theta/d\sigma-\sigma$曲线上出现最小值；临界应变与峰值应变之间具有一定的相关性，即$\varepsilon_c/\varepsilon_p=0.58$；临界应变与Z参数之间的函数关系为$\varepsilon_c=5.50903\times10^{-4}Z^{0.08061}$，EBSD分析显示即使是动态回复型曲线同样存在等轴的动态再结晶晶粒，这证实了加工硬化率推导是正确的。

关键词：加工硬化率；动态再结晶；临界条件；TC18钛合金

Critical Conditions of Dynamic Recrystallization during Deformation of TC18 Titanium Alloy

QU Feng–sheng, WANG Zhen–hong, ZHANG Xin–jian, LONG Bo, WEI Yi–yun

(Institute of Materials China Academy of Engineering Physics)

Abstract: The hot simulation compression tests of TC18 titanium alloy were conducted at deformation temperature of 750℃ –950 ℃ and strain rate of $10^{-2}s^{-1}$-$10s^{-1}$ with Gleeble1500D hot simulation test machine. The critical conditions of dynamic recrystallization for TC18 titanium were studied. there are two kinds of curves of stress-strain for TC18 titanium alloy. The θ-σ curves show inflection point and the minimal values appear on the curves of $-d\theta/d\sigma-\sigma$. There is linear relationship between critical strain and peak strain i.e. $\varepsilon_c/\varepsilon_p=0.58$. The function relationship between critical and Z parameters is $\varepsilon_c=5.50903\times10^{-4}Z^{0.08061}$. The equiaxed grains of dynamic recrystallization exist in TC18 titanium alloy of dynamic recovery through EBSD analysis, which prove that the reason of work hardening rate is right.

Key words: work hardening rate; dynamic recrystallization; critical condition; TC18 titanium alloy

---

第一作者：曲凤盛，男，高级工程师。qufengsheng@163.com。

电磁场作用下金属凝固的多尺度模拟

李应举，冯小辉，腾跃飞，杨院生

（中国科学院　金属研究所）

摘　要：本文采用有限元方法对脉冲磁场作用的凝固过程中进行电磁力、流场和温度场分布的耦合数值模拟，获得宏观场的变化规律，用改进的元胞自动机(CA)方法模拟凝固过程的微观组织，同时将宏观场的计算结果引入 CA 计算，实现了脉冲磁场下合金凝固过程和组织演化的宏—微观耦合模拟。模拟计算结果表明脉冲磁场作用于合金熔体中的电磁力和脉冲励磁电流同步呈周期性脉冲变化，合金熔体在其作用下产生电磁振荡和电磁对流，改变了熔体温度场的分布，减小了熔体中的温度梯度。组织模拟结果表明熔体流动对凝固组织演化存在显著影响，使凝固组织由未施加脉冲磁场的柱状晶和少量等轴晶的混合组织向细小等轴晶组织转变。

关键词：凝固；熔体流动；脉冲磁场；数值模拟

Multi-scale Simulation for Metal Solidification with Electromagnetic Fields

LI Ying-ju, FENG Xiao-hui, TENG Yue-fei, YANG Yuan-sheng

(Institute of Metal Research, Chinese Academy of Sciences)

Abstract: A macro - micro coupled simulation model was built to describe the effect of the pulsed magnetic field on the metal solidification coupling with FEM and cellular automata (CA) method. The macroscopic field was obtained by studying the effects of the pulsed magnetic field on the magnetic force distribution, fluid flow field and heat distribution. The macro simulation results were introduced to CA model to simulate the solidification process and microstructure evolution. The macro simulation results show that the magnetic force is periodically changed both in direction and quantity. Under the action of the magnetic force, Electromagnetic vibration and electromagnetic convection are produced in the melt, which reduces the temperature gradient in the melt and changes heat distribution in the melt. The CA simulation calculation results showed that the melt flow significant influences the solidification microstructure evolution, making solidification structure transition from mixture structure of columnar dendrites and equiaxed grains to fine equiaxed grain structure.

Key words: solidification; melt flow; pulse magnetic field; numerical simulation;

基金项目：国家重点基础研究发展计划资助项目(Project No. 2010CB631205)；
国家自然基金重点项目(Project No. 51034012)。
第一作者：李应举，男。外场控制材料制备。yjli@imr.ac.cn。
通讯作者：杨院生，男。先进材料凝固与制备。ysyang@imr.ac.cn。

健身自行车架焊接顺序优化的数值模拟分析

张 华，周广涛，苏礼季

(华侨大学 机电及自动化学院)

摘 要：针对健身自行车架焊接变形影响装配精度的问题，用 MARC 模拟分析车架的焊接变形。采用壳单元替代实体单元建立了车架有限元模型，分析了两者的误差并得到修正系数。对车架的 5 道焊缝拟定的 12 种焊接顺序进行了焊接模拟计算，得到不同焊接顺序下的温度场、焊后结构变形。分析结果表明，焊接变形主要表现为座垫套管面外偏转变形，并得到了优化的焊接顺序。在该顺序下，车架结构在冷却至室温时座垫套管上端偏转变形量最小，为 2.386mm，且该顺序下的变形模拟值与实测值吻合良好。

关键词：健身车架；焊接变形；焊接顺序；数值模拟

Numerical simulation analysis for welding sequence optimization of gymnastic bicycle frame

ZHANG Hua, ZHOU Guang-tao, SU Li-ji

(College of Mechanical Engineering and Automation, Huaqiao University)

Abstract: As to the problem that the welding deformation of gymnastic bicycle frame affects the assembly accuracy, the welding deformation of frame is simulated and analyzed by MARC. The finite element model of frame is built using shell elements instead of solid elements. The error between these two models is analyzed to obtain correction coefficient. Welding simulation calculation on twelve kinds of welding sequences from five welds of frame was proposed. Under different welding sequences, temperature field and structural deformation after welding were obtained. The analysis results showed that welding deformation mainly presented as deflection of cushion tube, optimal welding sequence is derived, in that sequence, when the frame welded structure cooled to room temperature, the deflection of the upper end of cushion tube is minimum, 2.386mm. Moreover, the simulated values of deformation are in good consistence with the measured values.

Key words: gymnastic bicycle frame; welding sequence; welding deformation; numerical simulation

基金项目：福建省科技平台项目(2013H2003)；厦门市科技计划(3502Z20133023)。

第一作者：张 华，男。焊接工艺方向。zhanghua19910920@163.com。

通讯作者：周广涛，男。焊接结构可靠性方向。zhouguangtao@hqu.edu.cn。